Structural Steelwork: Design to Limit State Theory

Second edition

T. J. MacGinley, BE, ME, C Eng, FI Struct E, MIE Aust
Formerly Associate Professor
Nanyang Technological University

T. C. Ang, BE, MSc, PE, MIES
Senior Lecturer
Nanyang Technological University

BUTTERWORTH
HEINEMANN

Butterworth-Heinemann Ltd
Linacre House, Jordan Hill, Oxford OX2 8DP

℞ A member of the Reed Elsevier group

OXFORD   LONDON   BOSTON
MUNICH   NEW DELHI   SINGAPORE   SYDNEY
TOKYO   TORONTO   WELLINGTON

First published 1987
Reprinted 1988 (with corrections), 1990, 1991
Second edition 1992
Reprinted 1993 (twice)

© Butterworth-Heinemann Ltd 1987, 1992

British Library Cataloguing in Publication Data
MacGinley, T. J. (Thomas Joseph)
   Structural steelwork: design to limit state theory.
   I. Title   II. Ang, T. C.
   624.1821

ISBN 0 7506 0440 9

Library of Congress Cataloging in Publication Data
MacGinley, T. J. (Thomas Joseph)
   Structural steelwork: design to limit state theory/T.J.
   MacGinley, T. C. Ang. – 2nd ed.
   p.   cm.
   Includes bibliographical references (p.   ) and index.
   ISBN 0 7506 0440 9
   1. Plastic analysis (Engineering)   2. Steel, Structural.   I. Ang,
   T. C.   II. Title.
   TA652.M28 1992
   624.1'821–dc20   91–42136
                                                               CIP

Printed and bound in Great Britain by
Biddles Ltd, Guildford and King's Lynn

Contents

Preface to the second edition

The book has been updated to comply with

BS 5950: Part 1: 1990
Structural Use of Steelwork in Building
Code of Practice for Design in Simple and Continuous Construction:
Hot Rolled Sections

A new chapter on portal design has been added to round out its contents. This type of structure is in constant demand for warehouses, factories and for many other purposes and is the most common single-storey building in use. The inclusion of this material introduces the reader to elastic and plastic rigid frame design, member stability problems and design of moment-transmitting joints.

T.J.M.
T.C.A.

Preface to the first edition

The purpose of this book is to show basic steel design to the new limit state code BS 5950. It has been written primarily for undergraduates who will now start learning steel design to the new code, and will also be of use to recent graduates and designers wishing to update their knowledge.

The book covers design of elements and joints in steel construction to the simple design method; its scheme is the same as that used in the previous book by the principal author, *Structural Steelwork Calculations and Detailing*, Butterworths, 1973. Design theory with some of the background to the code procedures is given and separate elements and a complete building frame are designed to show the use of the code.

The application of microcomputers in the design process is discussed and the listings for some programs are given. Recommendations for detailing are included with a mention of computer-aided drafting (CAD).

T.J.M.
T.C.A.

Acknowledgements

The authors gratefully wish to make the following acknowledgements:

The British Standards Institution: Reference has been made to British Standards, published by the British Standards Institution. Complete copies can be obtained from BSI at Linford Wood, Milton Keynes MK14 6LE, UK. The standard to which the most reference has been made is BS 5950: 1990, *The Structural Use of Steelwork in Building,* Part 1: Code of Practice for Design in Simple and Continuous Construction: Hot Rolled Sections. *Ward Building Components Sherburn, Malton, North Yorkshire YO17 8PQ, UK* : For permission to include a section on cold-rolled multibeam purlins and sheeting rails.

The principal author was very privileged to be associated with the project 'The New Approach to Teaching Structural Steel Design'. (This was funded by the British Steel Corporation and administered from Imperial College, London University.) He acknowledges a great debt to colleagues with whom he worked on the project.

Special thanks are due to Mr L. V. Leech for reading the manuscript and making many corrections and helpful comments. The authors would also like to thank Sancia for carefully tracing the diagrams and Cynthia for typing the manuscript.

The authors wish to record their special thanks to Dr D. J. Fraser of the School of Civil Engineering of the University of New South Wales for giving permission to include his work on effective lengths of members in portal frames.

We dedicate this book to Trudy and Cynthia

Introduction

1.1 Steel structures

Steel frame buildings consist of a skeletal framework which carries all the loads to which the building is subjected. The sections through three common types of buildings are shown in Figure 1.1. There are:

(1) Single-storey lattice roof building;
(2) Single-storey rigid portal; and
(3) Medium-rise braced multi-storey building.

These three types cover many of the uses of steel frame buildings such as factories, warehouses, offices, flats, schools, etc. A design for the lattice roof building (Figure 1.1(a)) is given and the design of the elements for the braced multi-storey building (Figure 1.1(c)) is also included. Design of rigid portals has been added in the new edition.

The building frame is made up of separate elements—the beams, columns, trusses and bracing—listed beside each section in Figure 1.1. These must be joined together and the building attached to the foundations. Elements are discussed more fully in Section 1.2.

Buildings are three dimensional and only the sectional frame has been shown in Figure 1.1. These frames must be propped and braced laterally so that they remain in position and carry the loads without buckling out of the plane of the section. Structural framing plans are shown in Figures 1.2 and 1.3 for the building types depicted in Figures 1.1(a) and (c).

Various methods for analysis and design have been developed over the years. In Figure 1.1 the single-storey structure in (a) and the multi-storey building in (c) are designed by the simple design method, while the rigid portal in (b) is designed by the continuous design method. All design is in accordance with the new limit state design code BS 5950: Part 1. Design theories are discussed briefly in Section 1.4 and design methods are set out in detail in Chapter 3.

1.2 Structural elements

As mentioned above, steel buildings are composed of distinct elements:

(1) Beams and girders—members carrying lateral loads in bending and shear;
(2) Ties—members carrying axial loads in tension;

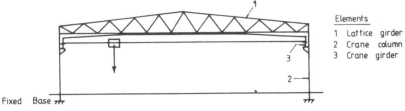

Elements

1 Lattice girder
2 Crane column
3 Crane girder

Fixed Base

a) Single Storey Lattice Roof Building with Crane

Elements

1 Portal rafter
2 Portal column

Haunched
Joint

Pinned Base

b) Single Storey Rigid Pinned Base Portal

Elements

1 Floor beam
2 Plate girder
3 Column
4 Bracing

c) Multi-storey   Building

Figure 1.1 Three common types of steel buildings

(3) Struts, columns or stanchions—members carrying axial loads in com-
    pression. These members are often subjected to bending as well as
    compression;
(4) Trusses and lattice girders—framed members carrying lateral loads. These
    are composed of struts and ties;
(5) Purlins—beam members carrying roof sheeting;
(6) Sheeting rails—beam members supporting wall cladding;
(7) Bracing—diagonal struts and ties that, with columns and roof trusses,
    form vertical and horizontal trusses to resist wind loads and stabilize the
    building.

Joints connect members together such as the joints in trusses, joints between
floor beams and columns or other floor beams. Bases transmit the loads from
the columns to the foundations.

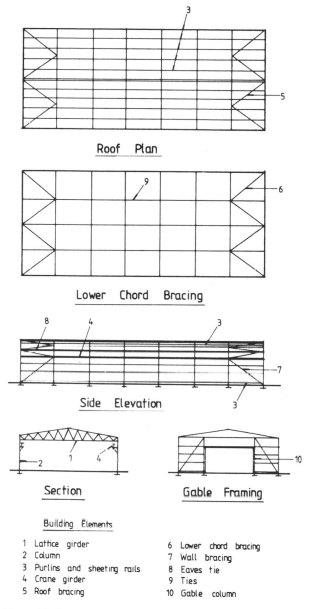

Roof   Plan

Lower   Chord   Bracing

Side   Elevation

Section   Gable   Framing

Building Elements

1 Lattice girder   6 Lower chord bracing
2 Column   7 Wall bracing
3 Purlins and sheeting rails 8 Eaves tie
4 Crane girder   9 Ties
5 Roof bracing   10 Gable column

Figure 1.2 Factory building

The structural elements are listed in Figures 1.1–1.3, and the types of members making up the various elements are discussed in Chapter 2. Some details for a factory and a multi-storey building are shown on Figure 1.4.

1.3 Structural design

An architect draws up plans for a building to meet the client's requirements.

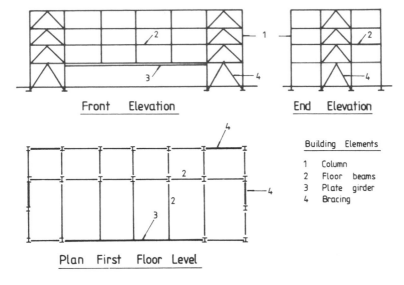

Front   Elevation   End   Elevation

Plan First Floor Level

Building Elements

1   Column
2   Floor beams
3   Plate girder
4   Bracing

Figure 1.3 Multi-storey office building

The structural engineer examines various alternative framing arrangements and may carry out preliminary designs to determine which is the most economical. This is termed the 'conceptual design stage'. For a given framing arrangement the problem in structural design consists of:

(1) Estimation of loading;
(2) Analysis of main frames, trusses or lattice girders, floor systems, bracing and connections to determine axial loads, shears and moments at critical points in all members;
(3) Design of the elements and connections using design data from step (2);
(4) Production of arrangement and detail drawings from the designer's sketches.

This book covers the design of elements first. Then, to show various elements in their true context in a building, the design for the basic single-storey structure with lattice roof shown in Figure 1.2 is given.

1.4 Design methods

Steel design may be based on three design theories:

(1) Elastic design;
(2) Plastic design; and
(3) Limit state design.

Elastic design is the traditional method. Steel is almost perfectly elastic up to the yield point and elastic theory is a very good method on which to base design. Structures are analysed by elastic theory and sections are sized so that the permissible stresses are not exceeded. Design is in accordance with BS 449:

The Use of Structural Steel in Building.

Plastic theory developed to take account of behaviour past the yield point is based on finding the load that causes the structure to collapse. Then the working load is the collapse load divided by a load factor. This too is permitted under BS 449.

Finally, limit state design has been developed to take account of all con-

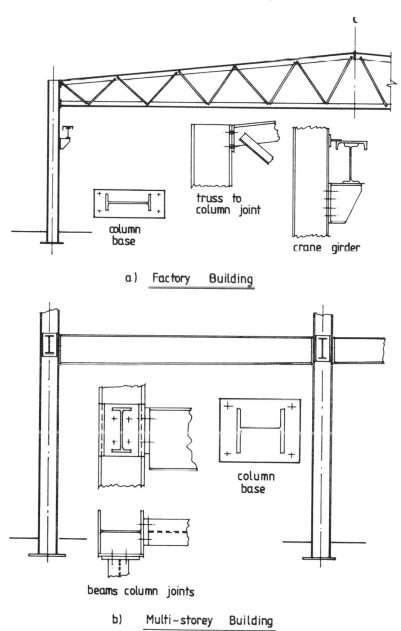

a)   Factory   Building

b)   Multi-storey   Building

Figure 1.4 Factory and multi-storey building

ditions that can make the structure become unfit for use. The design is based on the actual behaviour of materials and structures in use and is in accordance with BS 5950: *The Structural Use of Steelwork in Building*; Part 1—Code of Practice for Design in Simple and Continuous Construction: Hot Rolled Sections.

The code requirements relevant to the worked problems are noted and discussed and the complete code should be obtained and read in conjunction with this book.

The aim of structural design is to produce a safe and economical structure that fulfils its required purpose. Theoretical knowledge of structural analysis must be combined with a knowledge of design principles and theory and the constraints given in the standard to give a safe design. A thorough knowledge of materials properties, methods of fabrication and erection is essential for the experienced designer. The learner must start with the basics and gradually build up experience through doing coursework exercises in conjunction with a study of design principles and theory.

British Standards are drawn up by panels of experts from the professional institutions, and include engineers from educational and research institutions, consulting engineers, government authorities and the fabrication and construction industries. The standards give the design methods, factors of safety, design loads, design strengths, deflection limits and safe construction practices.

As well as the main design standard for steelwork in buildings, BS 5950: Part 1, reference must be made to other relevant standards, including:

(1) BS 4360: *Weldable Structural Steels*. This gives the mechanical properties for the various grades of structural steels.
(2) BS 6399: Part 1, Code of Practice for Dead and Imposed Loads. CP 3: Chapter V, Part 2, Wind Loads.

Representative loading may be taken for element design. Wind loading depends on the complete building and must be estimated using the wind code.

1.5 Design calculations and computing

Calculations are needed in the design process to determine the loading on the structure, carry out the analysis and design the elements and joints, and must be set out clearly in a standard form. Design sketches to illustrate and amplify the calculations are an integral part of the procedure and are used to produce the detail drawings.

Computing now forms an increasingly larger part of design work, and all routine calculations can be readily carried out on a microcomputer. The use of the computer speeds up calculation and enables alternative sections to be checked, giving the designer a wider choice than would be possible with manual working. (Computer applications are given in Chapter 11.) However, it is most important that students understand the design principles involved before using computer programs.

It is through doing exercises that the student consolidates the design theory given in lectures. Problems are given at the end of most chapters.

1.6 Detailing

The final chapter deals with the detailing of structural steelwork. In the earlier chapters sketches are made in design problems to show building arrangements, loading on frames, trusses, members, connections and other features pertinent to the design. It is often necessary to make a sketch showing the arrangement of a joint before the design can be carried out. At the end of the problem, sketches are made to show basic design information such as section size, span, plate sizes, drilling, welding, etc. These sketches are used by the draughtsman to produce the working drawings.

The general arrangement drawing and marking plans give the information for erection. The detail drawings show all the particulars for fabrication of the elements.

The draughtsman must know the conventions for making steelwork drawings, such as the scales to be used, the methods for specifying members, plates, bolts, welding, etc. He must be able to draw standard joint details and must also have a knowledge of methods of fabrication and erection.

Computer drafting suites are becoming generally available and the student should be given an appreciation of their use. A brief discussion of computer drafting software has been included in this chapter.

Materials

2.1 Structural steels—properties

Structural steel products are manufactured to conform to BS 4360: *Weldable Structural Steels*. Steel is composed of about 98 per cent of iron with the main alloying elements carbon, silicon and manganese. Copper and chromium are

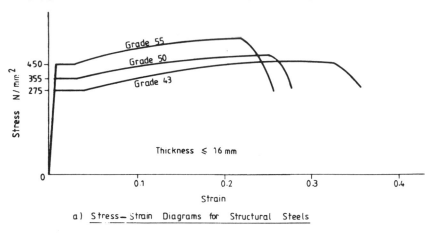

a) Stress—Strain Diagrams for Structural Steels

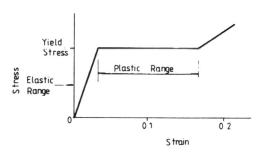

b) Stress—Strain Diagram for Plastic Design

Figure 2.1 Stress—strain diagrams for structural steels

added to produce the weather-resistant steels that do not require corrosion protection. Steel is produced in three strength grades—43, 50 and 55. The important properties are strength, ductility, impact resistance and weldability.

The stress-strain curves for the three grades of steel are shown in Figure 2.1(a) and these are the basis for the design methods used for steel. Elastic design is kept within the elastic region and because steel is almost perfectly elastic, design based on elastic theory is a very good method to use.

The stress–strain curves show a small plateau beyond the elastic limit and then an increase in strength due to strain hardening. Plastic design is based on the horizontal part of the stress strain shown in Figure 2.1(b).

The mechanical properties for steels are set out in BS 4360. The yield strengths for the various grades which vary with the thickness and other important properties are given in Section 3.7 of this book.

2.2 Design considerations

Special problems occur with steelwork and good practice must be followed to ensure satisfactory performance in service. These factors are discussed briefly below in order to bring them to the attention of students, although they are not generally of great importance in the design problems covered.

2.2.1 Fatigue

Fatigue failure can occur in members or structures subjected to fluctuating loads such as crane girders, bridges and offshore structures. Failure occurs through progressive growth of a crack that starts at a fault and the failure load may be well below its static value.

Welded connections have the greatest effect on the fatigue strength of steel structures. Tests show that butt welds give the best performance in service while continuous fillet welds are much superior to intermittent fillet welds. Bolted connections do not reduce the strength under fatigue loading. To help avoid fatigue failure, detail should be such that stress concentrations and abrupt changes of section are avoided in regions of tensile stress. Cases where fatigue could occur are noted in this book, and for further information the reader should consult reference (1).

2.2.2 Brittle fracture

Structural steel is ductile at temperatures above 10°C but it becomes more brittle as the temperature falls, and fracture can occur at low stresses below 0°C. The Charpy test is used to determine the resistance of steel to brittle fracture. In this test a small specimen is broken by a hammer and the energy to cause failure at a given temperature is measured.

In BS 4360 steels are graded A, B, C, D and E in order of increasing resistance to brittle fracture. The Charpy test fracture energy is specified for the various steel grades: for example, Grade 55 C steel is to have a minimum fracture energy of 27 Joules 0°C.

To reduce the likelihood of brittle fracture occurring it is necessary to take care in the selection of the steel to be used and to pay special attention to the

design detail. Thin plates are more resistant than thick ones. Abrupt changes
of section and stress concentration should be avoided. Fillets welds should not
be laid down across tension flanges and intermittent welding should not be
used.

Cases where brittle fracture may occur in elements designed in the book are
noted. For further information the reader should consult reference 2.

2.2.3 Fire protection

Structural steelwork performs badly in fires, with the strength decreasing with
increase in temperature. At 550°C the yield stress has fallen to approximately
0.7 of its value at normal temperatures; that is, it has reached its working stress
and failure occurs under working loads.

The statutory requirements for fire protection are set out in the Building
Regulations, Approved Documents.[3] These lay down the fire-resistance
period that any load-bearing element in a given building must have, and also
give the fire-resistance periods for different types of fire protection. Fire protection is provided by encasing the member in a fire-resistant material such as
concrete. The main types of fire protection for columns and beams are shown
in Figure 2.2.

All multi-storey steel buildings require fire protection. Single-storey factory
buildings normally do not require fire protection for the steel frame. Further
information is given in reference 4.

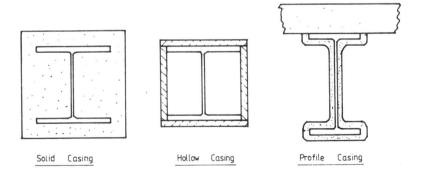

Solid Casing   Hollow Casing   Profile Casing

Figure 2.2 Fire protection for columns and beams

2.2.4 Corrosion protection

Exposed steelwork can be severely affected by corrosion in the atmosphere,
particularly if pollutants are present, and it is necessary to provide surface
protection in all cases. The type of protection depends on the surface conditions and length of life required.

The main types of protective coatings are:

(1) Metallic coatings. Either a sprayed-on coating of aluminium or zinc is
    used or the member is coated by dipping it in a bath of molten zinc in the
    galvanizing process.

(2) Painting, where various systems are used. One common system consists of using a primer of zinc chromate followed by finishing coats of micaceous iron oxide. Plastic and bituminous paints are used in special cases.

The single most important factor in achieving a sound corrosion-protection coating is surface preparation. Steel is covered with mill scale when it cools after rolling, and this must be removed before the protection is applied, otherwise the scale can subsequently loosen and break the film. Blast cleaning makes the best preparation prior to painting. Acid pickling is used in the galvanizing process.

Careful attention to design detail is also required (for example, upturned channels that form a cavity where water can collect should be avoided) and access for future maintenance should also be provided. For further information the reader should consult BS 5493: Code of Practice for Protective Coating of Iron and Steel Structures against Corrosion.

2.3 Steel sections

2.3.1 Rolled and formed sections

Hot-rolled sections are produced in steel mills from steel billets by passing them through a series of rolls. The main sections are shown on Figure 2.3 and their principal properties and uses are discussed briefly below:

(1) Universal beams. These are very efficient sections for resisting bending moment about the major axis.
(2) Universal columns. These are sections produced primarily to resist axial load with a high radius of gyration about the minor axis to prevent buckling in that plane.
(3) Channels. These are used for beams, bracing members, truss members and in compound members.
(4) Equal and unequal angles. These are used for bracing members, truss members and for purlins and sheeting rails.
(5) Structural tees. The sections shown are produced by cutting a universal beam or column into two parts. Tees are used for truss members, ties and light beams.
(6) Circular, square and rectangular hollow sections. These are produced from flat plate. The circular section is made first and then this is converted to the square or rectangular shape. These sections make very efficient compression members, and are used in a wide range of applications as members in lattice girders, in building frames, for purlins, sheeting rails, etc.

Note that the range in serial sizes is given for the members shown in Figure 2.3. A number of different members are produced in each serial size by varying the flange, web, leg or wall thicknesses. The sizes and properties of sections are given in the following publications:

BS 4: Structural Steel Sections
   Part 1: Hot Rolled Sections

BS 4848: *Hot Rolled Structural Steel Sections*
        Part 2: Hollow Sections
        Part 4: Equal and Unequal Angles
    Steelwork Design: Guide to BS 5950: Part 1: 1985, Volume 1. Section Properties, Member Capacities, Constrado, 1985.

2.3.2 Compound sections

Compound section are formed by the following means (Figure 2.4):

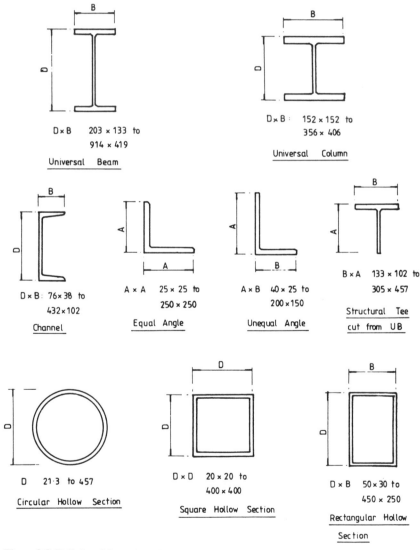

Figure 2.3 Rolled and formed sections

(1) Strengthening a rolled section such as a universal beam by welding on cover plates, as shown in Figure 2.4(a);

(2) Combining two separate rolled sections, as in the case of the crane girder in Figure 2.4(b). The two members carry loads from separate directions.

(3) Connecting two members together to form a strong combined member. Examples are the laced and battened members shown in Figures 2.4(c) and (d).

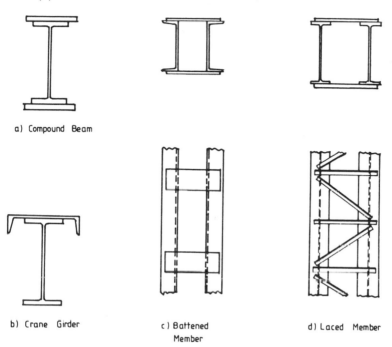

a) Compound Beam

b) Crane Girder   c) Battened   d) Laced Member
                                        Member

Figure 2.4 Compound sections

2.3.3 Built-up sections

Built-up sections are made by welding plates together to form I, H or box members which are termed plate girders, built-up columns box girders or columns, respectively. These members are used where heavy loads have to be carried and in the case of plate and box girders where long spans may be required. Examples of built-up sections are shown in Figure 2.5.

2.3.4 Cold-rolled sections

Thin steel plates can be formed into a wide range of sections by cold rolling. The most important uses for cold-rolled sections in steel structures are for purlins and sheeting rails. Three common sections—the zed, sigma and lipped channel—are shown in Figure 2.6. Reference should be made to the manufacturer's literature for the full range of sizes available and the section properties.[5] Some members and their properties are given in Sections 5.12.6 and 5.13.5 in design of purlins and sheeting rails.

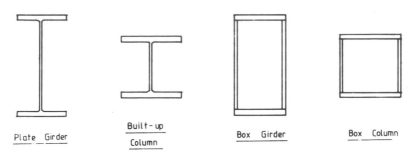

Figure 2.5 Built-up sections

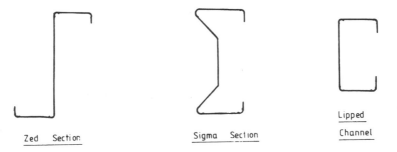

Figure 2.6 Cold-rolled sections

2.4 Section properties

For a given member serial size, the section properties are:

(1) The exact section dimensions;
(2) The location of the centroid if the section is asymmetrical about one or both axes;
(3) Area of cross section;
(4) Moments of inertia about various axes;
(5) Radii of gyration about various axes;
(6) Moduli of section for various axes, both elastic and plastic.

The section properties for hot rolled and formed sections are listed in *Steelwork Design*, Guide to BS 5950: Part 1: 1985, Volume 1, Section Properties, Member Capacities, Constrado, 1985.

For compound and built-up sections, the properties must be calculated from first principles. The section properties for the symmetrical I section dimensioned as shown in Figure 2.7(a) are as follows:

(1) Elastic properties:

$$A = 2\,BT + \mathrm{d}t$$

Moment of inertia XX axis $I_x = BD^3/12 - (B-t)\mathrm{d}^3/12$

Moment of inertia YY axis $I_y = 2\,TB^3/12 + \mathrm{d}t^3/12$

Radius of gyration XX axis $r_x = (I_x/A)^{0.5}$

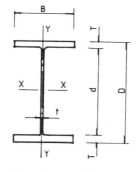

a) Symmetrical   I-Section

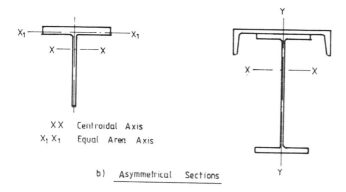

X X   Centroidal Axis
$X_1 X_1$ Equal Area Axis

b)   Asymmetrical   Sections

Figure 2.7 Beam sections

Radius of gyration YY axis $r_y = (Iy/A)^{0.5}$
Modulus of section XX axis $Z_x = 2 I_x/D$
Modulus of section YY axis $Z_y = 2 I_y/B$

(2) Plastic moduli of section:
The plastic modulus of section is equal to the algebraic sum of the first moments of area about the equal area axis. For the I section shown:

$$S_x = 2 B T(D-T)/2 + td^2/4$$
$$S_y = 2 T B^2/4 + dt^2/4$$

For asymmetrical sections such as those shown in Figure 2.7(b) the neutral axis must be located first. In elastic analysis the neutral axis is the centroidal axis while in plastic analysis it is the equal area axis. The other properties may then be calculated using procedures from strength of materials.[6] Calculations of properties for unsymmetrical sections are given in various parts of this book.

Other properties of universal beams, columns, joists and channels, used for determining the buckling resistance moment are:

Buckling parameter (u)
Torsional index (x)
Warping constant (H)
Torsional constant (J)

These properties may be calculated from formulae given in Appendix B of BS 5950: Part 1 or taken from tables in the *Guide* to BS 5950: Part 1, Volume 1, Constrado. (See Section 5.5.)

3

Limit state design

3.1 Limit state design principles

The central concepts of limit state design are that:

(1) All the separate conditions that make the structure unfit for use are taken into account. These are the separate limit states.
(2) The design is based on the actual behaviour of materials and performance of structures and members in service.
(3) Ideally, design should be based on statistical methods with a small probability of the structure reaching a limit state.

The three concepts are examined in more detail below.

Requirement (1) means that the structure should not overturn under applied loads and its members and joints should be strong enough to carry the forces to which they are subjected. In addition, other conditions such as excessive deflection of beams or unacceptable vibration, though not in fact causing collapse, should not make the structure unfit for use.

In concept (2) the strengths are calculated using plastic theory and post-buckling behaviour is taken into account. The effect of imperfections on design strength is also included. It is recognized that calculations cannot be made in all cases to ensure that limit states are not reached. In cases such as brittle fracture, good practice must be followed to ensure that damage or failure does not occur.

Concept (3) implies recognition of the fact that loads and material strengths vary, approximations are used in design and imperfections in fabrication and erection affect the strength in service. All these factors can only be realistically assessed in statistical terms. However, it is not yet possible to adopt a complete probability basis for design, and the method adopted is to ensure safety by using suitable factors. Partial factors of safety are introduced to take account of all the uncertainties in loads, materials strengths, etc. mentioned above. These are discussed more fully below.

3.2 Limit states for steel design

The limit states for which steelwork is to be designed are set out in Section 2 of BS 5950: Part 1. These are as follows.

3.2.1 Ultimate limit states

(1) Strength (including general yielding, rupture, buckling and transformation into a mechanism);
(2) Stability against overturning and sway;
(3) Fracture due to fatigue;
(4) Brittle fracture.

When the ultimate limit states are reached, the whole structure or part of it collapses.

3.2.2 Serviceability limit states

(5) Deflection;
(6) Vibration (for example, wind-induced oscillation);
(7) Repairable damage due to fatigue;
(8) Corrosion and durability.

The serviceability limit states, when reached, make the structure or part of it unfit for normal use but do not indicate that collapse has occurred.

All relevant limit states should be considered, but usually it will be appropriate to design on the basis of strength and stability at ultimate loading and then check that deflection is not excessive under serviceability loading. Some recommendations regarding the other limit states will be noted when appropriate, but detailed treatment of these topics is outside the scope of this book.

3.3 Working and factored loads

3.3.1 Working loads

The working loads (also known as the specified, characteristic or nominal loads) are the actual loads the structure is designed to carry. These are normally thought of as the maximum loads which will not be exceeded during the life of the structure. In statistical terms, characteristic loads have a 95 per cent probability of not being exceeded. The main loads on buildings may be classified as:

(1) Dead loads. These are due to the weights of floor slabs, roofs, walls, ceilings, partitions, finishes, services and self weight of steel. When sizes are known, dead loads can be calculated from weights of materials or from the manufacturer's literature. However, at the start of a design, sizes are not known accurately and dead loads must often be estimated from experience. The values used should be checked when the final design is complete. For examples on element design, representative loading has been chosen, but for the building design examples actual loads from BS 6399: Part 1 are used.
(2) Imposed loads. These take account of the loads caused by people, furniture, equipment, stock, etc. on the floors of buildings and snow on roofs. The values of the floor loads used depend on the use of the building. Imposed loads are given in BS 6399: Part 1 for various type of buildings.
(3) Wind loads. These loads depend on the location and building size. Wind loads are given in CP3: Chapter V: Part 2. Calculation of wind loads is given in the examples on building design.

(4) Dynamic loads. These are caused mainly by cranes. An allowance is made for impact by increasing the static vertical loads and the inertia effects are taken into account by applying a proportion of the vertical loads as horizontal loads. Dynamic loads from cranes are given in BS 6399: Part 1. Design examples show how these loads are calculated and applied to crane girders and columns.

Other loads on the structures are caused by waves, ice, seismic effects, etc., and these are outside the scope of this book.

3.3.2 Factored loads for the ultimate limit states

In accordance with Section 2.4.1 of BS 5950: Part 1, factored loads are used in design calculations for strength and stability.

Factored load = working or nominal load × relevant overall load factor γ_f

The overall load factor takes account of:

(1) The unfavourable deviation of loads from their nominal values; and
(2) The reduced probability that various loads will all be at their nominal value simultaneously.

It also allows for the uncertainties in the behaviour of materials and of the structure as opposed to those assumed in design.

The γ_f factors are given in Table 2 of BS 5950: Part 1 and some of the factors are given in Table 3.1.

Table 3.1   Overall load factors γ_f

Loading	Factors γ_f
Dead load	1.4
Dead load restraining uplift or overturning	1.0
Dead load, wind load and imposed load	1.2
Imposed load	1.6
Wind load	1.4
Crane loads	
Vertical load	1.6
Vertical and horizontal load	1.4
Horizontal load	1.6
Crane loads and wind load	1.2

Clause 2.4.1.1 of BS 5950 states that the factored loads should be applied in the most unfavourable manner and members and connections should not fail under these load conditions. Brief comments are given on some of the load combinations:

(1) The main load for design of most members and structures is dead plus imposed load.
(2) In light roof structures uplift and load reversal occurs and tall structures must be checked for overturning. The load combination of dead plus wind load is used in these cases with a load factor of 1.0 for dead and 1.4 for wind load.
(3) It is improbable that wind and imposed loads will simultaneously reach their maximum values and load factors are reduced accordingly.

(4) It is also unlikely that the impact and surge load from cranes will reach maximum values together and so the load factors are reduced. Again, when wind is considered with crane loads the factors are further reduced.

3.4 Stability limit states

To ensure stability, Clause 2.4.2 of BS 5950 states that structures must be checked using factored loads for the following two conditions:

(1) Overturning. The structure must not overturn or lift off its seat.
(2) Sway. To ensure adequate resistance two design checks are required:
 (a) Design to resist the applied horizontal loads.
 (b) A separate design for notional horizontal loads. These are to be taken as the greater of 1 per cent of the factored dead loads and 0.5 per cent of the factored dead plus imposed load, and are to be applied at the roof and each floor level. They are to act with 1.4 times the dead and 1.3 times the imposed load.

Sway resistance may be provided by bracing rigid-construction shear walls, stair wells or lift shafts. The designer should clearly indicate the system he is using. In examples in this book, stability against sway will be ensured by bracing and rigid portal action.

3.5 Structural integrity

The provisions of Section 2.4.5 of BS 5950 ensure that the structure complies with the Building Regulations and has the ability to resist progressive collapse following accidental damage. The main parts of the clause are summarized below:

(1) All structures must be effectively tied at all floors and roofs. Columns must be anchored in two directions approximately at right angles. The ties may be steel beams or reinforcement in slabs. End connections must be able to resist a factored tensile load of 75 kN for floors and 40 kN for roofs. Ties are not required in light roofs.
(2) Additional requirements are set out for certain multi-storey buildings where the extent of accidental damage must be limited. In general, tied buildings will be satisfactory if the following five conditions are met:
 (a) Sway resistance is distributed throughout the building;
 (b) Extra tying is to be provided as specified;
 (c) Column splices are designed to resist a specified tensile force;
 (d) Any beam carrying a column is checked as set out in (3) below; and
 (e) Precast floor units are tied and anchored.
(3) Where required in (2) the above damage must be localized by checking to see if at any storey any single column or beam carrying a column may be removed without causing more than a limited amount of damage. If the removal of a member causes more than the permissible limit it must be designed as a key element. These critical members are designed for accidental loads set out in the Building Regulations.

The complete section in the code and the Building Regulations should be consulted.

3.6 Serviceability limit states—deflection

Deflection is the main serviceability limit state that must be considered in design. The limit state of vibration is outside the scope of this book and fatigue was briefly discussed in Section 2.2.1 and, again, is not covered in detail. The protection for steel to prevent the limit state of corrosion being reached was mentioned in Section 2.2.4.

BS 5950: Part 1 states in Clause 2.5.1 that deflection under serviceability loads of a building or part should not impair the strength or efficiency of the structure or its components or cause damage to the finishings. The serviceability loads used are the unfactored imposed loads except in the following cases:

(1) Dead + imposed + wind. Apply 80 per cent of the imposed and wind load.
(2) Crane surge + wind. The greater effect of either only is considered.

The structure is considered to be elastic and the most adverse combination of loads is assumed. Deflection limitations are given in Table 5 of BS 5950: Part 1. These are given here in Table 3.2. These limitations cover beams and structures other than pitched-roof portal frames.

It should be noted that calculated deflections are seldom realized in the finished structure. The deflection is based on the beam or frame steel section only and composite action with slabs or sheeting is ignored. Again, the full value of the imposed load used in the calculations is rarely achieved in practice.

Table 3.2 Deflection limits

Deflection of beams due to unfactored imposed loads

Cantilevers	Length/180
Beams carrying plaster	Span/360
All other beams	Span/200

Horizontal deflection of columns due to unfactored imposed and wind loads

Tops of columns in single-storey buildings	Height/300
In each storey of a building with more than one storey	Story height/300

Crane gantry girders	
Vertical deflection due to static wheel loads	Span/600
Horizontal deflection (calculated on top flange properties alone) due to crane surge	Span/500

3.7 Design strength of materials

The design strengths for steel complying with BS 4360 are given in Section 3.1.1 of BS 5950: Part 1. Note that the material strength factor γ_m, part of the overall safety factor in limit state design, is taken as 1.0 in the code. The design strength may be taken as

$$p_y = 1.0 \ Y_s \text{ but not greater than } 0.84 \ U_s$$

where Y_s = minimum yield strength

U_s = minimum ultimate tensile strength

For the common types of steel values of p_y are given in Table 6 of the code and reproduced in Table 3.3.

Table 3.3 Design strengths p_y for steel to BS 4360

BS 4360 Grade	Thickness (mm) less than or equal to	Sections, plates and hollow sections p_y (N/mm^2)
43	16	275
A, B, C	40	265
	100	245
50	16	355
B, C	63	340
	100	325
55	16	450
C	25	430
	40	425

The code states that the following values for the elastic properties are to be used:

Modulus of elasticity $E = 205 \ kN/mm^2$

Shear modulus $G = 79 \ kN/mm^2$

Poisson's ratio $v = 0.3$

Coefficient of linear thermal expansion $\alpha = 12 \times 10^{-6}$ per °C

3.8 Design methods for buildings

The design of buildings must be carried out in accordance with one of the methods given in Clause 2.1.2 of BS 5950. The design methods are:

(1) *Simple design.* In this method the connections between members are assumed not to develop moments adversely affecting either the members or structure as a whole. The structure is assumed to be pin jointed for analysis. Bracing or shear walls are necessary to provide resistance to horizontal loading.
(2) *Rigid design.* The connections are assumed to be capable of developing the strength and/or stiffness required by an analysis assuming full continuity. The analysis may be made using either elastic or plastic methods.
(3) *Semi-rigid design.* Practical joints are capable of transmitting some moment and the method takes this partial fixity into account. Two approaches to semi-rigid design are given:
    (a) Experimental determination of joint behaviour. Some limited plasticity is permitted, but the ultimate tensile capacity of the fastener is not to be the failure criterion;

(b) Simplified empirical method. The inter-restraint of the beams and columns is taken as not more than 10 per cent of the free moments in the beams. The code sets out other provisions which must be met when using the method.

(4) *Experimental verification.* The code states that where the design of a structure or element by calculation in accordance with any of the above methods is not practicable, the strength and stiffness may be confirmed by loading tests. The test procedure is set out in Section 7 of the code.

In practice, structures are designed to either the simple or rigid methods of design. Semi-rigid design has never found general favour with designers. Examples in this book are generally of the simple method of design.

Connections

4.1 Types of connections

Connections are needed to join:

(1) Members together in trusses and lattice girders;
(2) Plates together to form built-up members;
(3) Beams to beams, beams, trusses, bracing, etc. to columns in structural frames; and
(4) Columns to foundations.

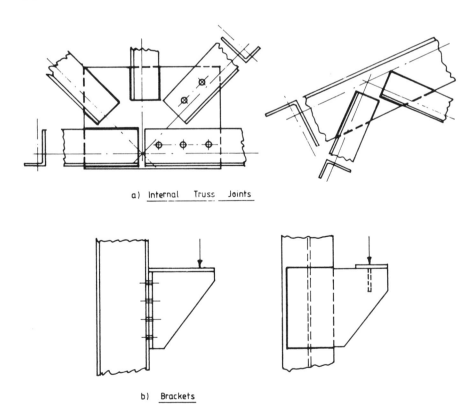

a) Internal Truss Joints

b) Brackets

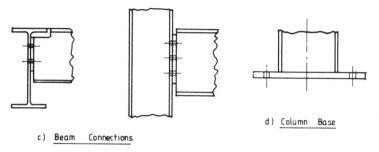

c) Beam   Connections

d) Column   Base

Figure 4.1 Typical connections

Some typical connections are shown on Figure 4.1. Basic connections are considered in this chapter and end connections for beams and column bases are treated in the chapters on beams and columns, respectively.

Connection are made by:

Bolting —ordinary bolts in clearance holes;
          —friction-grip bolts;
Welding—fillet and butt welds.

4.2 Ordinary bolts

4.2.1 Bolts, nuts and washers

The 'black' hexagon head bolt shown in Figure 4.2 with nut and washer is the most commonly used structural fastener. The bolts are in two strength grades as specified in BS 4190:

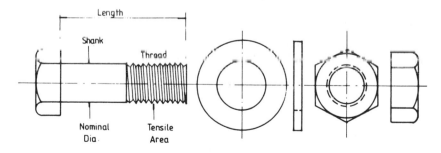

Figure 4.2 Hexagon head bolt, nut and washer

Grade 4.6: Mild steel   : Yield stress = 235 N/mm^2
Grade 8.8: High-strength steel : Yield stress = 627 N/mm^2

The main diameters used are:

16, 20, (22), 24, (27) and 30 mm

The sizes shown in brackets are not preferred.

4.2.2 Direct shear joints

Bolts may be arranged to act in single or double shear, as shown in Figure 4.3. Provisions governing spacing, edge and end distances are set out in Section 6.2 of BS 5950, Part 1. The principal provisions in normal conditions are:

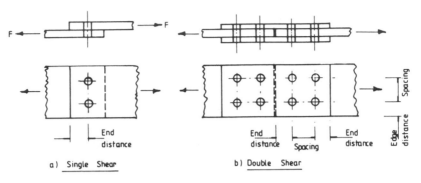

Figure 4.3 Bolts in single and double shear

(1) The minimum spacing is 2.5 times the bolt diameter;
(2) The maximum spacing in unstiffened plates in the direction of stress is $14t$, where t is the thickness of the thinner plate connected.
(3) The minimum edge and end distance as shown in Figure 4.3 from a rolled, machine-flame cut or planed edge is $1.25D$, where D is the hole diameter.
(4) The maximum edge distance is $11t\varepsilon$, where $\varepsilon = (275/p_y)^{0.5}$

The clause should be consulted for requirements in corrosive areas.

Note that the nominal diameters of holes for ordinary bolts are greater than the bolt diameter by:

2 mm for bolts up to 22 mm diameter
3 mm for larger-diameter bolts.

A shear joint can fail in the following four ways:

(1) By shear on the bolt shank;
(2) By bearing on the member or bolt;
(3) By tension in the member;
(4) By shear at the end of the member.

The failure modes are shown in Figure 4.4.

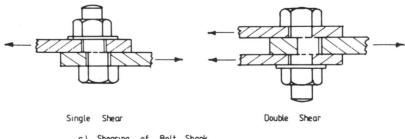

Single Shear   Double Shear

a) Shearing of Bolt Shank

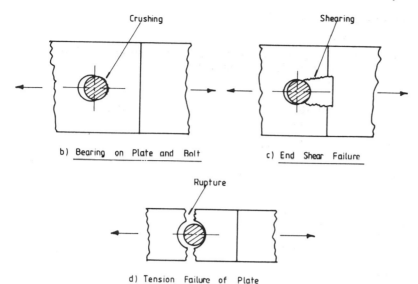

b) Bearing on Plate and Bolt   c) End Shear Failure

d) Tension Failure of Plate

Figure 4.4 Failure modes of a bolted joint

The failures noted above are prevented by taking the following measures:

(1) For modes 1 and 2, provide sufficient bolts of suitable diameter.
(2) For mode 3, design tension members for the effective area. (See Chapter 7.)
(3) Provide sufficient end distance for mode 4.

The design of bolted joints is set out in Section 6.3 of BS 5950: Part 1. The provisions in this section are:

(1) Effective area resisting shear A_S. Where threads occur in the shear plane:

$$A_S = A_T$$

where A_t = tensile stress area

Where threads do not occur in the shear plane:

$$A_S = A$$

where A = shank area based on the nominal diameter.

(2) Shear capacity P_S of a bolt:

$$P_S = p_S A_S$$

where p_s = shear strength given in Table 32 in the code. (See Table 4.1). (Note that long joints are not discussed. Refer to the code.)

(3) Bearing capacity should be taken as lesser of:

Capacity of the bolt $P_{bb} = dt p_{bb}$

where d = nominal diameter of the bolt

t = thickness of the connected play

p_{bb} = bearing strength of the bolt given in Table 32 in the code. (See Table 4.1.)

Capacity of the connected plate:

$$P_{bs} = dtp_{bs} \leqslant 1/2\ etp_{bs}$$

where p_{bs} = bearing strength of the connected parts given in Table 33 in the code. (See to Table 4.1.)

e = end distance

The second part of the bearing check ensures that the plate does not fail by end shear as shown in Figure 4.4(c).

Table 4.1 Ordinary bolts in clearance holes

Strength of bolts and bearing strength of bolts and connected ply (N/mm^2)

Strength of bolts	Bolt grade		
	4.6	8.8	
Shear strength p_s	160	375	
Bearing strength p_{bb}	435	970	
Bearing strength of	Steel grade		
connected parts	43	50	55
Bearing strength p_{bs}	460	550	650

Load-capacity tables can be made up for use in design and Table 4.2 is such a design aid.

With regard to this table it should be noted that the minimum end distance to ensure that the bearing capacity of the connected ply is controlled by the bearing on the plate is given by equating:

$$P_{bs} = dtp_{bs} = 0.5\ e\ t\ p_{bs}$$
End distance $e = 2d$

4.2.3 Direct tension joints

(1) Tension capacity of bolts

Joints with bolts in direct tension are shown in Figure 4.5(a). The tension capacity of an ordinary bolt is given in Clause 6.3.6 of **BS 5950**: Part 1. This states that:

Capacity $P_t = p_t\ A_t$
    where p_t = tension strength from Table 32 in the code
        = 195 N/mm^2 for Grade 4.6 bolts
        = 450 N/mm^2 for Grade 8.8 bolts

    A_t = tensile stress area.

Values for A_t are given in *Steelwork Design*, Guide to BS 5950: Part 1: 1985. Volume 1, Section Properties, Member Capacities, Constrado, 1985. Tension values for Grade 4.6 bolts are given in Table 4.2.

(2) Prying Forces

Prying forces are set up in tee connections with bolts in tension, as shown in Figure 4.5(b). This situation occurs in many standard joints in structures such as in bracket and moment connections to columns.

Table 4.2 Load capacity (Grade 4.6 bolts and Grade of 43 steel)

Nominal diameter (mm)	Shank area (mm²)	Tensile stress Area (mm²)	Tension Capacity KN	Shear capacity Single shear		Bearing capacity for bolts (kN) for plate thickness (mm)				Bearing capacity for connected part (kN): for plate thickness (mm)			
				Shank (kN)	Threads (kN)	6	8	10	12	6	8	10	12
16	201	157	30.6	32.1	25.1	41.7	55.6	69.6	83.5	44.1	58.8	73.6	88.3
20	314	245	47.7	50.2	39.2	52.2	69.6	87.0	104.4	55.2	73.6	92.0	110.4
22	380	303	59.1	60.8	48.4	57.4	76.5	95.7	114.8	60.7	80.9	101.2	121.4
24	452	353	68.8	72.3	56.4	62.6	83.5	104.4	125.2	66.2	88.3	110.4	132.4

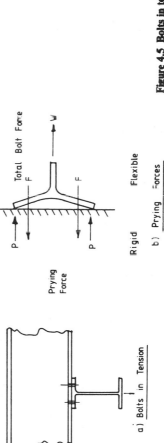

a) Bolts in Tension

Prying Force

Total Bolt Force

F

F

P

P

Rigid   Flexible

b) Prying Forces

Figure 4.5 Bolts in tension

The prying force P adds directly to the tension in the bolt. Referring to the figure,

Total bolt tension $F = W/2 + P$

where W = external tension on the joint

The magnitude of the prying force depends on the stiffnesses of the bolt and the flanges. Theoretical analyses based on elastic and plastic theory are available to determine values of prying forces. If the flanges are relatively thick, the bolt spacing not excessive and the edge distance sufficiently large, the prying forces are small and may be neglected. The reader should consult reference 7 for further particulars.

BS 5950: Part 1 states in Clause 6.3.6.2 that in connections subject to tension prying action need not be taken into account provided the stresses given in Table 32 and quoted in Section 4.2.3(1) above are used.

4.2.4 Eccentric connections

There are two principal types of eccentrically loaded connections:

(1) Bolt group in direct shear and torsion; and
(2) Bolt group in direct shear and tension.

These connections are shown in Figure 4.6.

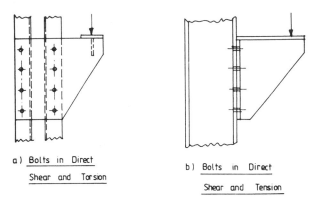

a) Bolts in Direct
    Shear and Torsion

b) Bolts in Direct
    Shear and Tension

Figure 4.6 Eccentrically loaded connections

4.2.5 Bolts in direct shear and torsion

In the connection shown in Figure 4.6(a) the moment is applied in the plane of the connection and the bolt group rotates about its centre of gravity. A linear variation of loading due to moment is assumed, with the bolt farthest from the centre of gravity of the group carrying the greatest load. The direct shear is divided equally between the bolts and the side plates are assumed to be rigid.

Consider the group of bolts shown in Figure 4.7(a), where the load P is applied at an eccentricity e. The bolts A, B, etc. are at distances r_1, r_2, etc. from the centroid of the group. The coordinates of each bolt are (x_1, y_1), (x_2, y_2), etc. Let the force due to the moment on bolt A be F_T. This is the force on the bolt farthest from the centre of rotation. Then the force on a bolt r_2 from the

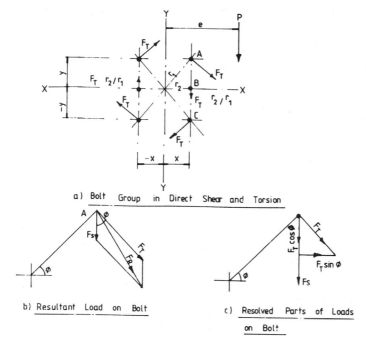

a) Bolt Group in Direct Shear and Torsion

b) Resultant Load on Bolt

c) Resolved Parts of Loads on Bolt

Figure 4.7 Bolt group in direct shear and torsion

centre of rotation is $F_T r_2/r_1$ and so on for all the other bolts in the group. The moment of resistance of the bolt group is given by Figure 4.7:

$$M_R = F_T.r_1 + F_T.r_2.r_1/r_1 + \ldots$$

$$= \frac{F_T}{r_1}. (r_1{}^2 + r_2{}^2 + \ldots)$$

$$= \frac{F_T}{r_1}. \Sigma r^2$$

$$= \frac{F_T}{r_1}. (\Sigma x^2 + \Sigma y^2)$$

$$= \text{applied moment} = P.e$$

The load F_T due to moment on the maximum loaded bolt A is given by

$$F_T = \frac{P.e.r_1}{\Sigma x^2 + \Sigma y^2}$$

The load F_s due to direct shear is given by

$$F_s = \frac{P}{\text{No. of bolts}}$$

The resultant load F_R on bolt A can be found graphically, as shown in

Figure 4.7(b). The algebraic formula can be derived by referring to Figure 4.7(c).

Resolve the load F_T vertically and horizontally to give

Vertical load on bolt $A = F_S + F_T \cos \varphi$

Horizontal load on bolt $A = F_T \sin \varphi$

Resultant load on bolt A

$$F_R = [(F_T \sin \varphi)^2) + (F_S + F_T \cos \varphi)^2]^{0.5}$$
$$= [F_S^2 + F_T^2 + 2F_S F_T \cos \varphi]^{0.5}$$

The size of bolt required can then be determined from the maximum load on the bolt.

4.2.6 Bolts in direct shear and tension

In the bracket-type connection shown in Figure 4.6(b) the bolts are in combined shear and tension. BS 5950: Part 1 gives the design procedure for these bolts in Clause 6.3.6.3. This is:

The factored applied shear F_s must not exceed the shear capacity P_S, where $P_S = p_S A_S$. The bearing capacity checks must also be satisfactory. The factored applied tension F_T must not exceed the tension capacity P_T, where $P_T = p_t A_T$.

In addition to the above the following relationship must be satisfied:

$$\frac{F_S}{P_S} + \frac{F_T}{P_T} \leqslant 1.4$$

The interaction diagram for this expression is shown on Figure 4.8. The shear F_S and tension F_T are found from an analysis of the joint.

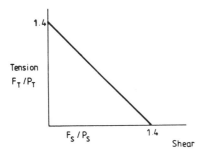

Figure 4.8 Interaction diagram: bolts in shear and tension

An approximate method of analysis that gives conservative results is described first. A bracket subjected to a factored load P at an eccentricity e is shown in Figure 4.9(a). The centre of rotation is assumed to be at the bottom bolt in the group. The loads vary linearly as shown on the figure, with maximum load F_T in the top bolt.

The moment of resistance of the bolt group is:

$$M_R = 2 [F_T \cdot y_1 + F_T \cdot y_2^2/y_1 + \ldots]$$
$$= 2 F_T/y_1 \cdot [y_1^2 + y_2^2 + \ldots]$$

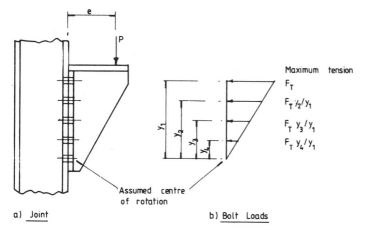

a) Joint

b) Bolt Loads

Figure 4.9 Bolts in direct shear and tension: approximate analysis

$$= \frac{2 F_T}{y_1} . \Sigma y^2$$
$$= P . e$$

The maximum bolt tension is:

$$F_T = P.e\, y_1 \,/\, 2\, \Sigma\, y^2$$

The vertical shear per bolt:

$$F_S = P/\text{No. of bolts}$$

A bolt size is assumed and checked for combined shear and tension as described above.

In a more accurate method of analysis the applied moment is assumed to be resisted by the bolts in tension with uniformly varying loads and an area at the bottom of the bracket in compression, as shown in Figure 4.10(b).

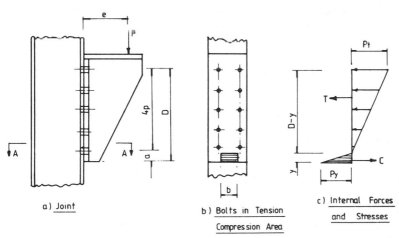

a) Joint

b) Bolts in Tension
Compression Area

c) Internal Forces
and Stresses

Figure 4.10 (contd. overleaf)

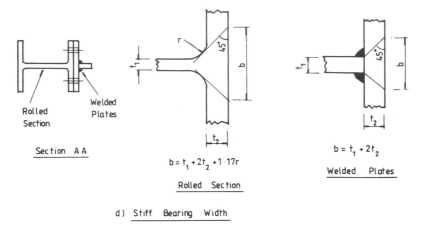

d) Stiff Bearing Width

Figure 4.10 Bolts in direct shear and tension: accurate analysis

For equilibrium, the total tension T must equal the total compression C. Consider the case where the top bolt is at maximum capacity P_t and the bearing stress is at its maximum value p_y. Referring to Figure 4.12(c), the total tension is given by:

$$T = P_t + \frac{(D - y - p)}{(D - y)} \cdot P_t + \frac{(D - y - 2p)}{(D - y)} \cdot P_t + \ldots$$

$P_t = p_t \cdot A_t$ (See Section 4.2.3 (1) above)

The total compression is given by:

$$C = 0.5 \, p_y \cdot b \cdot y$$

where b, the stiff bearing width, is obtained by spreading the load at 45 degrees, as shown in Figure 4.10(d). In the case of the rolled section the 45-degree line is tangent to the fillet radius.

The expressions for T and C can be equated to give a quadratic equation which can be solved to give y, the location of the neutral axis. The moment of resistance is then obtained by taking moments of T and C about the neutral axis. This gives

$$M_R = P_t (D - y) + P_t \frac{(D - y - p)^2}{(D - y)} + \ldots + 2/3.C.y$$

The actual maximum bolt tension F_T is then found by proportion, as follows:

Applied moment $M = P \cdot e$

Actual bolt tension $F_T = M \cdot P_t / M_R$

The direct shear per bolt $F_s = P/$No. of bolts and the bolts are checked for combined tension and shear.

Note that to use the method a bolt size must be selected first and the joint set out and analysed to obtain the forces on the maximum loaded bolt. The bolt can then be checked.

4.2.7 Examples of ordinary bolted connections

Example (1)

The joint shown in Figure 4.11 is subjected to a tensile dead load of 85 kN and a tensile imposed load of 95 kN. All data regarding the member and joint are shown in the figure. The steel is Grade 43 and the bolt Grade 4.6. Check that the joint is satisfactory.

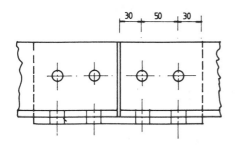
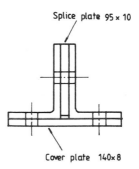

Splice plate 95 x 10

Cover plate 140x8

2 No. 100×65×8 $\rfloor\llcorner$

Bolts – 20mm dia. ordinary bolts

Holes – 22mm dia.

Figure 4.11 Double-angle splice

Using load factors from Table 2 of BS 5950, Part 1:

Factored load $= (1.4 \times 85) + (1.6 \times 95)$ $= 271$ kN

Strength of bolts. From Table 4.2 for 20-mm diameter bolts:

Single shear capacity on threads $= 39.2$ kN

Bearing capacity of bolts on 10-mm ply $= 87.0$ kN

Bearing capacity on 10-mm splice with 30-mm end distance

Bearing strength $p_{bs} = 460$ N/mm^2 from Table 4.1:

$P_{bs} = 1/2 \times 30 \times 10 \times 460/10^3$ $= 69$ kN

Bolt capacity—two bolts are in double shear and four in single

shear $= (4 \times 39.2) + (2 \times 39.2) + 69$ $= 304.2$ kN

Note that the capacity of the end bolt bearing on the 10 mm splice plate is controlled by the end distance (BS 5950: Part 1, Clause 6.3.3.3).

Strength of the angles. Gross area $= 12.7$ cm^2 per angle.

The angles are connected through both legs. Clause 4.6.3.3 of BS5950: Part 1 states that the net area defined in Clause 3.3.2 is to be used in design. The holes are 22-mm diameter:

Net area $= 2 (1270 - 2 \times 22 \times 8) = 1836$ mm^2

Design strength $p_y = 275$ N/mm (Table 6)

Capacity P_t $= 275 \times 1836/10^3 = 504.9$ kN

Splice plate and cover plate (see Clauses 3.3.3 of the code and Section 7.4.1 of this book):

Effective area $= 1.2\,[(95-22)10 + (140-44)8] = 1798\ mm^2$

$\quad\quad\quad\quad\quad\quad$ ≯ gross area $\quad\quad\quad\quad = 2070\ mm^2$

Capacity $P_t \quad\quad = 275 \times 1798 / 10^3 \quad\quad = 494.3\ kN$

The splice is adequate to resist the applied load.

Example (2)

Check that the joint shown in Figure 4.12 is adequate. All data required are given in the figure.

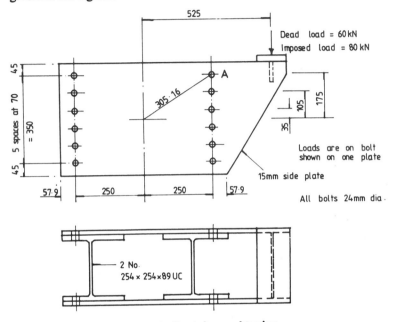

Figure 4.12 Example: bolt group in direct shear and torsion

Factored Load $= (1.4 \times 60) + (1.6 \times 80) \quad = 212\ kN$

Moment $M \quad\quad = (212 \times 525)/10^3 \quad\quad = 111.3\ kNm$

Bolt group $\Sigma\,x^2 = 12 \times 250^2 \quad\quad\quad\quad = 750 \times 10^3$

$\quad\quad\quad\quad \Sigma\,y^2 = 4(35^2 + 105^2 + 175^2) \quad = 171.5 \times 10^3$

$\quad\quad\quad\quad \Sigma x^2 + \Sigma y^2 \quad\quad\quad\quad\quad = 921.5 \times 10^3$

$\quad\quad\quad\quad Cos\ \varphi = 250/305.16 \quad\quad\quad = 0.819$

Bolt A is the bolt with the maximum load:

Load due to moment $= \dfrac{111.3 \times 10^3 \times 305.16}{921.5 \times 10^3} = 36.85\ kN$

Load due to shear $\quad = 212/12 \quad\quad\quad\quad = 17.67\ kN$

Resultant load on bolt
$\quad = [17.67^2 + 36.85^2 + (2 \times 17.67 \times 36.85 \times 0.819)]^{0.5}$
$\quad = 52.31\ kN$

Single-shear value of 24 mm diameter ordinary bolt on the threads

From Table 4.2	$= 56.4\,\text{kN}$
Universal column flange thickness	$= 17.3\,\text{mm}$
Side-plate thickness	$= 15\,\text{mm}$
Minimum end distance	$= 45\,\text{mm}$

Bearing capacity of the bolt

$$P_{bb} = 24 \times 15 \times 435/10^3 \qquad\qquad = 157\,\text{kN}$$

Bearing capacity of the plate

$$P_{bs} = 24 \times 15 \times 460/10^3 \qquad\qquad = 166\,\text{kN}$$
$$\leqslant 1/2 \times 45 \times 15 \times 460/10^3 \qquad = 155\,\text{kN}$$

The strength of the joint is controlled by the single shear value of the bolt. The joint is satisfactory.

Example (3)

Determine the diameter of ordinary bolt required for the bracket shown in Figure 4.13. The joint dimensions and loads are shown in the figure. Use Grade 4.6 bolts.

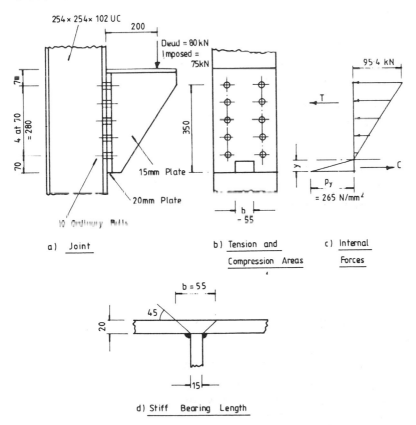

Figure 4.13 Bracket: bolts in direct shear and tension

Try 20-mm diameter ordinary bolts. From Table 4.2,

Tension capacity $P_t = 47.7\,\text{kN}$

The design strength from Table 6 of the code for plates:

20 mm thick $p_y = 265\,\text{N/mm}^2$

Referring to Figure 4.13(c), where the depth to the neutral axis is y, the total tension is:

$$T = 95.4 \left[1 + \frac{(280-y)}{(350-y)} + \frac{(210-y)}{(350-y)} + \frac{(140-y)}{(350-y)} + \frac{(70-y)}{(350-y)} \right]$$
$$= 95.4\,(1050-5y)/(350-y)\,\text{kN}\cdot$$

The width of stiff bearing is shown in Figure 4.13(d):

$b = 15 + 2 \times 20$ $= 55\,\text{mm}$

The total compression in terms of y is given by:

$C = 1/2 \times 265 \times 55\,y/10^3$ $= 7.29\,y\text{kN}$

Equate T and C and rearrange terms to give the quadratic equation:

$7.29\,y^2 - 3028\,y + 100\,170$ $= 0$

Solve to give $y = 36.25\,\text{mm}$

The moment of resistance is:

$$M_R = \frac{95.4}{10^3} \left[(350-36.25) + \frac{(280-36.25)^2}{(350-36.25)} + \frac{(210-36.25)^2}{(350-36.25)} \right.$$
$$\left. + \frac{(140-36.25)^2}{(350-36.25)} + \frac{(70-36.25)^2}{(350-36.25)} \right] + \frac{7.29 \times 36.25^2 \times 2}{3 \times 10^3}$$
$$= 67.17\,\text{kNm}$$

Factored load $= (1.4 \times 80) + (1.6 \times 75) = 232\,\text{kN}$

Factored moment $= 232 \times 200/10^3$ $= 46.4\,\text{kNm}$

Actual tension in top bolts

$F_T = 46.4 \times 47.7/67.17$ $= 32.95\,\text{kN}$

Direct shear per bolt $= 232/10$ $= 23.2\,\text{kN}$

Shear capacity on threads $P_S = 39.2\,\text{kN}$

Tension capacity $P_T = 47.7\,\text{kN}$

Combined shear and tension:

$$\frac{F_s}{P_s} + \frac{F_r}{P_t} = \frac{23.2}{39.2} + \frac{32.95}{47.7} = 1.28 < 1.4$$

Therefore 20-mm diameter bolts are satisfactory.

The reader can redesign the bolts using the approximate method. Note that only the bolts have been designed. The welds and bracket plates must be designed and the column checked for the bracket forces. These considerations are dealt with in Section 4.5.

4.3 Friction-grip bolts

4.3.1 General Considerations

Friction-grip bolts are made from high-strength steel so they can be tightened to give a high shank tension. The shear in the connected plates is transmitted by friction, as shown in Figure 4.14(a), and not by bolt shear, as in ordinary bolts. These bolts are used where strong joints are required, and a major use is in the joints in rigid frames.

The bolts are manufactured in three types to conform to BS 4395:

General grade. The strength is similar to Grade 8.8 ordinary bolts. This type is generally used.

Higher-grade: Parallel shank and waisted shank bolts are manufactured.

The use of general grade friction-grip bolts in structural steelwork is specified in BS 4604, Part 1. Parallel and waisted shank bolts are shown in Figure 4.14(b). Only general grade bolts will be discussed here.

The bolts must be used with hardened steel washers to prevent damage to the connected parts. The surfaces in contact must be free of mill scale, rust, paint, grease, etc. which would prevent solid contact between the surfaces and lower the slip factor (see below).

Care must be taken to ensure that bolts are tightened up to the required

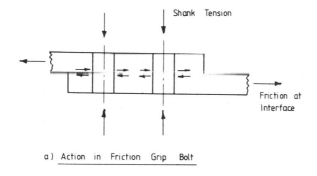

a) Action in Friction Grip Bolt

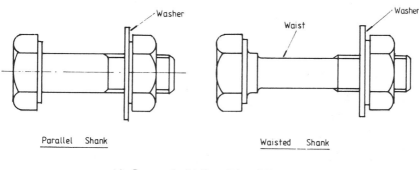

Parallel Shank   Waisted Shank

b) Types of Friction Grip Bolt

Figure 4.14 Friction-grip bolts

tension, otherwise slip will occur at service loads and the joint will act as an ordinary bolted joint. Methods used to achieve the correct shank tension are:

(1) Part-turning. The nut is tightened up and then forced a further half to three quarters of a turn, depending on the bolt length and diameter.
(2) Torque control. A power operated or hand-torque wrench is used to deliver a specified torque to the nut. Power wrenches must be calibrated at regular intervals.
(3) Load-indicating washers and bolts. These have projections which squash down as the bolt is tightened. A feeler gauge is used to measure when the gap has reached the required size.

Friction-grip bolts are generally used in clearance holes. The clearances are same as for ordinary bolts given in Section 4.2.2.

4.3.2. Design procedure

Friction-grip bolts can be used in shear, tension and combined shear and tension. The design procedure, given in Section 6.4 of BS 5950: Part 1 for general-grade parallel shank bolts, is set out below. Long joints are not discussed.

(1) Bolts in shear

The capacity is the lesser of the slip resistance and the bearing capacity. The code states that while the slip resistance is based on a serviceability criterion the design check is made for convenience, using factored loads. However, the joint can slip and go into bearing at loads greater than the working load, and hence the bearing capacity must be checked. The code further states that the shear capacity is automatically satisfied if the slip factor is not taken to be greater than 0.55.

The slip resistance is given by:

$$P_{SL} = 1.1 \, K_S \, \mu \, P_O$$

where P_O = minimum shank tension (i.e. the proof load given in Table 4.3);

K_S = 1.0 for bolts in clearance holes. It has lower values for bolts in oversized and slotted holes; and

μ = slip factor. For general-grade bolts and untreated surfaces $\mu = 0.45$.

The bearing resistance is given by:

$$P_{bg} = d \, t \, p_{bg}$$
$$\leqslant 1/3 \, e \, t \, p_{bg}$$

where d = nominal diameter of the bolt

t = thickness of the connected ply

e = end distance

p_{bg} = bearing strength of the parts connected.

The second part of the bearing check ensures that the joint does not fail by end shear in the plies as shown in Figure 4.4(c). The bearing strengths are given in Table 34 of BS 5950: Part 1. Values from the table are:

$$p_{bg} = 825 \text{ N/mm}^2 \text{ for steel Grade 43}$$
$$= 1065 \text{ N/mm}^2 \text{ for steel Grade 50}$$

The slip and bearing resistances for general grade bolts are given in Table 4.3. A minimum end distance of $e = 3d$ ensures that the bearing resistance is controlled by bearing on the plate. The factor K_s has been taken as 1.0 for bolts in clearance holes.

(2) Bolts in tension only

The tension capacity is given by

$$P_t = 0.9 \, P_O$$

The tension on capacities are given in Table 4.3.

(3) Bolts in combined shear and tension

The external capacity reduces the clamping action and slip resistance of the joint. Bolts in combined shear and tension must satisfy the following conditions:

(1) The transverse capacity in slip and bearing must be greater than the applied factored shear F_s.
(2) The tension capacity must be greater than the applied factored tension F_T.
(3) In addition, the interaction relationship must be satisfied:

$$\frac{F_s}{P_{3L}} + 0.8 \frac{F_t}{P_t} \leqq 1$$

This formula is shown graphically in Figure 4.15.

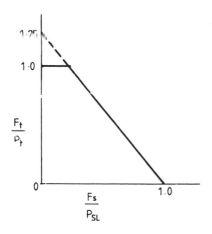

Figure 4.15 Friction-grip bolts in shear and tension

Table 4.3 Load capacities for friction bolts used in Grade 43 steel plate

Dia-meter (mm)	Minimum shank tension (kN)	Tension capacity (kN)	Slip resis-tance (kN)	Bearing capacity (kN) for plate thickness (mm)			
				6	8	10	12
16	92	82.8	45.5	79.2	105.6		
20	144	129.6	71.2	99.0	132.0	165.0	198.0
22	177	159.3	87.6	108.9	145.2	181.5	217.8
24	207	186.3	102.4	—	158.4	198.0	237.0

4.3.3 Examples of friction-grip bolt connections

Example (1)

Design the bolts for the moment and shear connection between the floor beam and column in a steel frame building. The following data are given:

Floor beam	610×229 UB 140
Column	254×254 UC 132
Moment	Dead load $= 180\,kNm$
	Imposed load $= 100\,kNm$
Shear	Dead load $= 300\,kN$
	Imposed load $= 160\,kN$

Set out the joint as shown in Figure 4.16

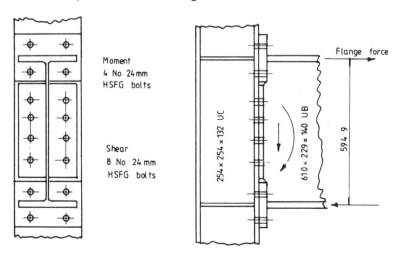

Figure 4.16 Beam-to-column connection

Moment Connection

The moment is taken by the flange bolts in tension:

Factored moment $= (1.4 \times 180) + (1.6 \times 100)$ $= 412\,kNm$
Flange force $= 412/0.594$ $= 693.6\,kN$

Provide 24-mm diameter friction-grip bolts:

Minimum shank tension $\qquad = 207\,kN$

Tension capacity of four bolts $= 4 \times 0.9 \times 207 = 745.2\,kN$

The joint is satisfactory for moment. Four bolts are also provided at the bottom of the joint but these are not loaded by the moment in the direction shown.

Shear Connection

The shear is resisted by the web bolts.

Factored shear $= (1.4 \times 300) + (1.6 \times 160) = 676\,kN$

Slip resistance of eight No. 24 mm diameter bolts in clearance
holes $= 8 \times 1.1 \times 0.45 \times 207 \qquad = 819.7\,kN$

If the end plate is 12 mm thick, the bearing resistance

$= 8 \times 825 \times 24 \times 12 \,/10^3 \qquad = 1901\,kN$

However, if the end distance of the two top bolts is 55 mm the bearing resistance must not exceed.

$= 6 \times 825 \times 24 \times 12/10^3 + 2 \times 1/3 \times 55 \times 12 \times$
$ 825/10^3 \qquad = 1788.6\,kN$

Therefore 24-mm diameter bolts are satisfactory.

Note that only the bolts have been designed. The welds, end plates and stiffeners must be designed and the column flange and web checked. These considerations are dealt with in Section 4.5.

Example (2)

Determine the bolt size required for the bracket loaded as shown in Figure 4.17(a).

Factored load $\quad = (1.4 \times 160) + (1.6 \times 110) = 400\,kN$

Factored moment $= 400 \times 0.28 \qquad = 112\,kNm$

Try 20-mm diameter friction-grip bolts:

Tension capacity from Table 4.3 $\qquad = 129.6\,kN$

The joint forces are shown in Figure 4.17(c). The stiff bearing width can be calculated from the bracket end plate (Figure 4.17(d)):

$b = 15 + 2 \times 20 \qquad = 55\,mm$

The total tension in terms of the maximum tension in the top bolts is

$$T = 259.2 \left[1 + \frac{370-y}{470-y} + \frac{270-y}{470-y} + \frac{170-y}{470-y} \right]$$

$$= 295.2\,(1280-4y)/(470-y)$$

The design strength from Table 6 in the code for plates 20 mm thick:

$P_y = 265\,N/mm^2$

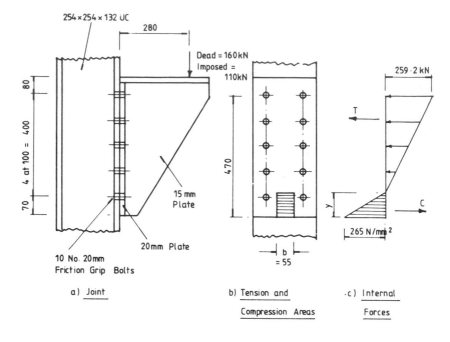

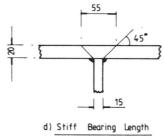

d) Stiff Bearing Length

Figure 4.17 Bracket: bolts in shear and tension

The total compression is given by:

$$C = 1/2 \times 265 \times 55\, y/10^3 \qquad\qquad = 7.29\, y$$

Equate T and C and rearrange to give:

$$y^2 - 631.9\, y + 51832 \qquad\qquad = 0$$

Solving gives $y = 96.87$ mm.

The moment of resistance is

$$M_R = \frac{259.2}{10^2}\left[(470-96.87) + \frac{(370-96.87)^2}{(470-96.87)} + \frac{(270-96.87)^2}{(470-96.87)} \right.$$
$$\left. + \frac{(170-96.87)^2}{(470-96.87)^2} \right] + \frac{7.29 \times 96.87^2 \times 2}{3 \times 10^3}$$
$$= 173.07 + 45.61 \qquad\qquad = 218.68 \text{ kNm}$$

The actual maximum tension in the top bolts is:

$$F_T = 112 \times 129.6/218.68 \qquad\qquad = 66.37 \, \text{kN}$$

The slip resistance from Table 4.3, $P_{SL} = 71.2 \, \text{kN}$

The bearing resistance is much greater than this value:

Applied shear per bolt $F_s = 400/10 \qquad = 40 \, \text{kN}$

Interaction criteria:

$$\frac{F_s}{P_{SL}} + \frac{0.8F_t}{P_t} = \frac{40}{71.2} - \frac{0.8 \times 66.37}{129.6} \qquad = 0.97 < 1.0$$

Therefore 20-mm diameter bolts are satisfactory

4.4 Welded connections

4.4.1 Welding

Welding is the process of joining metal parts by fusing them and filling in with molten metal from the electrode. The method is used extensively to join parts and members, attach cleats, stiffeners, end plates, etc. and to fabricate complete elements such as plate girders. Welding produces neat, strong and more efficient joints than are possible with bolting. However, it should be carried out under close supervision, and this is possible in the fabrication shop. Site joints are usually bolted. Though site welding can be done it is costly, and defects are more likely to occur.

Electric arc welding is the main system used, and the two main processes in structural steel welding are:

(1) Manual arc welding, using a hand-held electrode coated with a flux which melts and protects the molten metal. The weld quality depends very much on the skill of the welder.
(2) Automatic arc welding. A continuous wire electrode is fed to the weld pool. The wire may be coated with flux or the flux can be supplied from a hopper. In another process an inert gas is blown over the weld to give protection.

4.4.2 Types of welds, defects and testing

The two main types of welds, butt and fillet, are shown in Figures 4.18(a) and (b). Butt welds are named after the edge preparation used. Single and double U and V welds are shown in Figure 4.18(c). The double U welds require less weld metal than the V types. A 90-degree fillet weld is shown but other angles are used. The weld size is specified by the leg length. Some other types of welds—the partial butt, partial butt and fillet weld and deep penetration fillet weld—are shown in Figure 4.18(d). In the deep penetration fillet weld a higher current is used to fuse the plates beyond the limit of the weld metal.

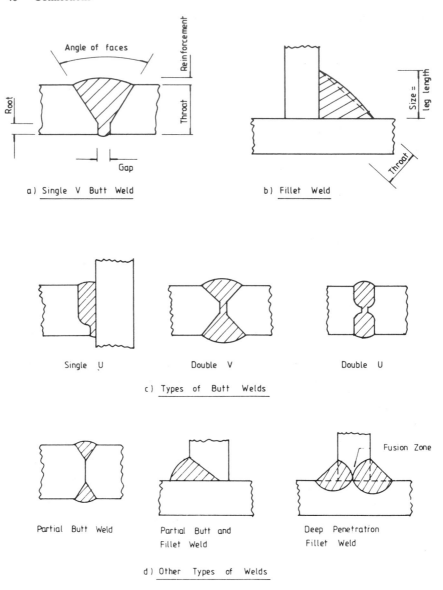

Figure 4.18 Weld types: fillet and butt welds

Cracks can occur in welds and adjacent parts of the members being joined. The main types are shown in Figure 4.19(a). Contraction on cooling causes cracking in the weld. Hydrogen absorption is the main cause of cracking in the heat-affected zone while lamellar tearing along a slag inclusion is the main problem in plates.

Faulty welding procedure can lead to the following defects in the welds, all of which reduce the strength (see Figure 4.19(b)):

(1) Over-reinforcement and undercutting;

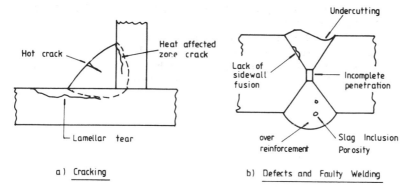

Figure 4.19 Cracks and defects in welds

(2) Incomplete penetration and lack of side-wall fusion;
(3) Slag inclusions and porosity

When the weld metal cools and solidifies it contracts and sets up residual stresses in members. It is not economic to relieve these stresses by heat treatment after fabrication, so allowance is made in design for residual stresses.

Welding also causes distortion, and special precautions have to be taken to ensure that fabricated members are square and free from twisting. Distortion effects can be minimized by good detailing and using correct welding procedure. Presetting, prebending and preheating are used to offset distortion.

All welded fabrication must be checked, tested and approved before being accepted. Tests applied to welding are given in reference 8:

(1) Visual inspection for uniformity of weld;
(2) Surface tests for cracks using dyes or magnetic particles;
(3) X-ray and ultrasonic tests to check for defects inside the weld.

Only visual and surface tests can be used on fillet welds. Butt welds can be checked internally, and such tests should be applied to important butt welds in tension.

When different thicknesses of plate are to be joined the thicker plate should be given a taper of 1 in 5 to meet the thinner one. Small fillet welds should not be made across members such as girder flanges in tension, particularly if the member is subjected to fluctuating loads, because this can lead to failure by fatigue or brittle fracture. With correct edge preparation if required, fit-up, electrode selection and a properly controlled welding process, welds are perfectly reliable.

4.4.3 Design of fillet welds

Important provisions regarding fillet welds are set out in Clause 6.6.2 of BS 5950: Part 1. Some of these are listed below (the complete clause in the code should be consulted):

(1) Fillet welds should be returned around corners for twice the leg length.
(2) In lap joints the lap length should not be less than four times the thickness of the thinner plate.

(3) In end connections the length of weld should not be less than the transverse spacing between the welds.
(4) Intermittent welds should not be used under fatigue conditions. The spacing between intermittent welds should not exceed 300 mm nor $16t$ for parts in compression nor $24t$ for parts in tension, where t is the thickness of the thinner plate. These provisions are shown on Figure 4.20.

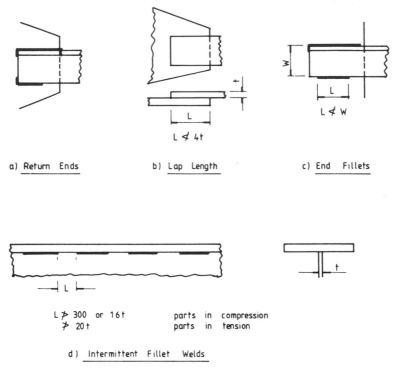

Figure 4.20 Design details for fillet welds

Some values of design strengths for fillet welds given in Table 36, Clause 6.6.5.1, BS 5950: Part 1 are given here in Table 4.4.

Table 4.4 Design strength of fillet welds p_w (N/mm²)

Steel grade	Electrodes Strength BS 639	
BS 4360	E 43	E 51
43	215	215
50	215	255
55	n.a.	255

The strength of a fillet weld is calculated using the throat thickness. For the 90-degree fillet weld shown in Figure 4.18(b) the throat thickness is taken as 0.7 times the size or leg length:

Strength of weld $= 0.7$ leg length $\times p_w/10^3$ kN/mm

Strengths of fillet welds are given in Table 4.5.

Table 4.5 Strength of fillet welds (kN/mm)

Weld size or leg length	Steel grade				
	43 Electrode		50 Electrode		55 Electrode
	E–43	E–51	E–43	E–51	E–51
5		0.75	0.75	0.89	0.89
6		0.90	0.90	1.07	1.07
8		1.20	1.20	1.42	1.42
10		1.50	1.50	1.78	1.78
12		1.80	1.80	2.14	2.14

In design the vector sum of the design stresses due to all the forces on the weld should not exceed the design strength. The code also notes that the strength of symmetrically loaded fillet welds can be taken as equal to the strength of the plate, provided that:

(1) The weld strength is not less than that of the plate;
(2) The sum of the throat thicknesses of the weld is greater than the plate thickness; and
(3) The weld is principally in direct compression or tension.

A weld meeting these provisions is shown in Figure 4.21.

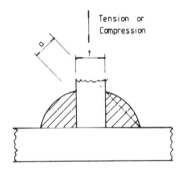

Tension or Compression

$2a \geqslant t$

Figure 4.21 Fillet welds as strong as plate

4.4.4 Design of butt welds

The design of butt welds is covered in Clause 6.6.6 of BS 5950: Part 1. This clause states that the design strength should be taken as equal to that of the parent metal provided that the strength of the weld metal is not less than that of the parent metal.

Full penetration depth is ensured if the weld is made from both sides or if a backing run is made on a butt weld made from one side (see Figures 4.18(a) and (c)). Full penetration is also achieved by using a backing plate (see Figure 4.22(a)).

If the weld is made from one side the throat thickness is reduced and the following values specified in Clause 6.6.6.2 are to be used:

(1) V weld—depth of preparation minus 3 mm (see Figure 4.22(b));
(2) J or U welds—depth of preparation unless it can be shown that greater penetration can be achieved (see Figure 4.18(c)).

The code also states in Clause 6.6.6.3 that if the weld is unsymmetrical relative to the parts joined the resulting eccentricity should be taken into account. The strength of a butt weld is then

= throat thickness × plate design strength.

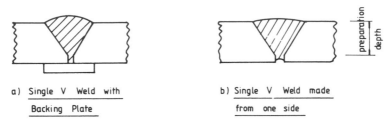

a) Single V Weld with
     Backing Plate

b) Single V Weld made
     from one side

Figure 4.22 Design details for butt welds

4.4.5 Eccentric connections

The two types of eccentrically loaded connections are shown in Figure 4.23. These are:

(1) The torsion joint with the load in the plane of the weld; and
(2) The bracket connection.

In both cases the fillet welds are in shear due to direct load and moment.

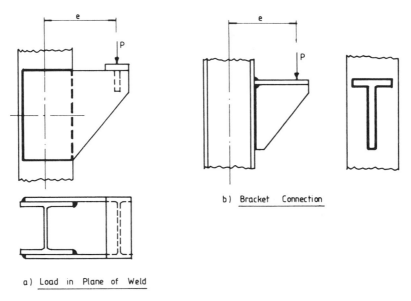

b) Bracket   Connection

a) Load in Plane of Weld

Figure 4.23 Eccentrically loaded connections

4.4.6 Torsion joint with load in plane of weld

The weld is in direct shear and torsion. The eccentric load causes rotation about the centre of gravity of the weld group. The force in the weld due to torsion is taken to be directly proportional to the distance from the centre of gravity and is found by a torsion formula. The direct shear is assumed to be uniform throughout the weld. The resultant shear is found by combining the shear due to moment and the direct shear, and the procedure is set out below. The side plate is assumed to be rigid.

A rectangular weld group is shown in Figure 4.24(a), where the eccentric load P is taken on one plate. The weld is of unit leg length throughout:

Direct shear $F_S = P/$length of weld
$$= P/[2(x+y)]$$

Shear due to torsion
$$F_T = Per/I_P$$
where I_P = polar moment of inertia of the weld group
$$= I_x + I_y$$
$$I_x = y^3/6 + xy^2/2$$
$$I_y = x^3/6 + x^2y/2$$
$$r = 0.5 \, (x^2 + y^2)^{0.5}$$

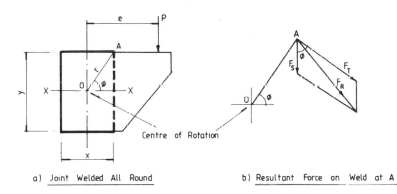

a) Joint Welded All Round   b) Resultant Force on Weld at A

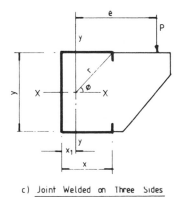

c) Joint Welded on Three Sides

Figure 4.24 Torsion joints load in plane of weld

The heaviest loaded length of weld is that at A, farthest from the centre of rotation O. The resultant shear on a unit length of weld at A is given by:

$$F_R = [F_S^2 + F_T^2 + 2F_S F_T \cos \varphi]^{0.5}$$

The resultant shear is shown on Figure 4.24(b). The weld size can be selected from Table 4.5.

If the weld is made on three sides only, as shown on Figure 4.24(c), the centre of gravity of the group is found first by taking moments about side BC:

$$x_1 = x^2/(2x+y)$$
$$I_x = y^3/12 + xy^2/2$$
$$I_y = x^3/6 + 2x(x/2 - x_1)^2 + yx_1^2$$

The above procedure can then be applied.

4.4.7 Bracket connection

Various assumptions are made for the analysis of forces in bracket connections. Consider the bracket shown in Figure 4.25(a), which is cut from a universal beam with a flange added to the web. The bracket is connected by fillet welds to the column flange. The flange welds have a throat thickness of unity and the web welds a throat thickness q, a fraction of unity. Assume rotation about the centroidal axis XX. Then:

Weld length	$L = 2b + 2a\,q$
Moment of inertia	$I_x = bd^2/2 + qa^3/6$
Direct shear	$F_S = P/L$
Load due to moment	$F_T = Ped/2I_x$
Resultant load	$F_R = (F_T^2 + F_S^2)^{0.5}$

Select the weld size from Table 4.5.

In a second assumption rotation takes place about the bottom flange $X_1 X_1$. The flange welds resist moment and web welds shear. In this case:

$$F_T = Pe/db$$
$$F_S = P/2a$$

The weld sizes can be selected from Table 4.5.

With heavily loaded brackets full-strength welds are required between the bracket and column flange.

The fabricated T-section bracket is shown in Figure 4.25(b). The moment is assumed to be resisted by the flange weld and a section of the web in compression of depth y, as shown in the figure. Shear is resisted by the web welds.

The bracket and weld dimensions and internal forces resisting moment are shown in the figure. The web area in compression is ty. Equating moments gives:

$$P.e = T(d - 2y/3) = C(d - 2y/3)$$
$$C = 1/2\,p_y\,ty = T$$
$$P.e = 1/2\,p_y\,ty\,(d - 2y/3)$$

Solve the quadratic equation for y and calculate C:

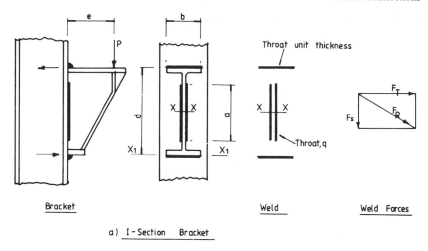

Bracket   Weld   Weld Forces

a) I-Section   Bracket

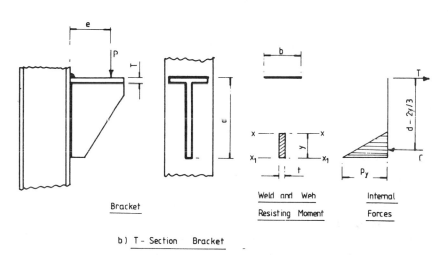

Bracket   Weld and Web   Internal
                          Resisting Moment   Forces

b) T - Section   Bracket

Figure 4.16 Bracket connections

The flange weld force $F_T = T/b$

The web weld force $F_S = P/2(d-y)$

Select welds from Table 4.5.

The calculations are simplified if the bracket is assumed to rotate about the $X_1 X_1$ axis, when

$F_T = Pe/db$

4.4.8 Examples on welded connections

(1) Direct shear connection

Design the fillet weld for the direct shear connection for the angle loaded as shown in Figure 4.26(a), where the load acts through the centroidal axis of the angle. The steel is Grade 43:

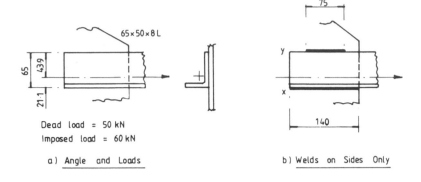

Dead load = 50 kN
Imposed load = 60 kN

a) Angle and Loads

b) Welds on Sides Only

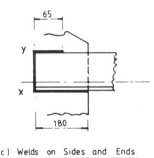

c) Welds on Sides and Ends

Figure 4.26 Direct shear connection

Factored load = $(1.4 \times 50) + (1.6 \times 60)$ = 166 kN
Use 6-mm fillet weld, strength from Table 4.5 = 0.9 kN/mm
Length required = 166/0.90   = 184.4 mm
Balance the weld on each side as shown in Figure 4.26(b):
Side X, length = $184.4 \times 43.9/65$ = 124.5 mm
    Add 12 mm, final length = 136.5 mm, say 140 mm
Side Y, length = 184.4 − 124.5   = 59.9 mm
    Add 12 mm, final length = 71.9, say 75 mm

Note that the length on side Y exceeds the distance between the welds, as required in Clause 6.6.2.2 of BS 5950: Part 1. A weld may also be placed across the end of the angle, as shown in Figure 4.26(c). The length of weld on side Y, L_y, may be found by taking moments about side X. In terms of weld lengths this gives:

$(L_y \times 65) + (65 \times 32.5)$ = (184.4×21.1)
$L_y = 27.4 + 6 = 33.4$, say 40 mm
Length on side X:
$L_x = 184.4 - 65 - 27.4 + 6 = 98$ mm, say 100 mm

Note that the leg length has been added at ends of all weld lengths calculated above to allow for craters at the ends. To comply with Clause 6.6.2.2 quoted above, weld L_y is increased to 65 mm. Weld L_x is also increased in proportion to 180 mm. In the above example the load is also eccentric to the plane of the gusset plate, as shown in Figure 4.26(a). It is customary to neglect this eccentricity.

(2) Torsion connection with load in plane of weld

One side plate of an eccentrically loaded connection is shown in Figure 4.27(a). The plate is welded on three sides only. Find the maximum shear force in the weld and select a suitable fillet weld from Table 4.5.

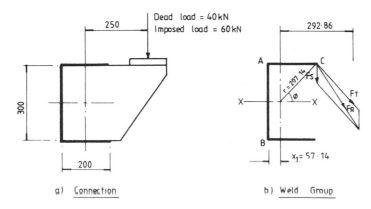

a) Connection   b) Weld Group

Figure 4.27 torsion Connection loaded in plane of weld

Find the position of the centre of gravity of the weld group by taking moments about side AB (see Figure 4.27(b)):

Length $\qquad\qquad L = 700$ mm
Distance to centroid $x_1 = 2 \times 200 \times 100/700 = 57.14$ mm
Eccentricity of load $e = 292.86$

Moments of inertia:

$I_x = (2 \times 200 \times 150^2) + 300^3/12 = 11.25 \times 10^6$ mm^3
$I_y = (300 \times 57.14^2) + (2 \times 200^3/12) + (2 \times 200 \times 42.86^2)$
$\qquad\qquad\qquad\qquad = 3.047 \times 10^6$ mm^3
$I_p = (11.25 + 3.047) 10^6 \qquad = 14.297 \times 10^6$ mm^2
Angle cos $\varphi = 142.86/207.14 \qquad = 0.689$
Factored load $= (1.4 \times 40) + (1.6 \times 60) = 152$ kN
Direct shear $F_s = 152/700 \qquad = 0.217$ kN/mm

Shear due to torsion on weld at C:

$$F_T = \frac{152 \times 292.86 \times 207.14}{14.297 \times 10^6} \qquad = 0.645 \text{ kN/mm}$$

Resultant shear:

$$F_R = [0.217^2 + 0.645^2 + 2 \times 0.217 \times 0.645 \times 0.689]^{0.5}$$
$$= 0.81 \, \text{kN/mm}$$

A 6-mm fillet weld, strength 0.90 kN/mm is required.

(3) Bracket connection

Determine the size of fillet weld required for the bracket connection shown in Figure 4.28. The web welds are to be taken as one half the leg length of the flange welds. All dimensions and loads are shown in the figure.

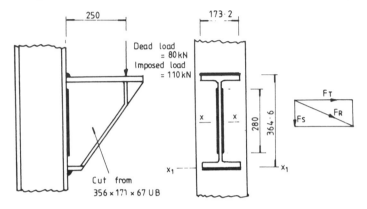

Figure 4.28 Bracket connection

Design assuming rotation about XX axis

Factored load $= (1.4 \times 80) + (1.6 \times 110)$	$= 288 \, \text{kN}$
Length $L = (2 \times 173.2) + 280$	$= 626.4 \, \text{mm}$
Inertia $I_X = (2 \times 173.2 \times 182^2) + 280^3/12$	$= 13.3 \times 10^6 \, \text{mm}^3$
Direct shear $F_S = 288/626.4$	$= 0.46 \, \text{kN/mm}$

$$\text{Shear from moment } F_T = \frac{288 \times 250 \times 182}{13.3 \times 10^6} \quad = 0.985 \, \text{kN/mm}$$

Resultant shear $F_R = [0.46^2 + 0.985^2]^{0.5}$ $= 1.09 \, \text{kN/mm}$

Provide 8-mm fillet welds for the flanges, strength 1.2 kN/mm. For the web welds provide 6-mm fillets (the minimum size recommended).

Design assuming rotation about $X_1 X_1$ axis

The flange weld resists the moment

$$F_T = \frac{288 \times 250}{364 \times 173.2} = 1.14 \, \text{kN/mm}$$

Provide 8-mm fillet welds, strength 1.2 kN/mm. The web welds resist the shear:

$$F_S = 288/(2 \times 280) \qquad\qquad = 0.514 \, \text{kN/mm}$$

Provide 6-mm fillet welds. The methods give the same results.

4.5 Further considerations in design of connections

4.5.1 Load paths and forces

The design of bolts and welds has been considered in the previous sections. Other checks which depend on the way the joint is fabricated are necessary to ensure that it is satisfactory. Consistent load paths through the joint must be adopted.

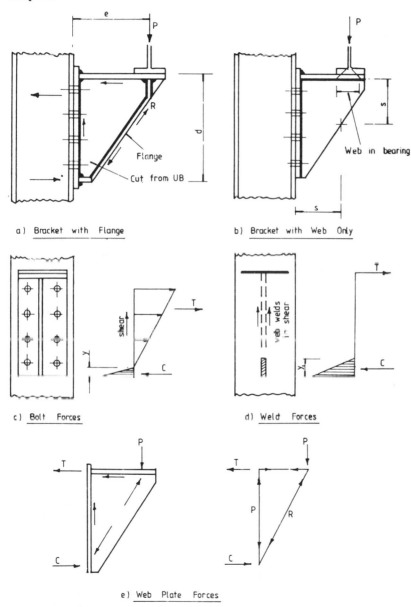

Figure 4.29 Brackets: load paths and forces

Consider the brackets shown in Figure 4.29. The design checks required are:

(1) The bolt group (see Section 4.3.3);
(2) The welds between the three plates (see Section 4.4.5);
(3) The bracket plates. These are in tension, bearing, buckling and local bending;
(4) The column in axial load, shear and moment. Local checks on the flange in bending and web in tension at the top and buckling and bearing at the bottom are also required.

4.5.2 Other design checks

Some points regarding the design checks are set out below.

(1) A direct force path is provided by the flange on the bracket in Figure 4.29(a). The flange can be designed to resist force

$$R = (d^2 + e^2)^{0.5} P/d$$
e = eccentricity of the load
d = depth of bracket

(2) Where the bracket has a web plate only, as shown in Figure 4.29(b), the maximum outstand should not exceed $13t\varepsilon$, where

t = thickness of plate
$\varepsilon = (275/p_y)^{0.5}$

This ensures that the plate can be stressed to the design strength p_y without buckling. (See Table 7 of BS 5950: Part 1.) Local buckling is dealt with in Section 5.3. If the web is satisfactory for bearing under the load P, the load R can be assumed to be carried on a strip of web of width equal to the length in bearing (see (3) below). The bolt, weld and web plate forces for this type of bracket are shown in Figures 4.29(c), (d) and (e), respectively. (See references 7 and 9 for a more rigorous treatment of this problem.)

(3) The bearing length at the top of the web is shown in Figure 4.30(a). For a universal beam the load is dispersed at 45 degrees from the beam fillets and at 1 in 2.5 through the top plate. (See Clauses 4.5.1.3 and 4.5.3 in BS 5950: Part 1. Bearing is dealt with in Section 5.8.2.)

(4) The end plate and column flange are checked for bending, as shown on Figure 4.30(b). The plates are in double curvature, produced by prying forces which are absorbed by the bolts. The length resisting bending is found by dispersing the load at 30 degrees, as shown in the figure. For a more rigorous approach using yield line analysis see references 10 and 11.

(5) The following checks are made on the column at the bottom of the bracket. The column web is checked for bearing at the end of the fillet between flange and web with load dispersed at 1 in 2.5 to give the length in bearing b_1, shown in Figure 4.30(c). Note that the compression force is assumed to spread over a depth $2y/3$, where y is the depth of compression area from the bolt force analysis (see Figure 4.29(c)). The column web is checked for buckling at the centre line of the column. The load is dispersed at 45 degrees to give the length b, shown in Figure 4.30(c), which is considered for buckling. (See Section 5.8.1 for details of this design check.) Stiffeners can be added to carry loads if the column web is overstressed.

(6) The column web is checked in tension at the top of the bracket, as shown in Figure 4.30(c). The length in tension is taken as the length *g* resisting bending, defined in (4) above.

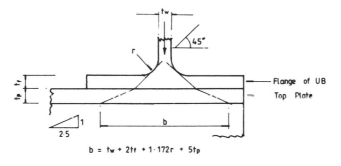

a) Bearing at Top of Web

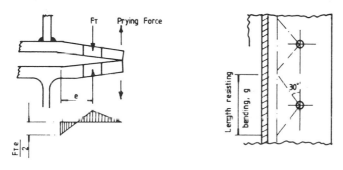

b) Column Flange and End Plate in Bending

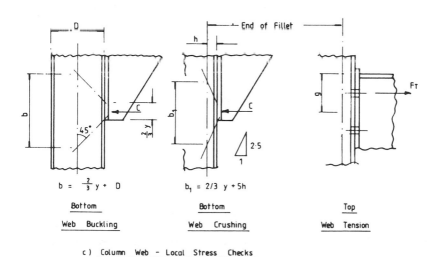

c) Column Web - Local Stress Checks

Figure 4.30 Brackets: local stress checks

Problems

4.1 A single-shear bolted lap joint (Figure 4.31) is subjected to an ultimate tensile load of 200 kN. Determine a suitable bolt diameter using Grade 4.6 bolts.

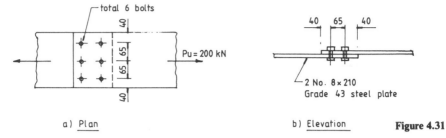

a) Plan   b) Elevation   **Figure 4.31**

4.2 A double-channel member carrying an ultimate tension load of 820 kN is to be spliced, as shown in Figure 4.32.
(1) Determine the number of 20-mm diameter Grade 4.6 bolts required to make the splice.
(2) Check the double-channel member in tension.
(3) Check the splice plates in tension.

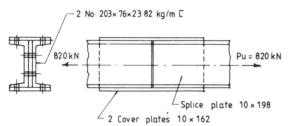

Figure 4.32

4.3 A bolted eccentric connection (illustrated in Figures 4.33(a) and (b)) is subjected to a vertical ultimate load of 120 kN. Determine the size of Grade 4.6 bolts required if the load is placed at an eccentricity of 300 mm.

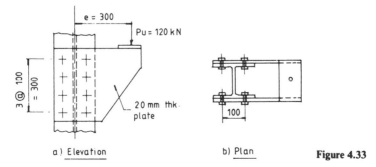

a) Elevation   b) Plan   **Figure 4.33**

4.4 The bolted bracket connection shown in Figure 4.34 carries a vertical ultimate load of 300 kN placed at an eccentricity of 250 mm. Check that 12 No. 24-mm diameter Grade 4.6 bolts are adequate. Use both approximate and accurate methods of analysis discussed in Section 4.2.6. Assume all plates to be 20 mm thick.

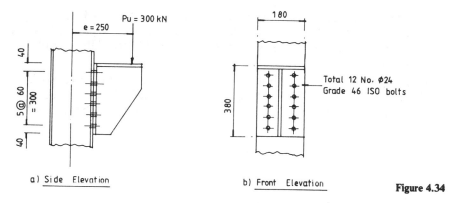

a) Side Elevation   b) Front Elevation

Figure 4.34

4.5 Design a beam-splice connection for a 533 × 210 UB 82. The ultimate moment and shear at the splice are 300 kNm and 175 kN, respectively. A sketch of the suggested arrangement is shown in Figure 4.35. Prepare the final connection detail drawing from your design results.

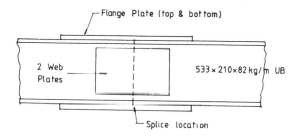

Figure 4.35

4.6 The arrangement for a friction-bolt grip connection provided for a tie carrying an ultimate force of 300 kN is shown in Figure 4.36. Check the adequacy if all the bolts provided are 20-mm diameter.

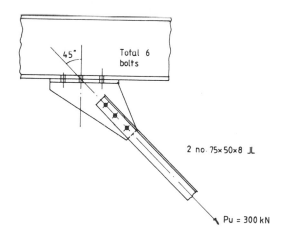

Figure 4.36

4.7 Redesign the bracked connection in Problem 4.4 using high-strength friction-grip bolts. What is the minimum bolt diameter required? Discuss the relative merits of using Grade 4.6, Grade 8.8 and HSFG bolt for connections.

4.8 The welded connection for a tension member in a roof truss is shown in Figure 4.37. Using an E43 electrode on Grade 43 plate, determine the minimum leg size of the welds if the ultimate tension in the member is 90 kN.

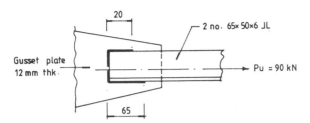

Figure 4.37

4.9 Determine the leg length of fillet weld required for the eccentric joint shown in Figure 4.38. The ultimate vertical load is 500 kN placed at 300 mm from the centre line. Use an E43 electrode on a Grade 43 plate.

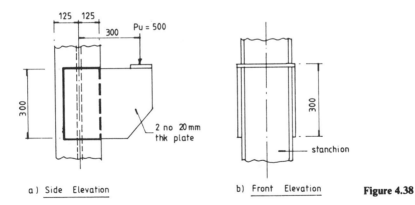

Figure 4.38

4.10 A bracked cut from a 533 × 210 UB 82 of Grade 43 steel is welded to a column, as shown in Figure 4.39. The ultimate vertical load on the bracket is 350 kN applied at an eccentricity of 250 mm. Design the welds between the bracket and column.

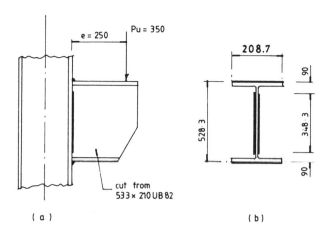

(a)   (b)   **Figure 4.39**

Beams

5.1 Types and uses

Beams span between supports to carry lateral loads which are resisted by bending and shear. However, deflections and local stresses are also important.

Beams may be cantilevered, simply supported, fixed ended or continuous, as shown in Figure 5.1(a). The main uses of beams are to support floors and columns, carry roof sheeting as purlins and side cladding as sheeting rails.

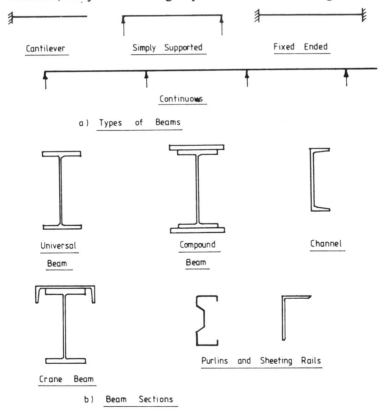

Cantilever   Simply Supported   Fixed Ended

Continuous

a) Types of Beams

Universal Beam   Compound Beam   Channel

Crane Beam   Purlins and Sheeting Rails

b) Beam Sections

Figure 5.1 Types of beams and beam sections

Any member may serve as a beam, and common beam sections are shown in Figure 5.1(b). Some comments on the different sections are given:

(1) The universal beam where the material is concentrated in the flanges is the most efficient section to resist uniaxial bending.
(2) The compound beam consisting of a universal beam and flange plates is used where the depth is limited and the universal beam itself is not strong enough to carry the load.
(3) The crane beam consists of a universal beam and channel. It must resist bending in two directions.

Beams may be of uniform or non-uniform section. Rolled beams may be strengthened in regions of maximum moment by adding cover plates or haunches. Some examples are shown in Figure 5.2.

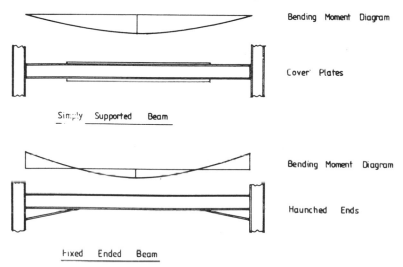

Figure 5.2 Non-uniform beams

5.2 Beam loads

Types of beam loads are:

(1) Concentrated loads from secondary beams and columns.
(2) Distributed loads from self weight and floor slabs.

The loads are further classified into:

(1) Dead loads from self weight, slabs, finishes, etc.
(2) Imposed loads from people, fittings, snow on roofs, etc.
(3) Wind loads, mainly on purlins and sheeting rails.

Loads on floor beams in a steel frame building are shown in Figure 5.3(a). The figure shows loads from a two-way spanning slab which give trapezoidal and triangular loads on the beams. One-way spanning floor slabs give uniform loads.

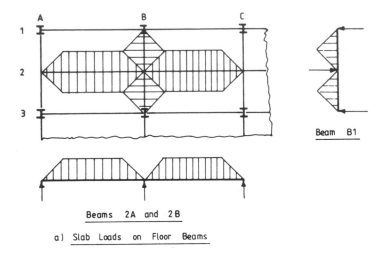

Beam B1

Beams 2A and 2B

a) Slab Loads on Floor Beams

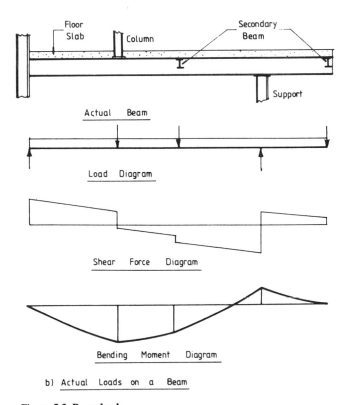

Floor
Slab   Column   Secondary
                             Beam

Support

Actual   Beam

Load   Diagram

Shear   Force   Diagram

Bending   Moment   Diagram

b) Actual Loads on a Beam

Figure 5.3 Beam loads

An actual beam with the floor slab and members it supports is shown in Figure 5.3(b). The load diagram and shear force and bending moment diagrams constructed from it are also shown.

5.3 Classification of beam cross sections

The projecting flange of an I beam will buckle prematurely if it is too thin. Webs will also buckle under compressive stress from bending and from shear. This problem is discussed in more detail in Section 6.2 in Chapter 6. (See also reference 12.)

To prevent local buckling occurring, limiting outstand/thickness ratios for flanges and depth/thickness ratios for webs are given in BS 5950: Part 1 in Section 3.5. Beam cross-sections are classified as follows in accordance with their behaviour in bending:

Class 1 Plastic cross section. This can develop a plastic hinge with sufficient rotation capacity to permit redistribution of moments in the structure. Only class 1 sections can be used for plastic design.

Class 2 Compact cross section. This can develop the plastic moment capacity but local buckling prevents rotation at constant moment.

Class 3 Semi-compact cross section. The stress in the extreme fibres should be limited to the yield stress because local buckling prevents development of the plastic moment capacity.

Class 4 Slender cross section. Premature buckling occurs before yield is reached.

Flat elements in a cross section are classified as:

Internal elements supported on both longitudinal edges.

Outside elements attached on one edge with the other free.

Elements are generally of uniform thickness but, if tapered, the average thickness is used. Elements are classified as plastic, compact or semi-compact if they meet limits given in Table 7 of the code. The limiting proportions for elements of rolled sections are shown in Figure 5.4. Note that the table in the figure does not include angles or tees.

5.4 Bending stresses and moment capacity

Both elastic and plastic theory are discussed here. Short or restrained beams are considered in this section. Plastic properties are used for plastic and compact sections and elastic properties for semi-compact sections to determine moment capacities.

5.4.1 Elastic theory

(1) Uniaxial bending

The bending stress distributions for an I-section beam subjected to uniaxial moment are shown in Figure 5.5(a). Define terms for the I-section:

M = applied bending moment;

I_x = moment of inertia about XX;

$Z_x = 2 I_x/D$ = modulus of section for XX axis;

D = overall depth of beam.

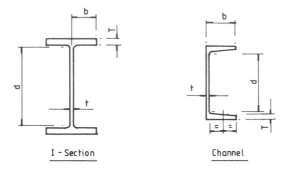

I – Section   Channel

Type of element		Class of Section		
		Plastic	Compact	Semi-Compact
Outstand element of compression flange	$\frac{b}{T} \leqslant$	$8.5\,\epsilon$	$9.5\,\epsilon$	$15\,\epsilon$
Web with neutral axis at mid-depth	$\frac{d}{T} \leqslant$	$79\,\epsilon$	$98\,\epsilon$	$120\,\epsilon$

$$\epsilon = (275/p_y)^{0.5}$$

Figure 5.4 Limiting proportions for rolled sections

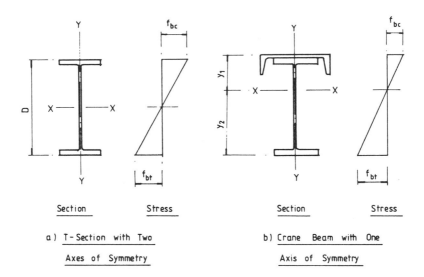

Section   Stress   Section   Stress

a) T–Section with Two   b) Crane Beam with One

   Axes of Symmetry   Axis of Symmetry

Figure 5.5 Beams in uniaxial bending

The maximum stress in the extreme fibres top and bottom is:

$$f_{bc} = f_{bt} = M_x/Z_x$$

The moment capacity $M_c = p_b Z_x$

where p_b = allowable stress

The moment capacity for a semi-compact section subjected to a moment due to factored loads is given in Clause 4.7.5 of BS 5950: Part 1 as

$M_c = p_y Z_x$

where p_y = design strength

For the asymmetrical crane beam section shown in Figure 5.5(b) the additional terms require definition:

$Z_1 = I_x/y_1$ = modulus of section for top flange
$Z_2 = I_x/y_2$ = modulus of section for bottom flange
y_1, y_2 = distance from centroid to top and bottom fibres.

The bending stresses are:

Top fibre in compression $f_{bc} = M_x/Z_1$
Bottom fibre in tension f_{bt} $= M_x/Z_2$

The moment capacity controlled by the stress in the bottom flange is $M_c = p_b Z_2$

(2) Biaxial bending

Consider that I section in Figure 5.6(a) which is subject to bending about both axes. Define the following terms:

M_x = moment about the XX axis
M_y = moment about the YY axis
Z_x = modulus of section for the XX axis
Z_y = modulus of section for the YY axis

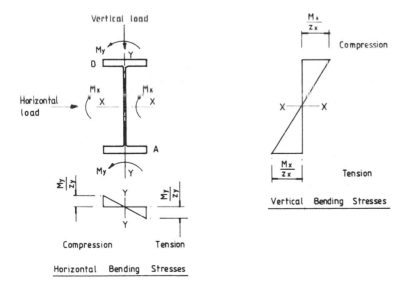

Figure 5.6 Biaxial bending. (a) Bending stresses

The maximum stress at A or B is:

$$f_A = f_B = M_x/Z_x + M_y/Z_y$$

If the allowable stress is p_b the moment capacities with respect to XX and YY axes are:

$$M_{cx} = p_b Z_x$$
$$M_{cy} = p_b Z_y$$

Putting the maximum stress equal to p_b and substituting for Z_x and Z_y in the expression above gives the interaction relationship

$$M_x/M_{cx} + M_y/M_{cy} = 1$$

This is shown graphically in Figure 5.6(b).

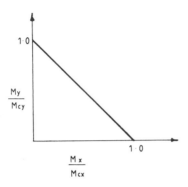

b) Interaction   Diagram

Figure 5.6 Biaxial bending: (b) Interaction diagram

(3) Asymmetrical sections

Note that with the channel section shown in Figure 5.7(a) the vertical load must be applied through the shear centre for bending in the free member to take place about the XX axis, otherwise twisting and biaxial bending occurs. However, a horizontal load applied through the centroid causes bending about the YY axis only.

For an asymmetrical section such as the unequal angle shown in Figure 5.7(b) bending takes place about the principle axes UU and VV in the free member when the load is applied through the shear centre. When the angle is used as a purlin the cladding restrains the member so that it bends about the XX axis.

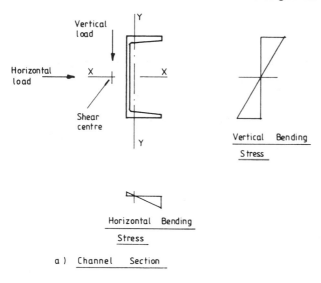

Vertical Bending Stress

Horizontal Bending Stress

a) Channel   Section

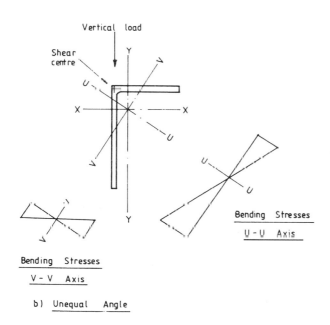

Bending Stresses U – U Axis

Bending Stresses V – V Axis

b) Unequal   Angle

Figure 5.7 Bending of asymmetrical sections

5.4.2 Plastic theory

(1) Uniaxial bending

The stress–strain curve for steel on which plastic theory is based is shown in Figure 5.8(a). In the plastic region after yield the strain increases without increase in stress. Consider the I section shown in **Figure 5.8(b)**. Under moment the stress first follows an elastic distribution. As the moment increases the stress at the extreme fibre reaches the yield stress and the plastic region proceeds inwards as shown until the full plastic moment is reached and a plastic hinge is formed.

For single axis bending the following terms are defined:

M_c = plastic moment capacity

S = plastic modulus of section

Z = elastic modulus of section

p_y = design strength

The moment capacity given in Clause 4.2.5 of BS 5950: Part 1 for sections with low shear load is:

$$M_c = p_y S$$
$$\leqslant 1.2\, p_y Z$$

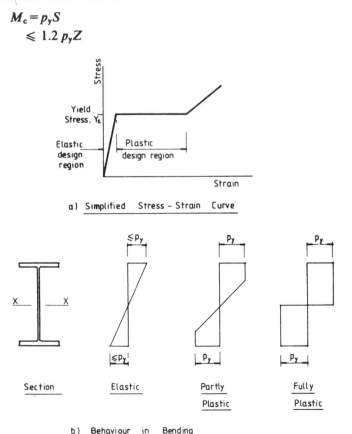

a) Simplified Stress – Strain Curve

b) Behaviour in Bending

Figure 5.8 Behaviour in bending

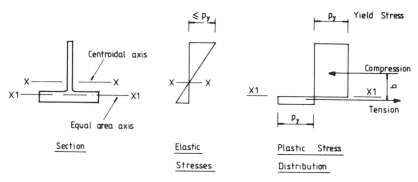

Figure 5.9 Section with one axis of symmetry

The first expression is the plastic moment capacity, the second ensures that yield does not occur at working loads in I sections bent about the YY axis.

For single-axis bending for a section with one axis of symmetry consider the T-section shown in Figure 5.9. In the elastic range bending takes place about the centroidal axis and there are two values for the elastic modulus of section.

In the plastic range bending takes place about the equal area axis and there is one value for the plastic modulus of section:

$$S = M_c/p_y$$
$$= Ab/2$$

where A = area of cross section

b = lever arm between the tension and compression forces.

(2) Biaxial bending

When a beam section is bent about both axes the neutral axis will lie at an angle to the rectangular axes which depends on the section properties and values of the moments. Solutions have been obtained for various cases and a relationship established between the ratios of the applied moments and the moment capacities about each axis. The relationship expressed in Sections 4.9 and 4.8.3 of BS 5950: Part 1 for plastic or compact cross sections is in the following form:

$$\left(\frac{M_x}{M_{cx}}\right)^{Z_1} + \left(\frac{M_y}{M_{cy}}\right)^{Z_2} < 1$$

where M_x = factored moment about the XX axis

M_y = factored moment about the YY axis

M_{cx} = moment capacity about the XX axis

M_{cy} = moment capacity about the YY axis

Z_1 = 2 for I and H sections and 1 for other open sections

Z_2 = 1 for all open sections

A conservative result is given if $Z_1 = Z_2 = 1$. The interaction diagram is shown in Figure 5.10.

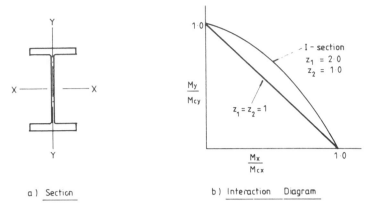

a) Section   b) Interaction Diagram

Figure 5.10 Biaxial bending

(3) Unsymmetrical sections

For sections with no axis of symmetry plastic analysis for bending is complicated but solutions have been obtained. In many cases where such sections are used the member is constrained to bend about the rectangular axis (see Section 5.4.1(3)). Such cases can also be treated by elastic theory using factored loads with the maximum stress limited to the design strength.

5.5 Lateral torsional buckling

5.5.1 General considerations

The compression flange of an I beam acts like a column and will buckle sideways if the beam is not sufficiently stiff or the flange is not restrained laterally. The load at which the beam buckles can be much less than that causing the full moment capacity to develop. Only a general description of the phenomenon and factors affecting it are set out here. The reader should consult references 12 and 13 for further information.

Consider the simply supported beam with ends free to rotate in plan but restrained against torsion and subjected to end moments, as shown in Figure 5.11. Initially the beam deflects in the vertical plane due to bending but, as the moment increases, it reaches a critical value M_E, less than the moment capacity, where it buckles sideways, twists and collapses.

Elastic theory is used to set up equilibrium equations to equate the disturbing effect to the lateral bending and torsional resistances of the beam. The solution of these equation gives the elastic critical moment:

$$M_E = \frac{\pi}{L}(EI_y GJ)^{0.5}\left(1 + \frac{\pi^2 EH}{L^2 GJ}\right)^{0.5}$$

where E = Young's modulus

G = shear modulus

J = torsion constant for the section

H = warping constant for the section

L = span

I_y = moment of inertia about the YY axis.

The theoretical solution applies to a beam subjected to a uniform moment. In other cases where the moment varies, the tendency to buckling is reduced. If the load is applied to the top flange and can move sideways it is destabilizing, and buckling occurs at lower loads than if the load were applied at the centroid or to the bottom flange.

In the theoretical analysis the beam was assumed to be straight. Practical beams have initial curvature and twisting, residual stresses, and the loads are applied eccentrically. The theory set out above requires modification to cover actual behaviour. Theoretical studies and tests show that slender beams fail at the elastic critical moment M_E and short or restrained beams fail at the plastic moment capacity M_c. A lower bound curve running between the two extremes can be drawn to contain the behaviour of intermediate beams. Beam behaviour as a function of slenderness is shown in Figure 5.11(b). Terms used in the curve are defined:

M = moment causing failure

M_c = moment capacity for a restrained beam

M_E = elastic critical moment

L_E/r_y = slenderness with respect to the YY axis (see the next section).

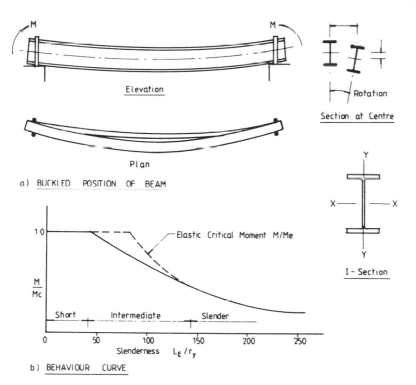

a) BUCKLED POSITION OF BEAM

b) BEHAVIOUR CURVE

Figure 5.11 Lateral torsional buckling

To summarize, factors influencing lateral torsional buckling are:

(1) The unrestrained length of compression flange. The longer this is, the weaker the beam. Lateral buckling is prevented by providing props at intermediate points.
(2) The end conditions. Rotational restraint in plan helps to prevent buckling.
(3) Section shape. Sections with greater lateral bending and torsional stiffnesses have greater resistance to buckling.
(4) Note that lateral restraint to the tension flange also helps to resist buckling. (See Figure 5.11.)
(5) The application of the loads and shape of the bending moment diagram between restraints.

A practical design procedure must take into account the effects noted above.

5.5.2 Lateral restraints and effective length

The code states in Clause 4.2.2 that full lateral restraint is provided by a floor slab if the friction or shear connection is capable of resisting a lateral force of 1 per cent of the maximum factored force in the compression flange. Other suitable construction can also be used. Members not provided with full lateral restraint must be checked for buckling.

The following two types of restraints are defined in Sections 4.3.2 and 4.3.3 of the code:

(1) Lateral restraint, which prevents sideways movement of the compression flange; and
(2) Torsional restraint, which prevents movement of one flange relative to the other.

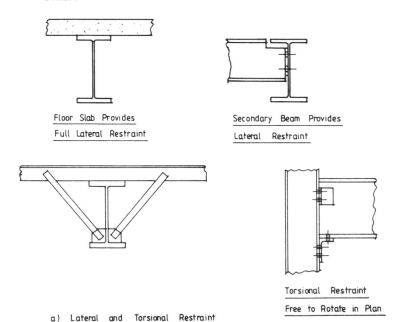

Floor Slab Provides
Full Lateral Restraint

Secondary Beam Provides
Lateral Restraint

Torsional Restraint
Free to Rotate in Plan

a) Lateral and Torsional Restraint

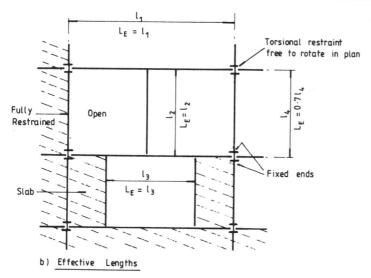

Figure 5.12 Restraints and effective lengths

Restraints are provided by floor slabs, end joints, secondary beams, stays, sheeting, etc., and some restraints are shown in Figure 5.12(a).

The effective length L_E is defined in Section 1 of the code as the length between points of effective restraint of a member multiplied by a factor to take account of the end conditions and loading. Note that a destabilizing load (where the load is applied to the top flange and can move with it) is taken account of by increasing the effective length of member under consideration.

The effective length for beams is discussed in Section 4.3.5 of BS 5950: Part 1. When the beam is restrained at the ends only the effective length should be obtained from Table 9 in the code. Some values from this table are given in Table 5.1.

Table 5.1 Effective lengths L_E—Beams

Support conditions	Loading conditions	
	Normal	Destabilizing
Beam torsionally unrestrained Compression flange laterally unrestrained Both flanges free to rotate on plan	$1.2(L+2D)$	$1.4(L+2 \,)$
Beam torsionally restrained Compression flange laterally restrained Compression flange only free to rotate on plan	$1.0\,L$	$1.2\,L$
Beam torsionally restrained Both flanges NOT free to rotate on plan	$0.7\,L$	$0.85\,L$

L = length of beam between restraints.
D = depth of beam.

Where the beam is restrained at intervals by other members the effective length L_E may be taken as L, the distance between restraints. Some effective lengths for floor beams are shown in Figure 5.12(b).

5.5.3 Code design procedure

(1) General procedure

The general procedure for checking the lateral torsional buckling resistance of a beam is set out in Section 4.3.7 of BS 5950: Part 1. This is:

(1) If the member or part being checked carries no loads between adjacent lateral restraints the equivalent uniform moment factor m is evaluated. Values of m which depend on the ratio and direction of the end moments are given in Table 18 in the code. Values for some common load cases are shown in Figure 5.13(a). A conservative result is given if m is taken as 1. Note that the slenderness factor n is 1 in this case. See (7) below.
(2) Calculate the equivalent uniform moment

$$\bar{M} = m\,M_A$$

M_A = maximum moment in the beam or part considered.

(3) Estimate the effective length L_E of the unrestrained compression flange using the rules from Section 5.5.2. Minor axis slenderness $\lambda = L_E/r_y$, where

r_y = radius of gyration for the YY axis

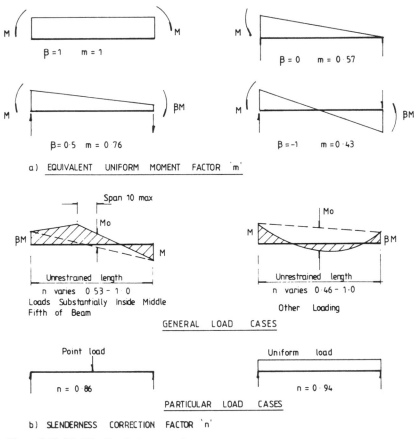

a) EQUIVALENT UNIFORM MOMENT FACTOR 'm'

b) SLENDERNESS CORRECTION FACTOR 'n'

Figure 5.13 Modification factors *m* and *n*

(4) Calculate the equivalent slenderness:

$$\lambda_{LT} = uv\lambda$$

where u = buckling parameter allowing for torsional resistance. This may be calculated from the formulae in Appendix B or taken from the published table in the *Guide* to BS 5950: Part 1: 1985, Vol. 1, Section properties, Member Capacities, Constrado

= 0.9 conservatively for a uniform rolled I section

v = slenderness factor which depends on values of N and λ/x

where $N = \dfrac{I_{cf}}{I_{cf} + I_{tf}}$

I_{cf} = moment of inertia of the compression flange about the minor axis of the section

I_{tf} = moment of inertia of the tension flange about the minor axis of the section

$N = 0.5$ for a symmetrical section

x = torsional index. This can be calculated from the formula in Appendix B or obtained from the published table in the *Guide* to BS 5950: Part 1

= D/T approximately

where D = overall depth of beam

T = thickness of the compression flange.

Values of v for uniform sections are given in Table 14 of the code. For other sections v can be calculated by the formulae in Appendix B.

(5) Read the bending strength p_b from Table 11 in the code. Values of p_b depend on the equivalent slenderness λ_{LT} and design strength p_y. Graphs giving values of p_b for rolled sections are shown in Figure 5.14.

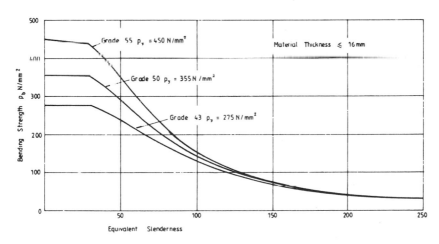

Figure 5.14 Bending strengths for rolled sections

(6) Calculate the buckling resistance moment:

$$M_b = S_x p_b$$

where S_x = plastic modulus for the major axis

For a safe design,

$$M_b > \bar{M}$$

Successive trials are needed to obtain an economic section.

(7) If loads are applied to the beam or part under consideration the slenderness factor n is calculated. This gives a more accurate method of correcting for non-uniform moment in this case than using the m factor as above. The m factor is taken as 1. Values of n given in Tables 15 and 16 of the code depend on how the load is applied and values of the end and mid-span moments. The general load cases, two particular beam loads and their n factors are shown in Figure 5.13(b). The equivalent slenderness:

$$\lambda_{LT} = nuv\lambda$$

The design proceeds from step (5) above.

(8) A conservative design results if $m = n = 1$. Values of m and n are given in Table 13 of BS 5950: Part 1. Note that if the loads are destabilizing $m = n = 1$.

(2) Conservative approach for rolled sections

The code gives a conservative approach for equal flanged rolled sections in Section 4.3.7.7. The buckling resistance moment

$$M_b = p_b S_x$$

where p_b = bending strength determined from Table 19 for values of λ and x

λ = slenderness L_E / r_y (see above)

x = torsional index = D/T (see above)

The following applies when using this approach:

(1) The equivalent uniform moment factor m is taken as 1.0.

(2) The slenderness correction factor n is taken from Table 20. This modifies the value of λ used.

5.5.4 Biaxial bending

Lateral torsional buckling affects the moment capacity with respect to the major axis only of I section beams. When the section is bent about the minor axis only it will reach the moment capacity given in Section 5.4.2(1).

Where biaxial bending occurs, BS 5950: Part 1 specifies in Section 4.9 that the following simplified interaction expressions must be satisfied for plastic or compact sections:

(1) Local capacity check at point of maximum combined moments:

$$\frac{M_x}{M_{cx}} + \frac{M_y}{M_{cy}} \leqslant 1$$

This design check was discussed in Section 5.4.2(2) above.

(2) Overall buckling check at the centre of the beam:

$$\frac{mM_x}{M_b} + \frac{mM_y}{p_yZ_y} \leqslant 1$$

where m = equivalent uniform moment factor from Table 18

M_x = applied moment about the major axis at the critical region

M_y = applied moment about the minor axis at the critical region

M_b = buckling resistance moment capacity about the major axis

Z_y = elastic modulus about the minor axis.

Note that a reduced moment capacity is specified for the YY axis.

More exact expressions are given in the code. Biaxial bending is discussed more fully in Chapter 8 of this book.

5.5.5 Computer programs

BS 5950: Part 1 gives formulae for calculating the parameters used in the check for lateral torsional buckling in the appendix. Formulae are given for m, x, v and p_b on which the tables in the code referred to above are based and the design process is easily programmed. A computer program is of great use in carrying out beam section checks. A beam design program is given in Chapter 11 of this book.

5.6 Shear in beams

5.6.1 Elastic theory

The value of shear stress at any point in a beam section is given by the following expression (see Figure 5.15(a)):

$$f_s = \frac{V.A.y}{I_x t}$$

where V = shear force at the section

A = area between the point where the shear stress is required and a free edge

y = distance from the centroid of the area A to the centroid of the section

I_x = moment of inertia about the XX axis

t = thickness of the section at the point where the shear stress is required.

Using this formula the shear stresses at various points in the beam section can be found. Thus the maximum shear stress at the centroid in terms of the beam dimensions shown in the figure is:

$$f_{max} = \frac{V}{I_x t}\left(\frac{BT(d+T)}{2} + \frac{td^2}{8}\right)$$

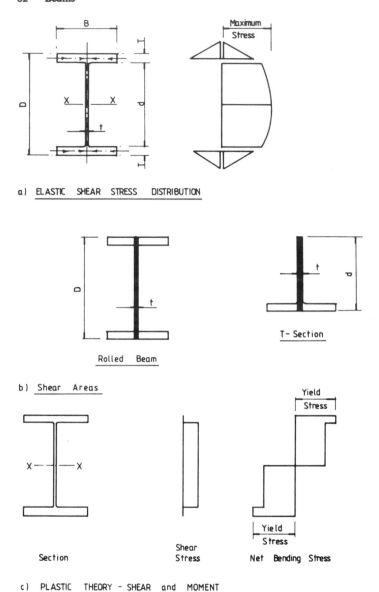

a) ELASTIC SHEAR STRESS DISTRIBUTION

Rolled Beam

T- Section

b) Shear Areas

Section

Shear
Stress

Yield
Stress

Net Bending Stress

c) PLASTIC THEORY - SHEAR and MOMENT

Figure 5.15 Shear in beams

Note that the distribution shows that the web carries the bulk of the shear. It has been customary in design to check the average shear stress in the web given by:

$$f_{av} = V/Dt$$

which should not exceed an allowable value.

5.6.2 Plastic theory

Shear is considered in BS 5950: Part 1 in Section 4.2.3. For a rolled member subjected to shear only, the shear force is assumed to be resisted by the web area A_v, shown in Figure 5.15(b), where:

A_v = web thickness × overall depth = tD

For the T section shown in the figure:

$A_v = 0.9\, A_o$

A_o = area of the rectilinear element which has the largest dimension in the direction parallel to the load

    = td

The shear area may be stressed to the yield stress in shear, that is, to $1/\sqrt{3}$ of the yield stress in tension. The capacity is given in the code as:

$P_v = 0.6\, p_y A_v$

The code in Table 7 also notes that when the depth to thickness ratio of the web exceeds 67ε it should be checked for buckling. (See Section 4.4.5 in the code.)

If moment as well as shear occurs at the section, the web is assumed to resist all the shear while the flanges are stressed to yield by bending. The section analysis is based on the shear stress and bending stress distributions shown in Figure 5.15(c). The web is at yield under the combined bending and shear stresses and von Mises' criterion is adopted for failure in the web. The shear reduces the moment capacity, but the reduction is small for all but high values of shear force. The analysis for shear and bending is given in reference 14.

BS 5950: Part 1 gives the following expression in Section 4.2.6 for the moment capacity for plastic or compact sections in the presence of high shear load.

When the average shear force F_v is less tha 0.6 of the shear capacity P_v no reduction in moment capacity is required. When F_v is greater than $0.6\, P_v$, the reduced moment capacity is given by:

$M_c = p_y\,(S - S_v \rho_1)$

    $\ngtr 1.2\, p_y Z$

$\rho_1 = (2.5\, F_v/P_v) - 1.5$

$S_v = tD^2/4$ for a rolled section with equal flanges

S_v is defined in the code for sections with unequal flanges. Other terms used have been defined previously.

5.7 Deflection of beams

The deflection limits for beams specified in Section 2.5.1 of BS 5950: Part 1 were set in in Section 3.6 of this book. The serviceability loads are the un-factored imposed loads.

Deflection formulae are given in design manuals.[15] Deflections for some common load cases for simply supported beams together with the maximum

Beam and Load	Maximum Moment	Deflection at Centre
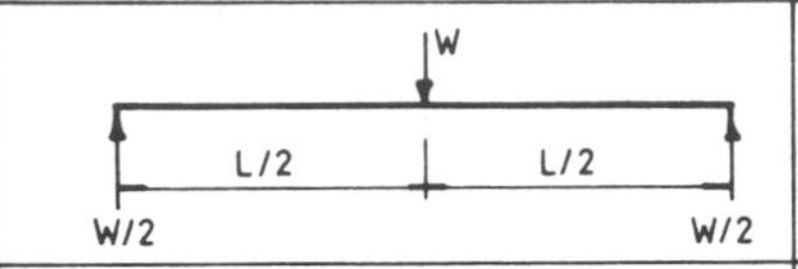 (simply supported beam, central point load W, spans $L/2$ and $L/2$, reactions $W/2$ and $W/2$)	$WL/4$	$\dfrac{WL^3}{48EI}$
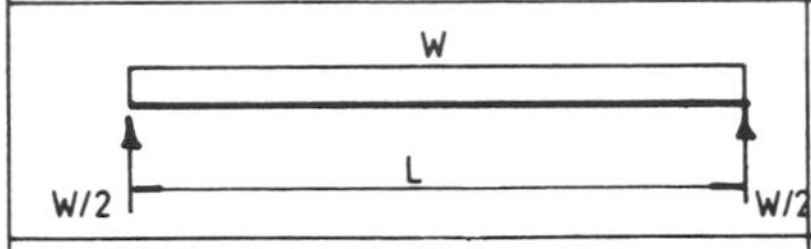 (uniformly distributed load W over span L, reactions $W/2$ and $W/2$)	$WL/8$	$\dfrac{5WL^3}{384EI}$
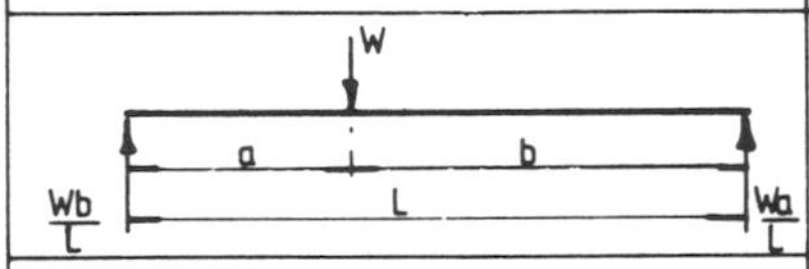 (point load W at distance a from one support, b from other, span L, reactions $\dfrac{Wb}{L}$ and $\dfrac{Wa}{L}$)	Wab/L	$\dfrac{WL^3}{48EI}\left[\dfrac{3a}{L}-4\left(\dfrac{a}{L}\right)^3\right]$
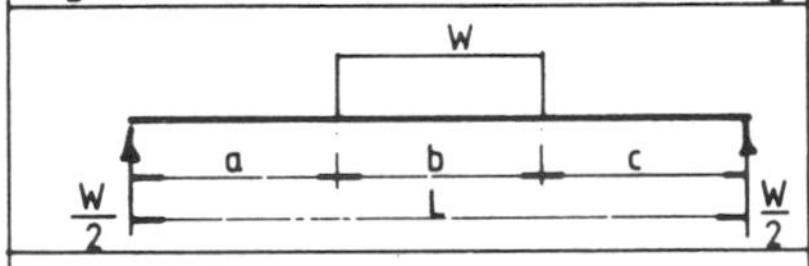 (distributed load W over central length b, with a and c on either side, span L, reactions $\dfrac{W}{2}$ and $\dfrac{W}{2}$)	$W\left(\dfrac{a}{2}+\dfrac{b}{8}\right)$	$\dfrac{W}{384EI}\left[8L^3-4Lb^2+b^3\right]$
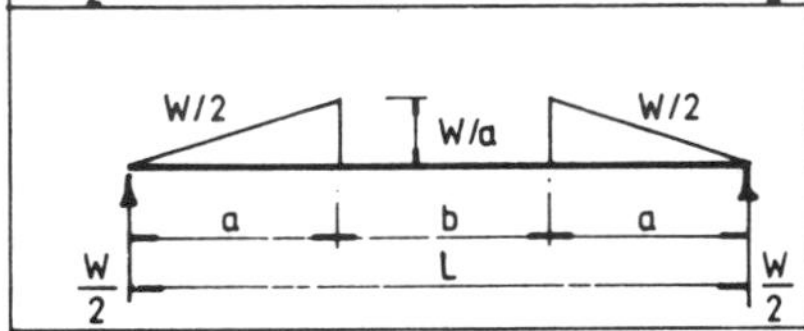 (triangular/trapezoidal load W/a, with lengths a, b, a, span L, reactions $\dfrac{W}{2}$ and $\dfrac{W}{2}$)	$Wa/3$	$\dfrac{Wa}{120EI}\left[16a^2+20ab+5b^2\right]$
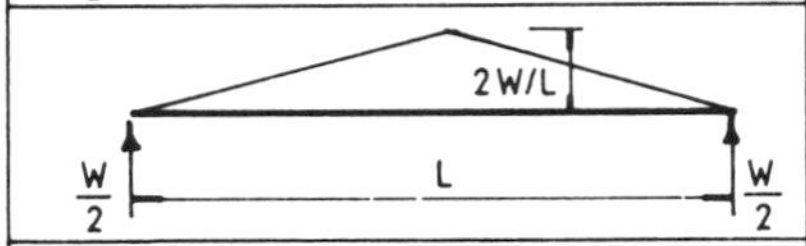 (triangular load $2W/L$ at centre, span L, reactions $\dfrac{W}{2}$ and $\dfrac{W}{2}$)	$WL/6$	$\dfrac{WL^3}{60EI}$
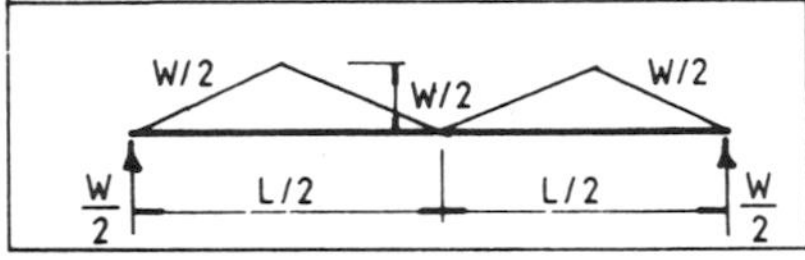 (double triangular load, $W/2$, $W/2$, $W/2$, spans $L/2$ and $L/2$, reactions $\dfrac{W}{2}$ and $\dfrac{W}{2}$)	$WL/8$	$\dfrac{WL^3}{73\cdot14EI}$

Figure 5.16 Simply supported beams: maximum moments and deflections

moments are given in Figure 5.16. For general-load cases deflections can be calculated by the moment area method.[15,16]

5.8 Beam connections

End connections to columns and other beams form an essential part of beam design. Checks for local failure are required at supports and points where concentrated loads are applied.

5.8.1 Buckling resistance of beam webs

Types of buckling caused by a load applied to the top flange are shown in Figure 5.17. The web buckles at the centre if the flanges are restrained, otherwise sideways movement or rotation of one flange relative to the other occurs.

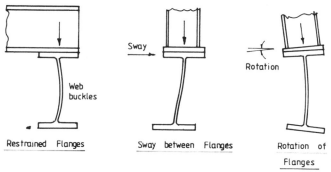

Restrained Flanges   Sway between Flanges   Rotation of

Flanges

Figure 5.17 Types of buckling

An end bracket support and an intermediate load on a beam are shown in Figures 5.18(a) and (b), respectively. The buckling resistance of a web to loads applied through the flange is given in Section 4.5.2.1 of BS 5950: Part 1. This is:

$$P_\mathrm{W} = (b_1 + n_1)\, t p_\mathrm{c}$$

where b_1 = stiff bearing length (see below)

n_1 = length assuming that the load is dispersed at 45 degrees through one half the depth of the beam

t = web thickness

p_c = compressive strength from Table 27(c) of the code for a value of slenderness specified below.

The stiff bearing is defined in Section 4.5.1.3 as the length which cannot deform appreciably in bending. The dispersion of the load is taken as 45 degrees through solid material. Stiff bearing lengths b_1 are shown in Figure 5.18(c). For the unstiffened angle a tangent is drawn at 45 degrees to the fillet and the length b_1 in terms of dimensions shown is:

$$b_1 = 2(t + 0.293\, r) - \text{clearance}$$

where r = radius of the fillet

t = thickness of the angle leg.

The slenderness for an unstiffened web:

$$\lambda = 2.5\, d/t$$

where d = clear depth of web between fillets

This applies when the flange where the load is applied is effectively restrained against rotation relative to the web and lateral movement relative to the other flange.

The code states that if the flange carrying the load is not restrained the slenderness of the web should be determined by considering the web as part of the compression member applying the load. If the load exceeds the buckling resistance of the web, stiffeners should be provided (see Section 6.3.7).

5.8.2 Bearing resistance of beam webs

The local bearing capacity of the web at its junction with the flange must be checked at supports and at points where loads are applied. The bearing capac-

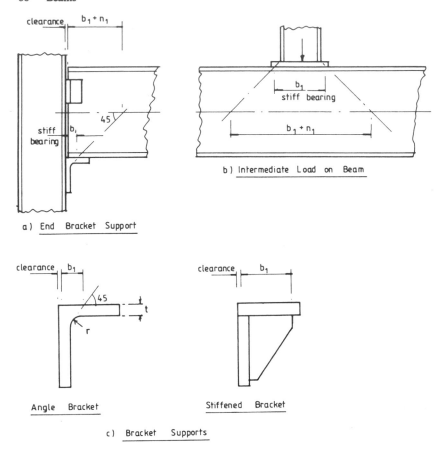

a) End   Bracket   Support

b) Intermediate   Load   on   Beam

Angle   Bracket

Stiffened   Bracket

c)   Bracket   Supports

Figure 5.18 Web buckling and bracket supports

ity is given in Section 4.5.3 of BS 5950: Part 1. An end bearing and an intermediate bearing are shown in Figures 5.19(a) and (b), respectively:

Bearing capacity $= (b_1 + n_2)\, tp_{yw}$

where b_1 = stiff bearing length

n_2 = length obtained by dispersing the load through the flange to the flange/web connection at a slope of 1 in 2.5

t = web thickness

p_{yw} = design strength of the web

For the beam supported on the angle bracket as shown in Figure 5.19(a) the bracket is checked in bearing at Section ZZ and the weld or bolts to connect it to the column are designed for direct shear only. If the bearing capacity of the beam web is exceeded, stiffeners must be provided to carry the load (see Section 6.3.7).

5.8.3 Beam-end shear connections

Design procedures for flexible end shear connections for simply supported

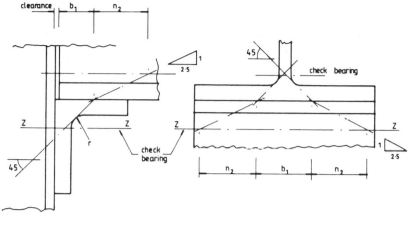

a) End   Bearing   on   an   Angle   Bracket

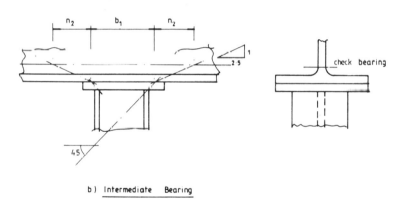

b) Intermediate   Bearing

Figure 5.19 Web and bracket bearing

beams are set out here. The recommendations are from the BCSA publication *Manual on Connections*.[10]

Two types of shear connections, beam to column and beam to beam, are shown in Figures 5.20(a) and (b), respectively. Design recommendations for the end plate are:

(1) Length—maximum = clear depth of web
    minimum = 0.6 of the beam depth
(2) Thickness— 8 mm for beams up to 457 × 191 serial size
    10 mm for larger beams
(3) Positioning—the upper edge should be near the compression flange.

Flexure of the end plate permits the beam end to rotate about its bottom edge, as shown in Figure 5.20(c). The end plate is arranged so that the beam flange at A does not bear on the column flange. The end rotation is taken as

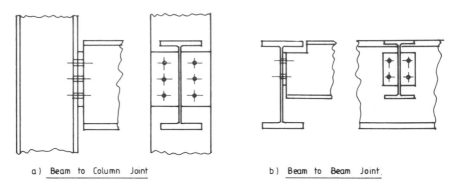

a) Beam to Column Joint   b) Beam to Beam Joint.

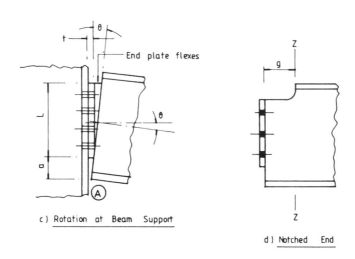

c) Rotation at Beam Support

d) Notched End

Figure 5.20 Flexible shear connection

0.03 radians, which represents the maximum slope likely to occur at the end of the beam. If the bottom flange just touches the column at A then

$$t/a = 0.03$$

or

a/t should be made $\leqslant 33$ to prevent contact.

The joint is subjected to shear only. The steps in the design are:

(1) Design the bolts for shear and bearing.
(2) Check the end plate in shear and bearing.
(3) Design the weld between the end plate and beam web.

If the beam is notched as shown in Figure 5.20(d) the beam web should be checked for shear and bending at Section Z–Z. To ensure that the web at the top of the notch does not buckle, the BCSA manual limits the maximum length of notch g to $24t$ for Grade 43 steel and $20t$ for Grade 50 steel, where t is the web thickness.

5.9 Examples of beam design

5.9.1 Floor beams for an office building

The steel beams for part of the floor of a library with book storage are shown in Figure 5.21(a). The floor is a reinforced concrete slab supported on universal beams. The design loading has been estimated as:

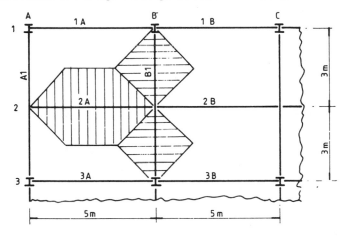

a) Part Floor Plan and Load Distribution

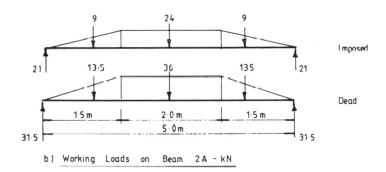

b) Working Loads on Beam 2A – kN

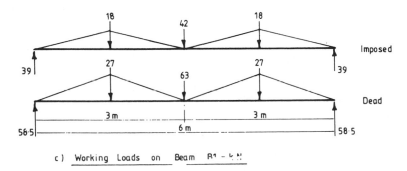

c) Working Loads on Beam B1 – kN

Figure 5.21 Library: part floor plan and beam loads

Dead load—slab, self weight of steel,
finishes, ceiling, partitions
services and fire protection: $= 6\,kN/m^2$
Imposed load from Table 1 of BS 6399: Part 1 $= 4\,kN/m^2$

Determine the section required for beams 2A and 1B and design the end connections. Use Grade 43 steel.

The distribution of the floor loads to the two beams assuming two-way spanning slabs is shown in Figure 5.21:

(1) Beam 2A

Service dead load	$= 6 \times 3$	$= 18\,kN/m$
Service imposed load	$= 4 \times 3$	$= 12\,kN/m$
Factored shear	$= (1.4 \times 31.5) + (1.6 \times 21)$	$= 77.7\,kN$

Factored moment $= 1.4\,[(31.5 \times 2.5) - (13.5 \times 1.5) - (18 \times 0.5)] +$
$\qquad\qquad\qquad 1.6\,[(21 \times 2.5) - (\ 9 \times 1.5) - (12 \times 0.5)]$
$\qquad\qquad\qquad\qquad\qquad\qquad\qquad = 122.1\,kNm$

Design strength, Grade 43 steel, thickness $\leqslant 16\,mm$

$p_y \qquad\qquad = 275\,N/mm^2$ (Table 6)

Plastic modulus $S = \dfrac{M}{p_y} = \dfrac{122.1 \times 10^3}{275} = 444\,cm^2$

Try 356×127 UB 33 $S_x = 539.8\,cm^3$
$\qquad Z_x = 470.6\,cm^3 \quad I_x = 8200\,cm^4$

The dimensions for the section are shown in Figure 5.22(a). The checks from Tables 7 BS 5950: Part 1 are:

$\varepsilon \quad = (275/p_y)^{0.5} = 1.0$
$b/T = 62.7/8.5 \quad = 7.37 < 8.5$
$d/t = 311.1/5.9 \ = 52.7 \ < 79$

This is a plastic section.

The moment capacity is $p_y S \leqslant 1.2\,p_y Z$

$p_y S_x = 275 \times 539.8/10^3 \qquad = 148.4$ kNm
$1.2\,p_y Z_x = 1.2 \times 275 \times 470.6/10^3 = 155.3$ kNm

The section is satisfactory for the moment.

The deflection due to the unfactored imposed load using formulae from Figure 5.16 is:

$$\delta = \frac{18 \times 10^3 \times 1500}{120 \times 205 \times 10^3 \times 8200 \times 10^4}[16 \times 1500^2 + 20 \times 1500 \times 2000 + 5 \times 2000^2]$$

$$+ \frac{24 \times 10^3}{384 \times 205 \times 10^3 \times 8200 \times 10^4}[8 \times 5000^3 - 4 \times 5000 \times 2000^2 + 2000^3]$$

$\qquad = 1.553 + 3.45$
$\qquad = 5.003$ mm
$\qquad \delta/span = 5.00/5000 = 1/1000 < 1/360$

The beam is satisfactory for deflection.

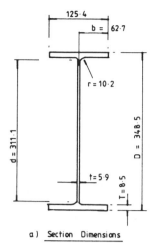

a) Section Dimensions

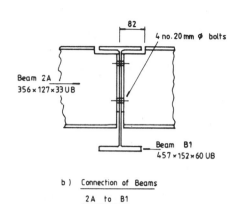

b) Connection of Beams

2A to B1

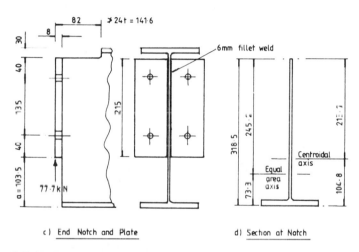

c) End Notch and Plate

d) Section at Notch

Figure 5.22 Section and end connection beam 2A

The end connection is shown in Figure 5.22(b) and the end shear is 77.7 kN. The notch required to clear the flange and fillet on beam B1 is shown in Figure 5.22(c). The end plate conforms to recommendations given in Section 5.8.3 above. To ensure end rotation:

$a/t = 103.5/8 = 12.93 < 33$

The bolts are 20 mm diameter, Grade 4.6

From Table 4.2:

Single shear value on threads	= 39.2 kN
Capacity of four bolts = 4 × 39.2	= 156.8 kN
Bearing capacity of a bolt on 8 mm thick end plate =	69.6 kN
Bearing on the end plate	= 73.6 kN

Bolts and end plate are satisfactory in bearing. The web of beam B1 is checked for bearing below:

Shear capacity of end plate in shear on both sides

$P_v = 2 \times 0.9 \times 0.6 \times 275 \times 8 \, (215 - 44)/10^3$ $= 406.2 \, \text{kN}$

Provide 6 mm fillet weld in two lengths of 215 mm each.

The strength at 0.9 kN/mm from Table 4.5 is

$= 2(215 - 12) \times 0.9$ $= 365 \, \text{kN}$

Check the beam end in shear at the notch (see Figure 5.22(d).

$P_v = 318 \times 5.9 \times 0.9 \times 0.6 \times 275/10^3$ $= 279.0 \, \text{kN}$

Check the beam end in bending at the notch. The locations of the centroid and equal area axes of the T section are shown in Figure 5.22(d). The elastic and plastic properties may be calculated from first principles (see Section 2.4). The properties are:

Elastic modulus top $Z = 148.5 \, \text{cm}^3$
Plastic modulus $S = 263.3 \, \text{cm}^3$

Moment capacity assuming a semi-compact section with the maximum stress limited to the design strength:

$M_c = 148.5 \times 275/10^3 = 40.8 \, \text{kNm}$

Factored moment at the end of the notch:

$M = 77.7 \times 82/10^3 = 5.37 \, \text{kNm}$

The beam end is satisfactory.

Note that the notch length 82 mm is taken from the *Guide* to BS 5950: Part 1: Vol. 1, Constrado.

(2) Beam B1

The beam loads are shown in Figure 5.21(c). The point load at the centre is twice the reaction of Beam 2A. The triangular loads are:

Dead $= 2 \times 1.5^2 \times 6 = 27 \, \text{kN}$
Imposed $= 2 \times 1.5^2 \times 4 = 18 \, \text{kN}$
Factored shear $= (1.4 \times 58.5) + (1.6 \times 39) = 144.3 \, \text{kN}$
Factored moment $= 1.4[(58.5 \times 3) - (27 \times 1.5)] + 1.6[(39 \times 3) - (18 \times 1.5)]$
                  $= 333 \, \text{kNm}$

Plastic modulus
$$S = 333 \times 10^3/275 = 1210.9 \, \text{cm}^3$$
Try 457 × 152 UB 60:
$$S_x = 1284 \, \text{cm}^3$$
$$Z_x = 1120 \, \text{cm}^3$$
$$I_x = 25464 \, \text{cm}^4$$

The section can be checked and will be found to be plastic. The moment capacity is:

$p_y S \quad \leqslant 1.2\, p_y Z$

$p_y S \quad = 275 \times 1284/10^3 = 353.1\ \text{kNm}$

$1.2\, p_y Z = 1.2 \times 1120 \times 275/10^3 = 369.6\ \text{kNm}$

The section is satisfactory.

Shear capacity $P_v = 0.6 \times 275 \times 454.7 \times 8.0/10^2$

$\qquad\qquad\qquad = 600.2\ \text{kN (satisfactory)}$

The deflection due to the unfactored imposed loads using formula from Figure 5.16 is:

$$\delta = \frac{42 \times 10^3 \times 6000^3}{48 \times 205 \times 10^3 \times 25\,464 \times 10^4} + \frac{36 \times 10^3 \times 6000^3}{73.14 \times 205 \times 10^3 \times 25\,464 \times 10^4}$$

$= 5.65\ \text{mm}$

$\delta/\text{span} = 5.65/6000 = 1/1062 < 1/360\ \text{(satisfactory)}$

The end connection is shown in Figure 5.23(a) with the beam supported on an angle bracket $150 \times 75 \times 10\text{L}$. Details for the various checks are shown below:

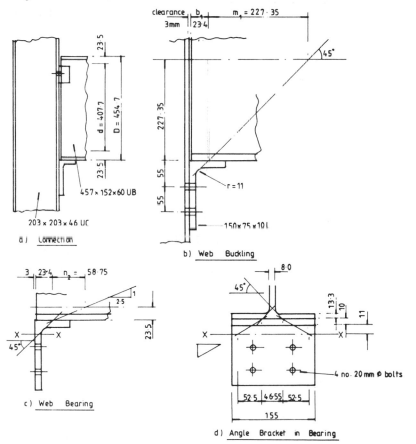

a) Connection

b) Web Buckling

c) Web Bearing

d) Angle Bracket in Bearing

Figure 5.23 End-connection beam B1

(1) **Buckling check** (see Figure 5.23(b)):

Stiff bearing b_1	$= 2(10 + 0.293 \times 11) - 3$	$= 23.4\,mm$
Web slenderness λ	$= 2.5 \times 407 \times 7/8.0$	$= 127.4$

Compressive strength (Table 27(c) $p_c = 88.6\,N/mm^2$

Buckling resistance $= P_w = (b_1 + n_1)\,tp_c$

$= (23.4 + 227.4)\,8.0 \times 88.6/10^3$ $= 177.7\,kN$

Satisfactory; reaction $= 144.3\,kN$

(2) **Bearing check** (see Figure 5.23(c)):

Bearing capacity $= (b_1 + n_2)t\,p_{yw}$

$= (23.4 + 58.75)\,8.0 \times 275/10^3$ $= 180.7\,kN$

(Satisfactory)

(3) Check bracket angle for bearing at Section XX (see Figures 5.23(d) and 4.10):

Stiff bearing b_1	$= 8.0 + 2 \times 13.3 + 1.172 \times 10.2 =$	$46.55\,mm$
Length in bearing $= 2 \times 52.5 + 46.55$		$= 151.6\ mm$
Bearing capacity $= 151.6 \times 10 \times 275/10^3$		$= 417\,kN$

(Satisfactory)

(4) **Bracket bolts:**

Provide four No. 20 mm diameter Grade 4.6 bolts

Shear capacity $= 4 \times 39.2$ $= 156.8\,kN$

The bolts are adequate.

(5) Check beam B1 for bolts from 2 No. beams 2A bearing on web 8.0 mm thick:

Reactions $= 2 \times 77.7$ $= 155.4\,kN$

Bearing capacity of bolts

$= 4 \times 435 \times 20 \times 8.0/10^3$ $= 278\,kN$

The joint is satisfactory.

5.9.2 Beam with unrestrained compression flange

Design the simply supported beam for the loading shown in Figure 5.24. The loads P are normal loads. The beam ends are restrained against torsion with the compression flange free to rotate in plan. The compression flange is unrestrained between supports. Use Grade 43 steel.

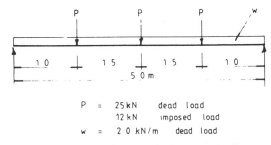

P =	25 kN	dead load
	12 kN	imposed load
w =	2·0 kN/m	dead load

Figure 5.24 Beam with unrestrained compression flange

Factored shear $= (1.4 \times 37.5) + (1.4 \times 5) + (1.6 \times 18) = 88.3 \text{ kN}$

Factored moment $= 1.4(37.5 \times 2.5 - 25 \times 1.5) + 1.4 \times 2 \times 5^2/8$
$$+ 1.6(18 \times 2.5 - 12 \times 1.5) \qquad = 130.7 \text{ kNm}$$

Try 457×152 UB 60. The properties are:

$r_y = 3.23 \text{ cm}; \ x = 37.5; \ u = 0.869$
$S_x = 1280 \text{ cm}^3$

Note that a check will confirm this is a plastic section.

Design strength $p_y = 275 \text{ N/mm}^2$ (Table 6, BS 5950)

The effective length L_E from Table 9 of BS 5950: Part 1:

$L_E = 5000 \text{ mm}$

The loads are applied within the unrestrained length. Hence the factor $m = 1$ and n is evaluated.

Equivalent slenderness $\lambda_{LT} = nuv\lambda$

λ	$= 5000/32.3$	$= 154.8$
N	$= 0.5$ and x	$= 37.5$
λ/x	$= 154.8/37.5$	$= 4.13$
v	$= 0.855$ from Table 14 of BS 5950: Part 1.	

The loads extend outside the middle fifth of the beam (use Table 16):

β	$=$ ratio of end moments	$= \beta M/M = 0$
M_0	$=$ midspan moment	$= 130.7 \text{ kNm}$
γ	$= M/M_o$	$= 0$
n	$= 0.94$	

$\lambda_{LT} = 0.94 \times 0.869 \times 0.855 \times 154.8 = 108.1$

Bending strength (Table 11) $p_b \ = 112 \text{ N/mm}^2$

Buckling resistance moment:

$M_b = 112 \times 1280/10^3 \qquad = 143.4 \text{ kNm}$

Shear capacity—overall depth $D = 454.7 \text{ mm}$
$\qquad\qquad$ —web thickness $t = 8.0 \text{ mm}$

$P_v = 0.6 \times 275 \times 454.7 \times 8/10^3 = 600.2 \text{ kN}$

The section is satisfactory.

The conservative approach in Section 4.3.7.7 gives:

$n = 0.94$—Table 20

Modified slenderness $\lambda = 0.94 \times 154.8 = 145.5$

For $x = 37.5$ and $\lambda = 145.5$

$p_b = 103.6 \text{ N/mm}^2$—Table 19(b)

$M_b = 103.6 \times 1280/10^3 = 132.6 \text{ kNm}$

The section is satisfactory.

5.9.3 Beam subjected to bending about two axes

A beam of span 5 m with simply supported ends not restrained against torsion has its major principal axis inclined at 30 degrees to the horizontal, as shown in Figure 5.25. The beam is supported at its ends on sloping roof girders. The unrestrained length of the compression flange is 5 m. If the beam is 457 × 152 UB 52, find the maximum factored load that can be carried at the centre. The load is applied by slings to the top flange.

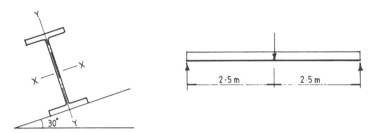

Figure 5.25 Beam in biaxial bending

Let the centre factored load $= W$ kN. The beam self weight is unfactored.

Moments $M_x = [W \times 5/4 + (52 \times 9.81 \times 5^2 \times 1.4/10^3 \times 8)] \cos 30°$
$$= 1.083W + 1.933$$
$$M_y = M_x \tan 30° = 0.625W + 1.116$$

Properties for 457 × 152 UB 52:

$$S_x = 1090 \text{ cm}^3$$
$$Z_y = 84.6 \text{ cm}^3 \qquad r_y = 3.11 \text{ cm}$$
$$x = 43.9 \qquad u = 0.859$$

The section is a plastic section. The design strength $p_y = 275 \text{ N/mm}^2$ from Table 6 of BS 5950.

(1) Moment capacity for XX axis

Effective length—The ends are torsionally unrestrained and free to rotate in plan and the load is destablizing. (Refer to Table 9 of BS 5950.)

$$L_E = 1.4 (5000 + 2 \times 449.8) = 8259.4 \text{ mm}$$

Slenderness $\lambda = 8259.4/31.1 = 265.57$

The load is destabilizing.

Factors $m = n = 1$ (Table 13 of BS 5950)

$N = 0.5$ uniform I section
$\lambda/x = 265.57/43.9 = 6.05$
$v = 0.768$ (Table 14 of BS 5950)

From Table 15 for $\gamma = \beta = 0$:

$n = 0.86$

Equivalent slenderness:

$$\lambda_{LT} = 0.859 \times 0.768 \times 265.57 = 175.2$$

Bending strength $p_b = 50.9\,\text{N/mm}^2$ (Table 11 of BS 5950: Part 1)
Buckling resistance moment

$$M_b = 50.9 \times 1090/10^3 = 55.5\,\text{kNm}$$

(2) Biaxial bending

The capacity in biaxial bending is determined by the buckling capacity at the centre of the beam (see Section 5.5.4). The interaction relationship to be satisfied is:

$$\frac{mM_x}{M_b} + \frac{mM_y}{p_y Z_y} \leqslant 1$$

The moment capacity for the YY axis

$$= p_y Z_y = 275 \times 84.6/10^3 = 23.3\,\text{kNm}$$

Factor $m = 1$

$$\frac{(1.083\,W + 1.933)}{55.5} + \frac{(0.625\,W + 1.116)}{23.3} = 1$$

and $W = 19.8\,\text{kN}$

5.10 Compound beams

5.10.1 Design considerations

A compound beam consisting of two equal flange plates welded to a universal beam is shown in Figure 5.26.

(1) Section classification

Compound sections are classified into plastic compact and semi-compact in the same way as discussed for universal beams in Section 5.3. However, the compound beam is treated as a section built up by welding. The limiting proportions from Table 7 of BS 5950: Part 1 for such sections are shown in Figure 5.26. The manner in which the checks are to be applied set out in Section 3.5.5 of the code is as follows:

(1) Whole flange consisting of flange plate and universal beam flange is checked using b_1/T, where

   b_1 = total outstand of the compound beam flange

   T = thickness of the universal beam flange

(2) The outstand b_2 of the flange plate from the universal beam flange is checked using b_2/T_f, where

   T_f = thickness of the flange plate

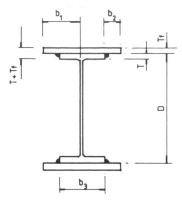

Compound   Beam

Element	Plastic	Compact	Semi - Compact
Outside element b_1 / T b_2 / T_f } $\leqslant$	7·5 ϵ	8·5 ϵ	13 ϵ
Internal element b_3 / T_f $\leqslant$	23ϵ	25ϵ	28ϵ

$$\epsilon = (275 / P_y)^{0.5}$$

Figure 5.26 Limiting proportions for flanges of compound beams fabricated by welding

(3) The width/thickness ratio of the flange plate between welds b_3/T_f is checked. The limits for an unrestrained element are given in Figure 5.26:

b_3 = width of the universal beam flange

(4) The universal beam flange itself and the web must be checked as set out in Section 5.3

(2) Moment capacity

The area of flange plates to be added to a given universal beam to increase the strength by a required amount may be determined as follows. This applies to a restrained beam (see Figure 5.27(a)):

Total plastic modulus required:

$$S_x = M/p_y$$

where M = applied factored moment

If S_{UB} is the plastic modulus for the universal beam, the additional plastic modulus required is:

$$S_{ax} = S_x - S_{UB}$$
$$= 2BT_f(D + T_f)/2$$

where $B \times T_f$ = flange area

$\qquad D$ = depth of universal beam

Suitable dimensions for the flange plates can be quickly established. If the beam is unrestrained, successive trials will be required.

(3) Curtailment of flange plates

For a restrained beam with a uniform load the theoretical cut-off points for the flange plates can be determined as follows (see Figure 5.27(b)):

The moment capacity of the universal beam:

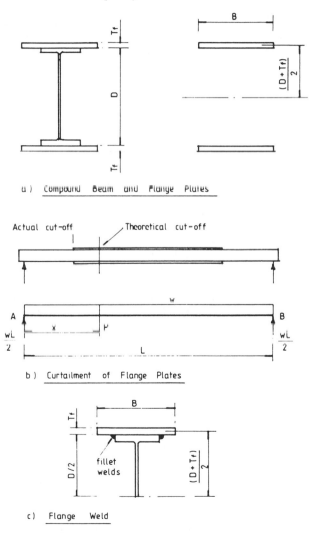

a) Compound Beam and Flange Plates

b) Curtailment of Flange Plates

c) Flange Weld

Figure 5.27 Compound beam design

$M_{UB} = p_y S_{UB} \leqslant 1.2 \, p_y Z_{UB}$

Z_{UB} = elastic modulus for the universal beam

Equate M_{UB} to the moment at P a distance x from the support:

$wLx/2 - wx^2/2 = M_{UB}$

where w = factored uniform load

L = span of the beam

Solve the equation for x. The flange plate should be carried beyond the theoretical cut-off point so that the weld on the extension can develop the load in the plate at the theoretical cut-off.

(4) Web

The universal beam web must be checked for shear. It must also be checked for buckling and crushing if the beam is supported on a bracket or column or if a point load is applied to the top flange.

(5) Flange plates to universal beam welds

The fillet welds between the flange plates and universal beam are designed to resist horizontal shear using elastic theory (see Figure 5.27(c)):

Horizontal shear in each fillet weld:

$$= \frac{F_v B T_f \, (D - T_f)}{4 \, I_x}$$

where F_v = factored shear

I_x = moment of inertia about XX axis

The other terms have been defined above.

The leg length can be selected from Table 4.5. In some cases a very small fillet weld is required, but the minimum recommended size of 6 mm should be used.

Intermittent welds may be specified but continuous welds made automatically are to be preferred. These welds considerably reduce the likelihood of failure due to fatigue or brittle fracture.

5.10.2 Design of a compound beam

A compound beam is to carry a uniformly distributed dead load of 400 kN and an imposed load of 600 kN. The beam is simply supported and has a span of 11 m. Allow 30 kN for the weight of the beam. The overall depth must not exceed 700 mm. The length of stiff bearing at the ends is 215.9 mm where the beam is supported on 203 × 203 UC 71 columns. Full lateral support is provided for the compression flange. Use Grade 43 steel.

(1) Design the beam section and check deflection assuming a uniform section throughout.
(2) Determine the theoretical and actual cut-off points for the flange plates and the possible saving in weight that would result if the flange plates were curtailed.

(3) Check the web for shear, buckling and bearing, assuming that plates are not curtailed.
(4) Design the flange plate to universal beam welds.

(1) Design of the beam section

The total factored load carried by the beam is:

$$= 1.4 \, (400 + 30) + (1.6 \times 600) = 1562 \, \text{kN (that is, } 142 \, \text{kN/m)}$$

Maximum moment $= 1562 \times 11/8 = 2147.8 \, \text{kNm}$

The loading, shear force and bending moment diagrams are shown in Figure 5.28(a).

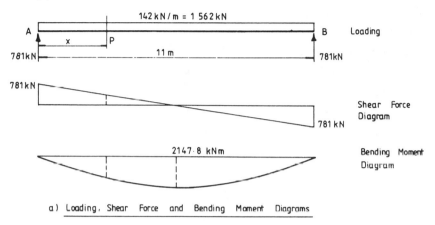

a) Loading, Shear Force and Bending Moment Diagrams

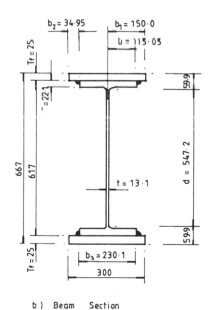

b) Beam   Section

Figure 5.28 Compound beam

Assume that the flanges of the universal beam are thicker than 16 mm:

$p_y = 265$ N/mm^2 (from Table 6)

Plastic modulus $S_x = 2147.8 \times 10^3/265 = 8104.9$ cm^3.

Try 610×229 UB 140, where $S_x = 4146$ cm^3

The beam section is shown in Figure 5.278(b). The additional plastic modulus required $= 8104.9 - 4146 = 3958.9$ cm^3

$$= 2 \times 300 \times T_f (617 + T_f)/(2 \times 10^3)$$

where the flange plate thickness T_f is to be determined for a width of 300 mm. This reduces to:

$$T_f^2 + 617\, T_f - 13\,196 = 0$$

Solving gives $T_f = 20.69$ mm.

Provide plates 300 mm $\times$ 25 mm.

The total depth is 667 mm (satisfactory).

Check the beam dimensions for local buckling:

$$\varepsilon = (275/265)^{0.5} = 1.02$$

Universal beam (see Figure 5.4):

Flange $b/T = 115.1/22.1 = 5.21 < 8.5 \times 1.02 = 8.67$

Web $d/t = 547.2/13.1 = 41.7 < 79 \times 1.02 = 80.58$

Compound beam flange (see Figure 5.26):

Flange $b_1/T = 150/22.1 = 6.79 < 1.02 \times 7.5 = 7.65$

$\qquad b_2/T_f = 34.95/25 = 1.40 < 7.65$

$\qquad b_3/T_f = 230.1/25 = 9.2 < 23 \times 1.02 = 23.46$

The section meets the requirements for a plastic section.

The moment of inertia about the XX axis for the compound section is calculated. Note for the universal beam:

$I_x = 111\,844$ cm^3

$I_x = 111\,844 + 2 \times 30 \times 2.5 \times 32.1^2 + 2 \times 30 \times 2.5^2/12$

$\quad = 266\,483$ cm^4

The deflection due to the unfactored imposed load is

$$\delta = \frac{5 \times 600 \times 10^3 \times 11\,000^3}{384 \times 205 \times 10^3 \times 266\,483 \times 10^4} = 19.03 \text{ mm}$$

$\delta/\text{span} = 19.03/11\,000 = 1/578 < 1/360$ (Satisfactory)

(2) Curtailment of flange plates

Moment capacity of the universal beam:

$$M_c = 4146 \times 265/10^3 = 1098.7 \text{ kNm}$$

Referring to Figure 5.28(a), determine the position of P where the bending moment in the beam is 1098.7 kNm from the following equation:

$$781x - 142x^2/2 = 1098.7$$

This reduces to

$$x^2 - 11x + 15.47 = 0$$
$$x = 1.656\,\text{m from each end}$$

The compound section will be the elastic range at this point with an average stress in the plate for the factored loads:

$$= \frac{1098.7 \times 10^6 \times 321}{266\,483 \times 10^4} = 132.4\;\text{N/mm}^2$$

Force in the flange plate:

$$= 132.4 \times 300 \times 25/10^3 = 993\,\text{kN}$$

Assume 6 mm fillet weld, strength 0.9 kN/mm from Table 4.5 (see (4) below). Length of weld to develop the force in the plate:

$$= [993/(2 \times 0.9)] + 6 = 557.7\,\text{mm}$$

Actual cut-off length = 1656 − 557.5 = 1098.5 mm

Cut plates off at 1000 mm from each end.

Saving in material from curtailment.

Area of universal beam = 178.4 cm²

Area of flange plates   = 150   cm²

Volume of the compound beam with no curtailment of plates

= 328.4 × 1100   = 36.12 × 10⁴ cm³

Volume of material saved = 200 × 150 = 3.0 × 10⁴ cm³

Saving in material   = 8.3%

(3) Web in shear, buckling and bearing

(1) Shear capacity (see Figure 5.28(b)). This is checked on the web of the universal beam:

$$P_v = 0.6 \times 265 \times 617 \times 13.1/10^3 = 1285\,\text{kN}$$
$$\text{Factored shear } F_v \qquad\quad = 781\,\text{kN}$$

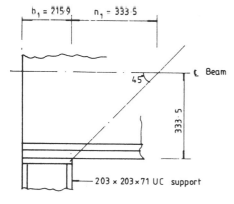

a)   Web   Buckling

Figure 5.29 (contd. overleaf)

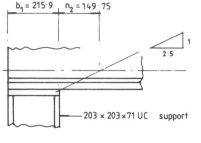
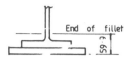

b) Web Crushing

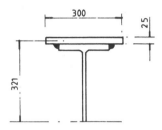

c) Flange Plate to Universal Beam Weld Design

Figure 5.29 Buckling and bearing check and flange weld design

(2) Web buckling (see Figures 5.28(b) and 5.29(a)):

Web slenderness $\lambda = 2.5 \times 547.2/13.1 = 104.4$

Compressive strength (Table 27(c)) for

$p_y = 265\,\text{N/mm}^2$, $p_c = 116.4\,\text{N/mm}^2$

Buckling resistance:

$P_w = (215.9 + 333.5)\,116.4 \times 13.1/10^3 = 838.7\,\text{kN}$ (Satisfactory)

(3) Web buckling (see Figures 5.28(b) and 5.29(b)). Bearing capacity:

$= (215.9 + 149.75)\,13.1 \times 265/10^3 = 1269.3\,\text{kN}$ (Satisfactory)

(4) Flange plate to universal beam weld
See Figure 5.29(c)

Factored shear at support $F_v = 781\,\text{kN}$

Horizontal shear on two fillet welds

$$= \frac{781 \times 300 \times 25 \times 321}{2 \times 266\,483 \times 10^4} = 0.353\,\text{kN/mm}$$

Provide 6 mm fillet welds, strength 0.9 kN/mm. This is the minimum size weld to be used.

5.11 Crane beams

5.11.1 Types and uses

Crane beams carry hand-operated or electric overhead cranes in industrial buildings such as factories, workshops, steelworks, etc. Types of beams used are shown in Figures 5.30(a) and (b). These beams are subjected to vertical and horizontal loads due to the weight of the crane, the hook load and dynamic loads. Because the beams are subjected to horizontal loading a larger flange or horizontal beam is provided at the top on all but beams for very light cranes.

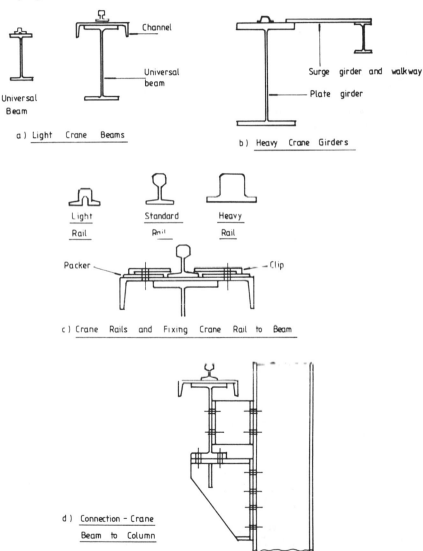

Figure 5.30 Type of crane beams and rails and connection to column

Light crane beams consist of a universal beam only or of a universal beam and channel, as shown in Figure 5.30(a). Heavy cranes require a plate girder with surge girder, as shown in Figure 5.30(b). Only light crane beams are considered in this book.

Some typical crane rails and the fixing of a rail to the top flange are shown in Figure 5.30(c). The connection of a crane girder to the bracket and column is shown in Figure 5.30(d). The size of crane rails depends on the capacity and use of the crane.

5.11.2 Crane data

Crane data can be obtained from the manufacturer's literature. The data required for crane beam design are:

Crane capacity
Span
Weight of crane
Weight of the crab
End carriage wheel centres
Minimum hook approach
Maximum static wheel load

The data are shown in Figure 5.31.

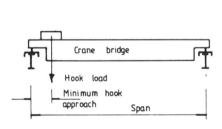

Figure 5.31 Crane design data

(1) Loads on crane beams

Crane beams are subjected to:

Vertical loads from self weight, the weight of the crane, the hook load and impact; and
Horizontal loads from crane surge.

Cranes are classified into four classes in BS 2573: *Rules for Design of Cranes*, Part 1: Specification for Classification, Stress Calculations and Design Criteria for Structures. The classes are:

Class 1—light. The safe working load is rarely hoisted;
Class 2—moderate. The safe working load is hoisted fairly frequently;
Classes 3 and 4 are heavy and very heavy cranes.

Only beams for cranes of classes 1 and 2 are considered in this book. The dynamic loads caused by these classes of cranes are given in BS 6399: Part 1, Section 7. The loading specified in the code is set out below.

The following allowances shall be deemed to cover all forces set up by vibration, shock from slipping of slings, kinetic action of acceleration and retardation and impact of wheel loads:

(1) For loads acting vertically, the maximum static wheel loads shall be increased by the following percentages:
    For electric overhead cranes   25%
    For hand-operated cranes   10%
(2) The horizontal force acting transverse to the rails shall be taken as a percentage of the combined weight of the crab and the load lifted as follows:
    For electric overhead cranes   10%
    For hand-operated cranes   5%
(3) The horizontal force acting along the rails shall be taken as 5 per cent of the static wheel loads for either electric or hand-operated cranes.

The forces specified in (2) or (3) may be considered as acting at rail level. Either of these forces may act at the same time as the vertical load. The load factors to be used with crane loads given in Table 2 in the code are:

Vertical or horizontal crane loads considered sepearately:

$$\gamma_f = 1.6$$

Vertical and horizontal crane loads acting together:

$$\gamma_f = 1.4$$

The application of these clauses will be shown in an example.

(2) Maximum shear and moment

The wheel loads are rolling loads, and must be placed in position to give maximum shear and moment. For two equal wheel loads:

(1) The maximum shear occurs when one load is nearly over a support;
(2) The maximum moment occurs when the centre of gravity of the loads and one load are placed equidistant about the centre line of the girder. The maximum moment occurs under the wheel load nearest the centre of the girder.[6]

The load cases are shown in Figure 5.32.

Note that if the spacing between the loads is greater than 0.586 of the span of the beam, the maximum moment will be given by placing one wheel load at the centre of the beam.

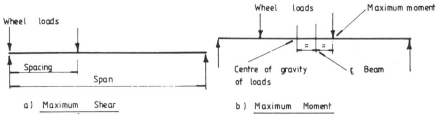

Figure 5.32 Rolling loads: maximum shear and moment

5.11.3 Crane beam design

(1) Buckling resistance moment for XX axis

Section properties

A crane girder section consisting of a universal beam and channel is shown in Figure 5.33(a) and the elastic properties for a range of sections are given in the *Structural Steelwork Handbook*. The plastic properties are calculated as fol-

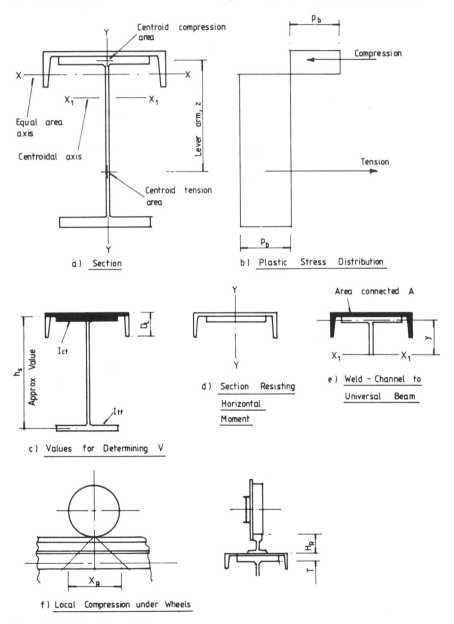

a) Section

b) Plastic Stress Distribution

c) Values for Determining V

d) Section Resisting

Horizontal

Moment

e) Weld – Channel to

Universal Beam

f) Local Compression under Wheels

Figure 5.33 Crane beam design

lows for the plastic stress distribution with bending about the equal area axis shown in Figure 5.33(b).

First locate the equal area axis by trial and error and then calculate the positions of the centroids of the tension and compression areas. If z is the lever arm between these centroids, the plastic modulus:

$$S_x = Az/2$$

where A = total area of cross section

The plastic modulus may also be calculated from the definition. This is the algebraic sum of the first moments of area about the equal area axis.

Lateral torsional buckling

The code specifies in Section 4.11.3 that no reduction is to be made for moment gradient: that is, $m = n = 1.0$. The effective length L_E = span for a simply supported beam with the ends torsionally restrained and the compression flange laterally restrained but free to rotate on plan.

The slenderness $\lambda = L_E/r_y$, where r_y = radius of gyration for the whole section about the YY axis.

The factors modifying the slenderness are set out in Appendix B.2.5 of the code. The buckling parameter $u = 1.0$. This may also be calculated from a formula in Appendix B. The torsional index x for a flanged section symmetrical about the minor axis is:

$$x = 0.566 \, h_s \, (A/J)^{0.5}$$

where h_s = distance between the shear centres of the flanges.

As a conservative approximation, h_s may be taken as the distance from the centre of the bottom flange to the centroid of the channel web and universal beam flange, as shown in Figure 5.33(c):

A = area of cross section
J = torsion constant
    = $1/3 \, (\Sigma b t^3 + h_w t_w^3)$
b = flange width
t = flange thickness
h_w = web depth
t_w = web thickness

Note that the top flange of the universal beam and channel web act together, so t is the sum of the thicknesses. The width may be taken as the average of the widths of the universal beam flange and the depth of web of the channel:

$$N = \frac{T_{cf}}{T_{cf} + T_{tf}} > 0.5$$

where T_{cf} = moment of inertia of the top flange about the YY axis

    = I_x (channel) + $1/2 \, I_y$ (universal beam)

T_{tf} = moment of inertia of the bottom flange about the YY axis

    = $1/2 \, I_y$ (universal beam)

The monosymmetry index for an I or T section with lipped flange is:

$$\psi = 0.8 \, (2N-1) \, (1+\frac{D_L}{2D}) \text{ for } N > 0.5$$

where D = overall depth of section

$\qquad D_L$ = depth of lip

$\qquad$ = breadth of channel flange

The slenderness factor:

$$v = \{[4N(1-N)+(\lambda'/x)^2/20 + \psi^2]^{0.5} + \psi\}^{-0.5}$$

The modified slenderness:

$$\lambda_{LT} = u.v. \, \lambda$$

The bending strength p_b is obtained from Table 12 for welded sections.
The buckling resistance moment:

$$M_b = S_x p_b$$

This must exceed the factored moment for the vertical loads only including impact with load factor 1.6.

(2) Moment capacity for the YY axis (see Figure 5.33(d))

The horizontal bending moment is assumed to be taken by the channel and top flange of the universal beam. The elastic modulus Z_y for this section is given in the *Structural Steelwork Handbook*. The moment capacity:

$$M_{cy} = Z_y p_y$$

(3) Biaxial bending check

The overall buckling check using the simplified approach is given in Section 4.8.3 of BS 5950: Part 1. This is:

$$\frac{M_x}{M_b} + \frac{M_y}{M_{cy}} \leqslant 1$$

Two checks are required.

(1) Vertical crane loads with no impact and horizontal loads only with load factor 1.6; and
(2) Vertical crane loads with impact and horizontal loads both with load factor 1.4.

(4) Shear capacity

The vertical shear capacity is checked as for a normal beam (see Section 5.6.2). The horizontal shear load is small and is usually not checked.

(5) Weld between channel and universal beam (see Figure 5.33(e))

The horizontal shear force in each weld:

$$\frac{FAy}{2I_x}$$

where F = factored shear

A = area connected by the weld

= area of the channel

y = distance from the centroid of the channel to the centroid of the crane beam

I_x = moment of inertia of the crane beam about the XX axis.

The elastic properties are given in the *Structural Steelwork Handbook*.

(6) Web buckling and bearing

The web is to be checked for buckling and bearing as set out in Section 5.8. The length to be taken for stiff·bearing depends on the bracket construction or other support for the crane beam (for example, if it is carried on a crane column).

(7) Local compression under wheels (see Figure 5.33(f))

BS 5950: Part 1 specifies in Section 4.11.5 that the local compression on the web may be obtained by distributing the crane wheel load over a length:

$x_R = 2(H_R + T)$

H_R = rail height

T = flange thickness

Bearing stress = $p/(t\,x_R)$

p = crane wheel load

t = web thickness

This stress should not exceed the design strength of the web p_{yw}.

5.11.4 Crane beam deflection

The deflection limitations for crane beams given in Table 5 of BS 5950: Part 1 are quoted in Table 3.2 in this book. These are:

(1) Vertical deflection due to static wheel loads = span/600
(2) Horizontal deflection due to crane surge calculated using the top flange properties alone   = span/500

The formula for deflection at the centre of the beam is given in Figure 5.34 for crane wheel loads placed in the position to give the maximum moment. The deflection should also be checked with the loads placed equidistant about the centre of the beam, when $a = c$ in the formula given.

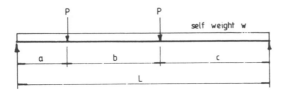

Deflection at centre

$$\frac{PL^3}{48EI}\left[\frac{3(a+c)}{L}-\frac{4(a^3+c^3)}{L^3}\right]+\frac{5wL^3}{384EI}$$

Figure 5.34 Crane beam deflection

5.11.5 Design of a crane beam

Design a simply supported beam to carry an electric overhead crane. The design data are as follows:

Crane capacity	$= 100\,\text{kN}$
Span between crane rails	$= 20\,\text{m}$
Weight of crane	$= 90\,\text{kN}$
Weight of crab	$= 20\,\text{kN}$
Minimum hook approach	$= 1.1\,\text{m}$
End carriage wheel centres	$= 2.5\,\text{m}$
Span of crane girder	$= 5.5\,\text{m}$
Self weight of crane girder	$= 8\,\text{kN}$

Use Grade 43 steel.

(1) Maximum wheel loads, moments and shear

The crane loads are shown in Figure 5.35(a). The maximum static wheel loads at A

$$=\frac{90}{4}+\frac{120\times18.9}{20\times2}=79.2\,\text{kN}$$

The vertical wheel load, including impact:

$$=79.2+25\%\qquad=99\,\text{kN}$$

The horizontal surge load transmitted by friction to the rail through four wheels:

$$=10\%\,(100+20)/4=3\,\text{kN}$$

Load factors from Table 3.1

Dead load—self weight $\gamma_f = 1.4$

Vertical and horizontal crane loads considered separately

$$\gamma_f = 1.6$$

Vertical and horizontal crane loads acting together

$$\gamma_f = 1.4$$

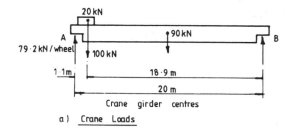

a) Crane Loads

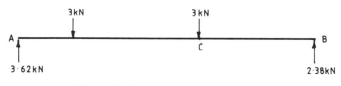

Vertical   Loads - Maximum   Moment

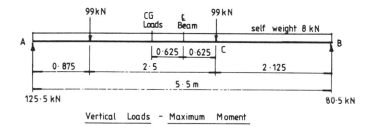

Horizontal Loads - Maximum   Moment

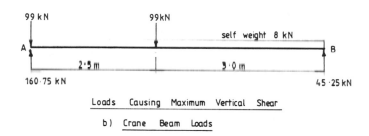

Loads   Causing   Maximum   Vertical   Shear

b) Crane   Beam   Loads

Figure 5.35 Crane and crane beam loads

The crane loads in a position to give maximum vertical and horizontal moments and maximum vertical shear are shown in Figure 5.35(b). The maximum vertical moments due to dead load and crane loads are calculated separately:
Dead load:

$$R_B = \qquad\qquad\qquad\qquad = \quad 4\,kN$$
$$M_c = (4 \times 2.125) - (8 \times 2.125^2/5.5 \times 2) = 5.22\,kNm$$

Crane load, including impact:

$$R_B = 99(0.875 + 3.375)55.5 \qquad = 76.5\,\text{kN}$$
$$M_c = 76.5 \times 2.125 \qquad\qquad = 162.6\,\text{kNm}$$

Crane loads with no impact:

$$M_c = 162.6 \times 79.2/99 \qquad\qquad = 130.1\,\text{kNm}$$

The maximum horizontal moment due to crane surge:

$$R_B = 3(0.875 + 3.375)/5.5 \qquad = 2.32\,\text{kN}$$
$$M_c = 2.32 \times 2.125 \qquad\qquad = 4.93\,\text{kNm}$$

The maximum vertical shear:

$$\text{Dead load } R_A \qquad\qquad = 4\,\text{kN}$$

Crane loads, including impact:

$$R_A = 99 + 99 \times 3.0/5.5 \qquad\qquad = 153.0\,\text{kN}$$

The load factors are introduced to calculate the design moments and shear for the various load combinations:

(1) Vertical crane loads with impact and no horizontal crane load.
Maximum moment:

$$M_c = (1.4 \times 5.22) + (1.6 \times 162.6) \qquad = 267.5\,\text{kNm}$$

Maximum shear:

$$F_A = (1.4 = 4) + (1.6 \times 153.0) \qquad = 250.4\,\text{kN}$$

(2) Horizontal crane loads and vertical crane loads with no impact.
Maximum horizontal moment:

$$M_c = 1.6 \times 4.93 \qquad\qquad = 7.89\,\text{kNm}$$

Maximum vertical moment:

$$M_c = (1.4 \times 5.22) + (1.6 \times 130.1) \qquad = 215.47\,\text{kNm}$$

(3) Vertical crane loads with impact and horizontal crane loads acting together.
Maximum vertical moment:

$$M_c = (1.4 \times 5.22) + (1.4 \times 162.6) \qquad = 234.95\,\text{kNm}$$

Maximum horizontal moment:

$$M_c = 1.4 \times 4.93 \qquad\qquad = 6.9\,\text{kNm}$$

(2) Buckling resistance moment for the XX axis

The following trial section is selected:

457×191 UB $74 + 254 \times 76$ channel

Referring to Figure 5.36, the equal area axis XX and centroids of the tension and compression areas are located and the plastic modulus calculated. Computations are shown in the figure. The elastic properties for this crane

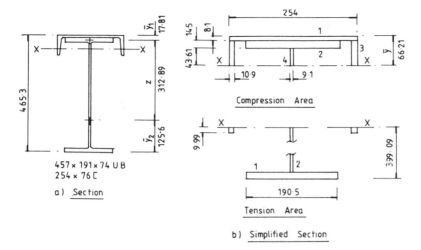

a) Section

457 × 191 × 74 U B
254 × 76 C

Compression Area

Tension Area

b) Simplified Section

1. Locate Equal Area Axis

Total area = $(25.4 \times 0.81) + (6.81 \times 2 \times 1.09) + (2 \times 19.05 \times 1.45)$
$+ (42.82 \times 0.91) = 129.64 \, cm^2$
$64.82 = (25.4 \times 0.82) + (19.05 \times 1.45) + 2 \times 1.09(y - 0.81)$
$+ (0.91(y - 1.45 - 0.81)$
$\bar{y} = 6.618 \, cm$

2. Locate Centroids of Compression and Tension Areas

Compression Area Area moments about top				Tension Area Area moments about bottom			
No	Area	y	Ay	No	Area	y	Ay
1	20.54	0.405	8.32	1	27.62	0.725	20.02
2	27.62	1.535	42.39	2	35.00	20.27	709.5
3	12.67	3.175	47.07	3	2.18	38.59	84.12
4	3.97	4.441	17.63				
Sum	64.83		115.41		64.80		813.59
	$y_1 = 1.78$				$y_2 = 12.56$		

3. Lever arm $Z = 46.53 - 1.78 - 12.56 = 31.19$ cm
4. Plastic modulus Sx = 64.83×32.19 $= 2086.8 \, cm^3$

Figure 5.36 Crane beam: plastic properties

beam are taken from the *Structural Steelwork Handbook*, and these are shown in Figure 5.37.

The bending strength p_b, taking lateral torsional buckling into account, is determined:

Effective length—L_E = span = 5500
Slenderness $\lambda = L_E/r_y = 5500/62 = 88.7$
Factors modifying slenderness:
Buckling parameter $u = 1.0$

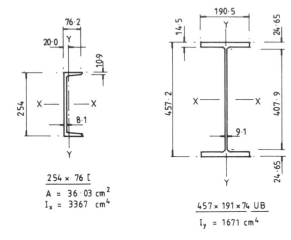

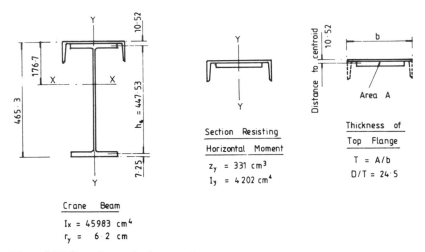

Figure 5.37 Crane beam: elastic properties

This is conservative: the value is 0.81 is calculated from the formula in Appendix B of the code.

The slenderness factor v is calculated from the formulae in Appendix B:

$$I_{cf} = I_x \text{ (channel)} + 1/2\ I_y \text{ (U.B.)}$$
$$= 3367 + 835.5 = 4202.5\ cm^4$$
$$I_{tf} = 1/2\ I_y \text{ (UB)} = 835.5\ cm^4$$
$$N = \frac{4202.5}{4202.5 + 835.5} = 0.834$$

The distance between the shear centres of the flanges:

h_s = distance from centre of bottom flange to centroid of channel web and universal beam flange = 447.53 mm approximately (see Figure 5.37)

Torsion constant:

$$J = 1/3 [(14.5^2 \times 190.5) + (9.1^2 \times 428.2) + (22.6^2 \times 211.35)$$
$$+ (2 \times 10.9^3 \times 76.2)] \qquad = 1.18 \times 10^6 \, \text{mm}^4$$

Area $A = 12964 \, \text{mm}^2$

The torsional index:

$$x = 0.566 \times 447.53 \, (13103/1.18 \times 10^6)^{0.5} = 26.5$$

This compares with $D/T = 24.5$ from the *Handbook*.

The monosymmetry index ψ for a T section with lipped flanges, where:

D_L = depth of the lip $\qquad\qquad\qquad\qquad$ = 76.2 mm
D = overall depth $\qquad\qquad\qquad\qquad$ = 465.3 mm
ψ = $0.8[(2 \times 0.834) - 1] [1 + (76.2/2 \times 465.3)] = 0.578$

The slenderness factor:

$$v = \{4 \times 0.834(1 - 0.834) + \frac{1}{20}\left(\frac{88.7}{26.5}\right)^2 + 0.578^2]^{0.5} + 0.578\}^{-0.5}$$
$$= 0.75$$

Table 14 gives:

$v = 0.769$ for $N = 0.834$ and $\lambda/x = 88.7/26.5 = 3.34$

The equivalent slenderness:

$\lambda_{LT} = 1.0 \times 0.769 \times 88.7$ $\qquad\qquad\qquad$ $= 68.2$

From Table 12 for welded sections for p_y $\quad = 265 \, \text{N/mm}^2$
Top flange thickness $= 23.6$ mm total:

$p_b = 154.3 \, \text{N/mm}^2$

The buckling resistance:

$M_b = 2086.8 \times 154.3/10^3$ $\qquad\qquad\qquad$ $= 321.9 \, \text{kNm}$

(3) Moment capacity for the top section for the YY axis

$M_{cy} = 265 \times 331/10^3 = 87.7 \, \text{kNm}$
$Z_y = 331 \, \text{cm}^3$ (from the *Structural Steelwork Handbook*)

(4) Check beam in bending

(1) Vertical moment, no horizontal moment:
$M_x = 267.5 \, \text{kNm} < M_b = 321.9 \, \text{kNm}$

(2) Vertical moment no impact + horizontal moment:

$$\frac{M_x}{M_b} + \frac{M_y}{M_{cy}} = \frac{215.5}{321.9} + \frac{7.89}{87.7} = 0.76 < 1$$

(3) Vertical moment with impact + horizontal moment:

$$\frac{M_x}{M_b} + \frac{M_y}{M_{cy}} = \frac{234.9}{321.9} + \frac{6.9}{87,7} = 0.81 < 1$$

The crane girder is satisfactory in bending.

(5) Shear capacity (see Section 5.62)

$P_y = 0.6 \times 457.2 \times 9.1 \times 265/10^3 = 611.5\,\text{kN}$

Maximum factored shear $= 250.4\,\text{kN}$

(6) Weld between channel and universal beam

The dimensions for determining the horizontal shear are shown in Figure 5.38(a). The location of the centroidal axis is taken from the *Structural Steelwork Handbook*.

Horizontal shear force in each weld:

$$= \frac{250.4 \times 3603 \times 158.1}{45\,983 \times 10^4} \qquad = 0.31\,\text{kN/mm}$$

Provide 6 mm continuous fillet (weld strength from Table 4.5 is 0.9 kN/mm).

(7) Web buckling and bearing

Assume a stiffened bracket 150 mm wide. The stiff bearing length allowing 3 mm clearance between the beams is 73.5 mm (see Section 5.8 and Figure 5.38(b)).

Slenderness $\lambda = 2.5 \times 407.9/9.1$ $= 112.1$

Compressive strength from Table 27(c) for

$p_y = 265\,\text{N/mm}^2$

$p_c = 105.9\,\text{N/mm}^2$

Buckling resistance:

$P_w = 302.1 \times 9.1 \times 105.9/10^3$ $= 291.1\,\text{kN}$

Factored reaction: $= 250.4\,\text{kN}$

The dimensions for checking web bearing are shown in Figure 5.38(c).

Bearing capacity $= 135.1 \times 9.1 \times 265/10^3 = 325.8\,\text{kN}$

(Satisfactory)

(8) Local compression under wheels

A 25 kg/m crane rail is used, and the height H_R is 65 mm. The length in bearing is shown in Figure 5.38(d):

Bearing capacity $= 265 \times 9.1 \times 175.2/10^3 = 422.4\,\text{kN}$

Factored crane wheel load
$= 99 \times 1.6$ $= 158.4\,\text{kN}$

(Satisfactory)

(9) Deflection

The vertical deflection due to the static wheel load must not exceed:

Span/600 = 5500/600 $= 9.17\,\text{mm}$

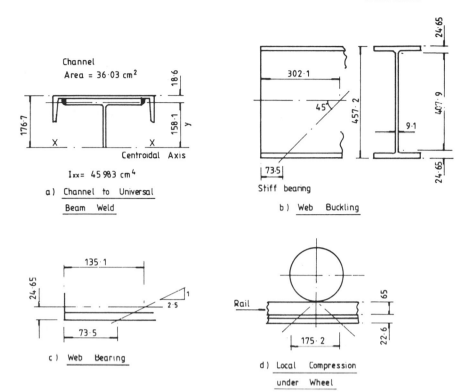

Figure 5.38 Diagrams for crane beam design

See Figures 5.34 and 5.35. The horizontal deflection due to crane surge must not exceed:

Span/500 = 5500/500   = 11 mm

The vertical deflection at the centre with the loads in position for maximum moment is

$$\delta = \frac{79.2 \times 10^3 \times 5500^3}{48 \times 205 \times 10^3 \times 45\,983 \times 10^4} \left(\frac{3(875 + 2125)}{5500} - \frac{4(875^3 + 2125^3)}{5500^3} \right)$$

$$+ \frac{5 \times 8000 \times 5500^3}{384 \times 205 \times 10^3 \times 45\,983 \times 10^4} = 4.03 + 0.18 = 4.21 \text{ mm}$$

If the loads are placed equidistant about the centre line of beam $a = c$ = 1500 mm:

$\delta = 4.29 + 0.18$ = 4.47 mmm

This gives the maximum deflection.

The horizontal deflection due to the surge loads

$$= \frac{4.29 \times 3 \times 45\,983}{79.2 \times 4202.5}$$ = 1.77 mm

The crane girder is satisfactory with respect to deflection.

5.12 Purlins

5.12.1 Types and uses

The purlin is a beam and it supports roof decking on flat roofs or cladding on sloping roofs on industrial buildings.

Members used for purlins are shown in Figure 5.39. These are cold-rolled sections, angles, channels joists and structural hollow sections. Cold-rolled sections are now used on most industrial buildings.

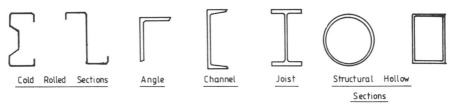

| Cold Rolled Sections | Angle | Channel | Joist | Structural Hollow Sections |

Figure 5.39 Sections used for purlins and sheeting rails

5.12.2 Loading

Roof loads are due to the weight of the roof material and the imposed load. The sheeting may be steel or aluminium corrugated or profile sheets or decking. On sloping roofs sheeting is placed over insulation board or glass wool. On flat roofs, insulation board, felt and bitumen are laid over the steel decking. Typical roof cladding and roof construction for flat and sloping roofs are shown in Figure 5.40.

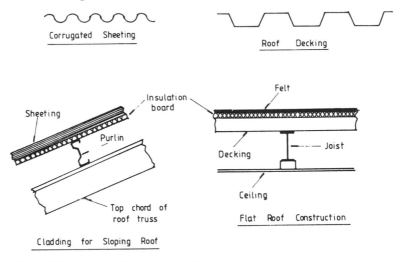

Figure 5.40 Roof materials and constructions

The weight of roofing varies from 0.3 to 1.0 kN/m², including the weight of purlins or joists, and the manufacturer's literature should be consulted. Purlins carrying sheeting are usually spaced at from 1.4 to 2.0 m centres. Joists carrying roof decking can be spaced at larger centres up to 6 m or more, depending on thickness of decking sheet and depth of profile.

Imposed loading for roofs is specified in BS 6399: Part 1 in Section 6. The code states:

(1) *Flat roofs*. On flat roofs and sloping roofs up to and including 10 degrees, where access in addition to that necessary for cleaning and repair is provided to the roof, allowance shall be made for an imposed load, including snow of $1.5 \, kN/m^2$ measured on plan or a load of $1.8 \, kN$ concentrated. On flat roofs and sloping roofs up to and including 10 degrees, where no access is provided to the roof other than that necessary for cleaning and repair, allowance shall be made for an imposed load, including snow of $0.75 \, kN/m^2$ measured on plan or a load of $0.9 \, kN$ concentrated.
(2) *Sloping roofs*. On roofs with a slope greater than 10 degrees and with no access provided to the roof other than that necessary for cleaning and repair the following imposed loads, including snow, shall be allowed for:
    (a) For a roof slope of 30 degrees or less, $0.75 \, kN/m^2$ measured on plan or a vertical load of $0.9 \, kN$ concentrated;
    (b) For a roof slope of 75 degrees or more, no allowance is necessary.
    For roof slopes between 30 and 75 degrees, the imposed load to be allowed for may be obtained by linear interpolation between $0.75 \, kN$ /m² for a 30 degree roof slope and nil for a 75 degree roof slope.

Wind loads are generally upward, or cause suction on all but steeply sloping roofs. In some instances the design may be controlled by the dead-wind load cases. Wind loads are estimated in accordance with CP 3: Chapter V: Part 2. The calculation of wind loads on a roof is given in Chapter 9 of this book.

5.12.3 Purlins for a flat roof

These members are designed as beams with the decking providing full lateral restraint to the top flange. If the ceiling is directly connected to the bottom flange the deflection due to imposed load may need to be limited to span/360, in accordance with Table 5 of BS 5950: Part 1. In other cases the code states in Section 4.12.2 that the deflection should be limited to suit the characteristics of the cladding system.

5.12.4 Purlins for a sloping roof

Consider a purlin on a sloping roof as shown in Figure 5.41(a). The load on an interior purlin is from a width of roof equal to the purlin spacing S. The load is made up of dead and imposed load acting vertically downwards.

A conservative method of design is to neglect the in-plane strength of the roof, resolve the load normal and tangential to the roof surface and design the purlin for moments about the XX and YY axes (see Figure 5.41(c)). If a section such as a channel is used where the strength about the YY axis is much less than that about the XX axis, a system of sag rods to support the purlin about the weak axis may be introduced, as shown in Figure 5.4.1(b). The purlin is then designed as a simply supported beam for bending about the XX axis and a continuous beam for bending about the YY axis.

A more realistic and economic design results if the in-plane strength of the cladding is taken into account. The purlin is designed for bending about the XX axis with the whole vertical load assumed to cause moment.

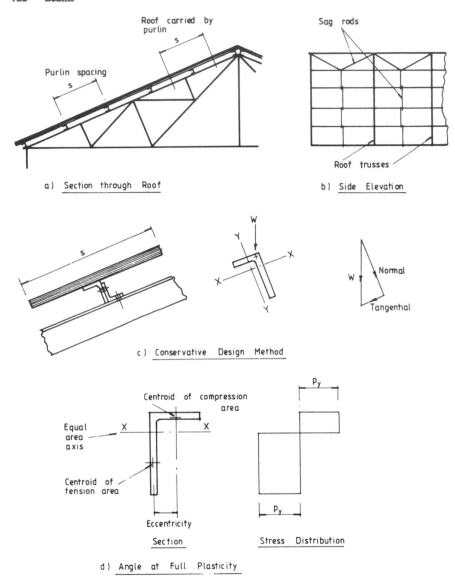

a) Section through Roof

b) Side Elevation

c) Conservative Design Method

d) Angle at Full Plasticity

Figure 5.41 Design of purlins for a sloping roof

An angle purlin bent at the full plastic moment about the XX axis is shown in Figure 5.41(d). Note that the internal resultant forces act at the centroids of the tension and compression areas. These forces cause a secondary moment about the YY axis. It is assumed in design that the sheeting absorbs this moment.

BS 5950: Part 1 gives the classification for angles in Table 7, where limiting width/thickness ratios are given for legs. The sheeting restrains the angle member so that bending take place about the XX axis. The unsupported downward leg is in tension in simply supported purlins, but it would be in compression under uplift from wind load or near the supports in continuous purlins.

The moment capacity for semi-compact outstand elements and a conservative value for plastic and compact sections is:

$M_c = p_y Z_x$

Z_x = elastic modulus for the XX axis

5.12.5 Design of purlins to BS 5950: Part 1, Section 4.12

The code states that the cladding may be assumed to provide restraint to an angle section or to the face against which it is connected in the case of other sections. Deflections as mentioned above are to be limited to suit the characteristics of the cladding used.

The empirical design method is set out in Section 4.12.4 of the code, and the general requirements are:

(1) Unfactored loads are used in the design;
(2) The span is not to exceed 6.5 m;
(3) If the purlin spans one bay it must be connected by two fasteners at each end;
(4) If the purlins are continuous over two or more bays with staggered joints in adjacent lines, at least one end of any single-bay member should be connected by not less than two fasteners.

The rules for empirical design of angle purlins are:

(1) The roof slope should not exceed 30 degrees.
(2) The load should be substantially uniformly distributed. Not more than 10 per cent of the total load should be due to other types of load;
(3) The imposed load should not be less than $0.75 \, kN/m^2$;
(4) The elastic modulus about the axis parallel to the plane of cladding should not be less than $W_p L/1800 \, cm^3$, where W_p is the total unfactored load on one span (kN) due to dead and imposed load or dead minus wind load and L is the span (mm).
(5) Dimension D perpendicular to the plane of the cladding is not to be less than $L/45$. Dimension B parallel to the plane of the cladding is not to be less than $L/60$.

The code notes that where sag rods are provided the sag rod spacing may be used to determine B. The empirical formula allows for partial fixity at supports of single-span purlins due to end connections and cladding.

5.12.6 Cold-rolled purlins

Cold-rolled purlins are almost exclusively adopted for industrial buildings. The design is to conform to BS 5950: Part 5, Code of Practice for Design of Cold Formed Sections. Detailed design of these sections is outside the scope of this book.

The purlin section for a given roof may be selected from manufacturer's data. Ward Building Components Ltd have kindly given permission for some of their design data to be reproduced in this book. This firm produces complete systems for purlins and cladding rails based on their cold-rolled 'Multibeam' section. Full information including fixing methods and accessories is given in their Technical Handbook. In addition, they have produced the Ward SIGMA software system for optimum design of purlins and side rails.

The Multibeam cold-rolled section and safe loads for double-span purlins for a limited range of purlins are shown in Table 5.2. Notes for use of the table are:

(1) The safe loads given are service loads that can actually be applied. The purlin weight has been deducted. Ultimate loads are also given in the Handbook.
(2) The loads given are based on lateral restraint being provided to the top flange by the cladding.
(3) The values given are also the permissible reversal loads due to wind uplift if the anti-sag tie arrangements specified in the Handbook are provided.

Table 5.2 Ward Building Components cold-rolled purlins—design data

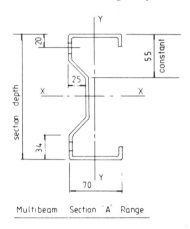

Multibeam   Section 'A' Range

Example—Section A 170/160
Depth = 170 mm; thickness = 1.6 mm

Safe loads for double-span purlins (kN/m²) including single-span end-bay members with sleeved connections

Span (m)	Section	Safe load for purlin centres (mm)				Purlin weight (kg/m)
		1500	1675	1800	2000	
5.0	A 140/155	1.14	1.02	0.95	0.86	3.75
	A 140/165	1.28	1.14	1.06	0.96	4.00
	A 170/160	1.52	1.36	1.27	1.14	4.24
	A 170/170	1.70	1.52	1.41	1.27	4.51
6.0	A 170/160	1.07	0.96	0.89	0.80	4.24
	A 170/170	1.19	1.07	0.99	0.89	4.51
	A 200/160	1.25	1.20	1.12	1.01	4.61
	A 200/180	1.50	1.34	1.25	1.13	5.20

5.12.7 Purlin design examples

Example 1. Design of a purlin for a flat roof

The roof consists of steel decking with insulation board, felt and rolled-steel joist purlins with a ceiling on the underside. The total dead load is $0.9\,kN/m^2$ and the imposed load is $1.5\,kN/m^2$. The purlins span 4 m and are at 2.5 m centres. The roof arrangement and loading are shown in Figure 5.42. Use Grade 43 steel.

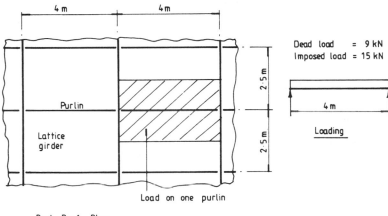

Figure 5.42 Purlin for a flat roof

Dead load $\quad = 0.9 \times 4 \times 2.5 \qquad = 9 \quad$ kN
Imposed load $= 1.5 \times 4 \times 2.5 \qquad = 15 \quad$ kN
Design load $\quad = (1.4 \times 9) + (1.6 \times 15) = 36.6\,kN$
Moment $\quad\quad - 36.6 \times 4/8 \qquad\qquad = 18.3\,kN$
Design strength (Table 6 of BS 5950) $p_y = 275\,N/mm^2$
Modulus $Z \quad = 18.3 \times 10^3/275 = 66.54\,cm^3$
Try 127×76 joist $13.36\,kg/m$ $Z = 74.94\,cm^3$
$$I_x = 475.9\,cm^4$$

Deflection due to imposed load:

$$\delta = \frac{5 \times 15 \times 10^3 \times 4000^3}{384 \times 205 \times 10^3 \times 475.9 \times 10^4} = 12.81\,mm$$

$\delta/span = 12.81/4000 = 1/312 > 1/360$
Increase section to 127×76 joist $16.37\,kg/m$, $I_x = 569.4\,cm^4$
$S/span = 1/373$ (Satisfactory)
Purlin 127×76 joist $16.37\,kg/m$

Example 2. Design of an angle purlin for a sloping roof

Design an angle purlin for a roof with slope 1 in 2.5. The purlins are simply supported and span 5.0 m between roof trusses at a spacing of 1.6 m. The total dead load, including purlin weight, is $0.32\,\text{kN/m}^2$ on the slope and the imposed load is $0.75\,\text{kN/m}^2$ on plan. Use Grade 43 steel.

The arrangement of purlins on the roof slope and loading are shown in Figure 5.43:

Dead load on slope $= 0.32 \times 5 \times 1.6$ $= 2.56\,\text{kN}$

Imposed load on plan $= 0.75 \times 5 \times 1.6 \times 2.5/2.69 = 5.58\,\text{kN}$

Design load $= (1.4 \times 2.56) + (1.6 \times 5.58) = 12.51\,\text{kN}$

Moment $= 12.51 \times 5/8$ $= 7.82\,\text{kNm}$

Assume that the angle bending about the XX axis resists the vertical load. The horizontal component is taken by the sheeting.

Design strength $p_y = 275$ N/mm^2

Applied moment = moment capacity of a single angle

7.82×10^3 $= 275 \times Z_x$

Elastic modulus $Z_x = 28.4\,\text{cm}^3$

Provide $125 \times 75 \times 10\ L \times 15\,\text{kg/m}$, $Z_x = 36.5\,\text{cm}^3$

Referring to Table 7; d/t for the downward leg $= 12.5$
  (i.e. semi-compact)

Deflection need not be checked in this case.

Example 3. Design using empirical method from BS 5950

Redesign the angle purlin above using the empirical method from BS 5950. The purlin specified meets the requirements for the design rules.

$W_p = $ total unfactored load $= 8.14\,\text{kN}$

$Z = $ elastic modulus $= \dfrac{8.14 \times 5000}{1800}$ $= 22.6\,\text{cm}^3$

Leg length perpendicular to plane of cladding
  $= 5000/45$ $= 111.1\,\text{mm}$

Leg length parallel to plane of cladding
  $= 5000/60$ $= 83.3\,\text{mm}$

Provide $120 \times 120 \times 8\ L \times 14.7\,\text{kg/m}$. Z_x $= 29.1\,\text{cm}^2$

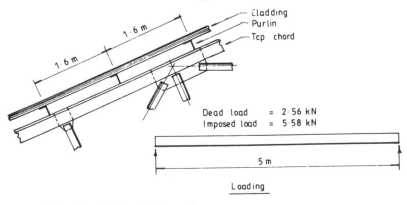

Dead load $= 2\cdot56\,\text{kN}$
Imposed load $= 5\cdot58\,\text{kN}$

Loading

Figure 5.43 Angle purlin for a sloping roof

Example 4. Select a cold-rolled purlin to meet the above requirements

Try purlin section A 140/155 from Table 5.2. Safe load for purlin span 5 m and spacing 1675 mm = 1.02 kN/m^2.

Purlin weight = $3.75 \times 9.81/103 \times 1.6 = 0.02$ kN/m^2
Imposed load normal to the roof slope = $0.75 \times 2.5/2.69 = 0.7$ kN/m^2
Net load on purlin = $0.7 + 0.32 - 0.02 = 1.0$ kN/m^2

The section is satisfactory and is much lighter than angle section.

5.13 Sheeting rails

5.13.1 Types of uses

Sheeting rails support cladding on walls and the sections used are the same as those for the purlins shown in Figure 5.39.

5.13.2 Loading

Sheeting rails carry a horizontal load from the wind and a vertical one from self weight and the weight of the cladding. The cladding materials are the same as used for sloping roofs (metal sheeting on insulation board). Wind loads are estimated using CP3: Chapter V: Part 2. For design examples in this section suitable values for wind loads will be assumed.

The arrangement of sheeting rails on the side of a building is shown in Figure 5.44(a) and the loading on the rails is shown in Figure 5.44(b). The wind may act in either direction due to pressure or suction on the building walls.

5.13.3 Design of angle sheeting rails

Sheeting rails may be designed as beams bending about two axes. It is assumed

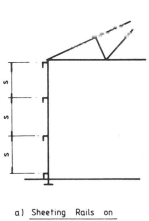

a) Sheeting Rails on
   Side of Building

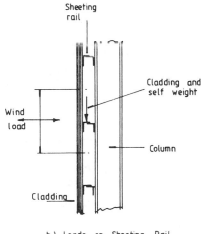

b) Loads on Sheeting Rail

Figure 5.44 (contd. overleaf)

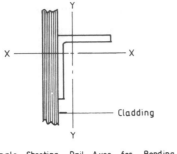

c) Angle Sheeting Rail Axes for Bending

Figure 5.44 Sheeting rails: arrangement and loading

for angle sheeting rails that the sheeting restrains the member and bending takes place about the vertical and horizontal axes. Eccentricity of the vertical loading (shown in Figure 5.44(b) is not taken into account.

The sheeting rail is fully supported on the downward leg. The outstand leg for simply supported sheeting rails is in compression from dead load and tension or compression from wind load.

The moment capacity is (see Section 5.12.4):

$M_c = p_y Z$

Z = elastic modulus for the appropriate axis

For biaxial bending:

$$\frac{M_x}{M_{bx}} + \frac{M_y}{M_{by}} < 1$$

M_x = moment about the XX axis

M_y = moment about the YY axis

M_{bx} = buckling resistance moment for the XX axis

M_{by} = buckling resistance moment for the YY axis

The design method is illustrated in the following example.

5.13.4 Design of angle sheeting rails to BS 5950: Part 1

The general requirements from Section 4.12.4 of the code set out for purlins in Section 5.12(4) above must be satisfied. Empirical rules for design of sheeting rails are given in Section 4.12.4.3 of the code. These state that:

(1) The loading should generally be due to wind load and weight of cladding. Not more than 10 per cent should be due to other loads or due to loads not uniformly distributed.
(2) The elastic moduli for the two axes of the sheeting rail from Table 30 in the code should not be less than the following values (see Figure 5.44(c):

(a) YY axis—parallel to plane of the cladding:

$Z_1 \geqslant W_1 L_1 / 1800 \, \text{cm}^3$

W_1 = unfactored load on one rail acting perpendicular to the plane of the cladding in kN. This is the wind load

L_1 = span in millimetres, centre to centre of columns

(b) XX axis—perpendicular to the plane of the cladding:

$Z_2 \geqslant W_2L_2/1200\,\mathrm{cm^3}$

W_2 = unfactored load on one railing acting parallel to the plane of the cladding (kN). This is the weight of the cladding and rail

L_2 = span centre to centre of columns or spacing of sag rods where these are provided and properly supported.

(3) The dimensions of the angle should not be less than the following:
   D—perpendicular to the cladding $< L_1/45$
   B—parallel to the cladding $< L_2/60$
   L_1 and L_2 were defined above.

5.13.5 Cold-rolled sheeting rails

The system using cold-rolled sheeting rails designed and marketed by Ward Building Components is described briefly with their kind permission.

The rail member is the Multibeam section placed with the major axis vertical. For bay widths up to 6.1 m a single tubular steel strut is provided to support the rails at mid-span. The strut is supported by diagonal wire rope ties and the cladding system can be levelled before sheeting by adjusting the ties. The system is shown in Figure 5.45. For larger width bays two struts are provided. The support system can be omitted when

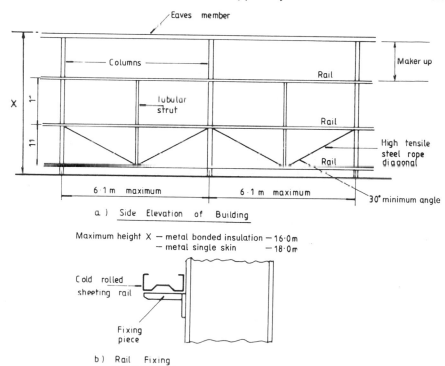

Figure 5.45 Cold-rolled sheeting rail system

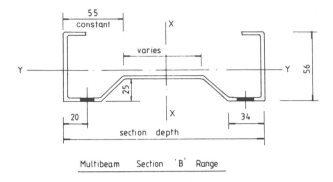

Multibeam   Section 'B' Range

Example—Section B 140/180
Depth = 140 mm; thickness = 1.8 mm

Table 5.3   Ward Building Components
Cold-rolled Sheeting Rail System

Safe loads for double-span sheeting rails (kN/m²)
including single-span end-bay members with sleeved connection

Span (m)	Section	Safe load for rail centres/(mm)				Rail weight (kg/m)
		1550	1700	1850	2000	
5.0	B120/150	1.08	0.99	0.91	0.84	3.03
	B140/150	1.32	1.20	1.10	1.02	3.26
	B140/165	1.51	1.38	1.26	1.17	3.59
6.0	B140/150	0.90	0.82	0.75	0.69	3.26
	B140/165	0.99	0.90	0.83	0.77	3.59
	B170/155	1.24	1.13	1.04	0.96	3.73

using 120 mm rails on bays up to 5 m in width, provided metal cladding fixed with self-tapping fasteners is used.

The allowable applied wind loads for a limited selection of sheeting rail spans and spacings are given in Table 5.3. Notes regarding use of the table are given below. The manufacturer's Technical Handbook should be consulted for full particulars regarding safe wind loads and fixing details for rails, support system and cladding.

Notes relating to Table 5.3 are:

(1) The loads shown are valid only when the rails and cladding are fixed exactly as indicated by the manufacturer.
(2) The loads shown are for positive external wind loads only. For negative suction loads multiply loads shown by a factor of 0.80.
(3) For intermediate spacings interpolation is acceptable.

5.13.6 Sheeting rail design examples

Example 1: Design of an angle sheeting rail

A simply supported sheeting rail spans 5 m. The rails are at 1.5-m centres. The total weight of cladding and self weight of rail is 0.32 kN/m². The wind loading on the wall is ±0.5 kN/m². The wind load would have to be carefully estimated for the particular building and the maximum suction and pressure may be different. The sheeting rail arrangement is shown in Figure 5.46(a). Use Grade 43 steel.

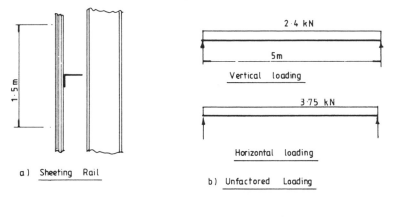

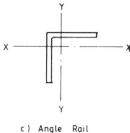

c) Angle   Rail

Figure 5.46 Angle sheeting rail

Vertical load $= 0.32 \times 1.5 \times 5 = 2.4$ kN
Horizontal load $= 0.5$ $\times 1.5 \times 5 = 3.75$ kN

The loading is shown in Figure 5.46(b).

The load factor $\gamma_f = 1.4$ for a wind load acting with dead load only. (Refer to Table 2 of BS 5950: Part 1.)

Factored vertical moment $M_x = 1.4 \times 2.4 \times 5/8 = 2.10$ kNm
Factored horizontal moment $M_y = 1.4 \times 3.75 \times 5/8 = 3.28$ kNm
Design strength $p_y = 275$ N/mm
Try $100 \times 100 \times 8$ L where $Z = 19.9$ cm^3
The moment capacity:
$M_{cx} = M_{cy} = 275 \times 19.9/10^3 = 5.47$ kNm

The biaxial bending interaction relationship:

$$\frac{M_x}{M_{cx}} + \frac{M_y}{M_{cy}} = \frac{2.1}{5.47} + \frac{3.28}{5.47} = 0.98 < 1$$

Provide $100 \times 100 \times 8$ $L \times 12.2$ kg/m.
For the outstand leg $b/t = 12.5$—semi-compact (Table 7).

Example 2: Design using empirical method from BS 5950

Redesign the angle sheeting rail above using the empirical method from BS 5950.

Unfactored wind load W_1	$= 3.75\,kN$
Elastic modulus	
$\quad Z_1 = Z_y = 3.75 \times 5000/1800$	$= 10.42\,cm^3$
Unfactored dead load W_2	$= 2.4\ \ kN$
Elastic modulus	
$\quad Z_2 = Z_x = 2.4 \times 5000/1200$	$= 10\ \ \ cm^3$

Dimensions specified are to be

D—perpendicular to cladding $< 5000/45 = 111.1\,mm$

B—parallel to cladding $< 5000/60 = \ \ 83.3\,mm$

$120 \times 120 \times 8\ L$ is the smallest angle to meet all the requirements.

Example 3: Select a cold-rolled sheeting rail to meet the following requirements

Wind load $= \pm\ 0.5\ kN/m^2$

Span $= 5\ m$

Spacing $= 1.5\ m$

Refer to Table 5.3. Rail section B120/150 will carry 1.08 kN/m² on a span of 5 m at 1.55 m spacing. This is the safe external pressure.
The safe suction load is $= 0.8 \times 1.08$ $= 0.86\ kN/m^2$
This section is satisfactory. (See Figure 5.45 for the rail support system.)

Problems

5.1 A simply supported steel beam of 6.0 m span is required to carry a uniform dead load of 40 kN/m and an imposed load of 20 kN/m. The floor slab system provides full lateral restraint to the beam. If a 457 × 191 UB 67 of Grade 43 steel is available for this purpose, check its adequacy in terms of bending, shear and deflection.

5.2 The beam carries the same loads as in Problem 5.1 but no lateral restraint is provided along the span of the beam. Determine the new size of universal beam required.

5.3 A steel beam of 8.0 m span carries the loading as shown in Figure 5.47. Lateral restraint is provided at the supports and the point of concentrated load (by cross beams). Using Grade 43 steel, select a suitable universal beam section to satisfy bending, shear and the code's serviceability requirements.

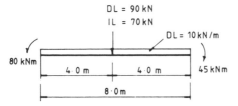

Figure 5.47

5.4 It is required to design a beam with an overhanging end. The dimension and loading are shown in Figure 5.48. The beam has torsional restraints at the supports but no intermediate lateral support. Select a suitable universal beam using Grade 43 steel.

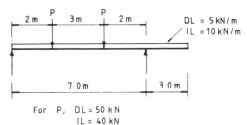

For P,   DL = 50 kN
            IL = 40 kN

Figure 5.48

5.5 A 610 × 229 UB 125 is used as a roof beam. The arrangement is shown in Figure 5.49 and the beam is of Grade 43 steel and fully restrained by the roof decking. Check the adequacy of the section in bending and shear and the web in buckling and crushing.

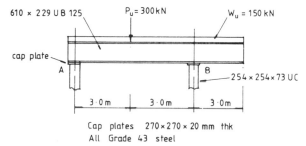

Cap plates   270 × 270 × 20 mm thk
All Grade 43 steel

Figure 5.49

5.6 The part floor plan for the internal panel of an office building is shown in Figure 5.50. The floor is precast concrete slabs 125 mm thick supported on steel beams. The following loading data may be used:

125 mm concrete slab = 3.0 kN/m²
Screed finishes   = 1.0 kN/m²
Partition   = 1.0 kN/m²
Imposed load   = 3.0 kN/m²

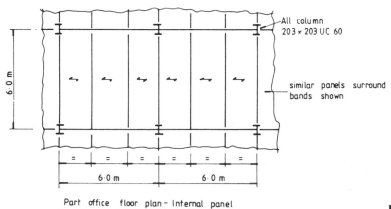

Part office floor plan – Internal panel

Figure 5.50

Design the floor beams, assuming that the self weight of main beams and secondary beams may be taken as 0.5 and 1.0 kN/m run, respectively.

5.7 A simply supported girder is required to span 7.0 m. The total load, including self weight of girder, is 130 kN/m uniformly distributed. The overall depth of the girder must not exceed 500 mm and a compound girder is proposed. If the compression flange has adequate lateral restraint and the two flange plates are not curtailed, carry out the following work:

(a) Check that a section consisting of 457 × 191 UB 98 and two No. 15 × 250 flange plates is satisfactory;
(b) Determine the weld size required for the plate-to-flange weld at the point of maximum shear;
(c) If the girder is supported on brackets at each end with a stiff bearing length of 80 mm, check the web shear, buckling and crushing.

5.8 A simply supported crane girder for a 200 kN (working load) capacity electric overhead crane spans 7 m. The maximum static wheel loads from the end carriage are shown in Figure 5.51. It is proposed to use a crane girder consisting of

533 × 210 UB 122 and
305 × 89 × 42 kg/m channel

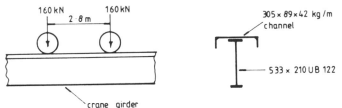

Figure 5.51

The weight of the crab is 40 kN and the self weight of the girder may be taken as 15 kN. Check the adequacy of the girder section.

5.9 A factory building has combined roof and crane columns at 8-m centres. It is required to install an electric overhead travelling crane. The crane data are as follows:

Hook load	= 150 kN
Span of crane	= 15 m
Weight of crane bridge	= 180 kN
Weight of crab	= 40 kN
No. of wheels in end carriage	= 2
Wheels centres in end carriage	= 3 m
Minimum hook approach	= 1 m

Design the crane girder using simply supported spans between columns.

5.10 Select a suitable size for a simply supported cold-rolled purlin. The purlin span is 5.0 m and the spacing is 1.8 m. The total dead load and imposed load on plan are 0.22 and 0.75 kN/m^2, respectively. Use Table 5.2 in the design. Redesign the purlin using the rules from BS 5950: Part 1.

6

Plate girders

6.1 Design considerations

6.1.1 Uses and construction

Plate girders are used to carry larger loads over longer spans than are possible with universal or compound beams. They are used in buildings and industrial structures for long-span floor girders, for heavy crane girders and in bridges.

Plate girders are constructed by welding steel plates together to form I sections. A closed section is termed a 'box girder'. Typical sections, including a heavy fabricated crane girder, are shown in Figure 6.1(a).

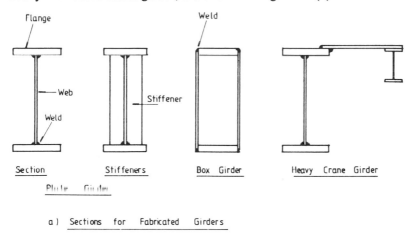

Flange   Weld

Web   Stiffener

Weld

Section   Stiffeners   Box Girder   Heavy Crane Girder

Plate Girder

a) Sections for Fabricated Girders

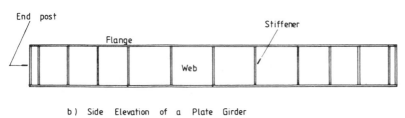

End post   Stiffener

Flange

Web

b) Side Elevation of a Plate Girder

Figure 6.1 Plate girder construction

The web of a plate girder is relatively thin, and stiffeners are required either to prevent buckling due to compression from bending and shear or to promote tension field action, depending on the design method used. Stiffeners are also required at load points and supports. Thus the side elevation of a plate girder has an array of stiffeners as shown in Figure 6.1(b).

6.1.2 Depth and flange breadth

The depth of a plate girder may be fixed by headroom requirements but it can often be selected by the designer. The depth is usually made from one tenth to one twelfth of the span. The breadth of flange plate is made about one third of the depth.

The deeper the girder is made, the smaller the flange plates required. However, the web plate must then be made thicker or additional stiffeners provided to meet particular design requirements. A simplified method to obtain the optimum depth is given in Section 6.3.3. A shallow girder can be very much heavier than a deeper girder carrying the same loads.

6.1.3 Variation in girder sections

Flange cover plates can be curtailed or single flange plates can be reduced in thickness when reduction in bending moment permits. This is shown in Figure 6.2(a). In the second case mentioned the girder depth is kept constant throughout.

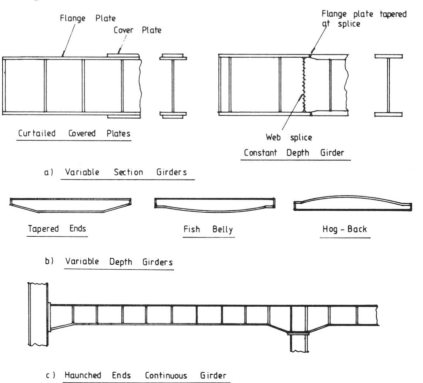

Figure 6.2 Variation in plate girder sections

For simply supported girders, where the bending moment is a maximum at the centre, the depth may be varied, as shown in Figure 6.2(b). In the past, hog-back or fish-belly girders were commonly used. In modern practice with automatic methods of fabrication it is more economical to make girders of uniform depth and section throughout.

In rigid frame construction and in continuous girders the maximum moment occurs at the supports. The girders may be haunched to resist these moments, as shown in Figure 6.2(c).

6.1.4 Plate girder loads

Loads are applied to plate girders through floor slabs, floor beams framing into the girder, columns carried on the girder or loads suspended from it through hangers. Some examples of loads applied to plate girders through secondary beams, a column and hanger are shown in Figure 6.3.

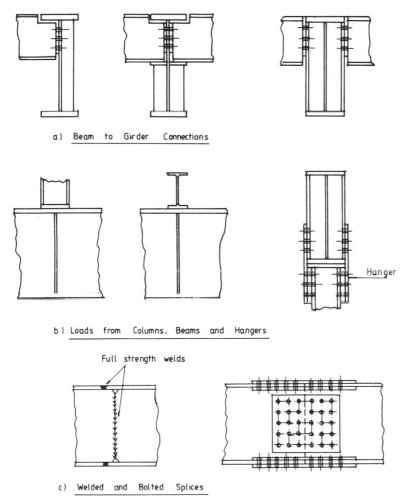

a) Beam to Girder Connections

b) Loads from Columns, Beams and Hangers

c) Welded and Bolted Splices

Figure 6.3 (contd. overleaf)

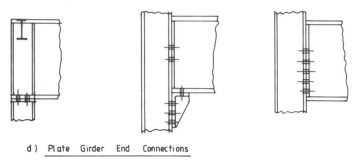

d) Plate Girder End Connections

Figure 6.3 Plate girder connections and splices

6.1.5 Plate girder connections and splices

Typical connections of beams and columns to plate girders are shown in Figures 6.3(a) and (b). Splices are necessary in long girders. Bolted and welded splices are shown in Figure 6.3(c) and end supports in Figure 6.3(d).

6.2 Behaviour of a plate girder

6.2.1 Girder stresses

The stresses from moment and shear for a plate and box girder in the elastic state are shown in Figure 6.4. The flanges have uniform direct stresses and the web shear and varying direct stress.

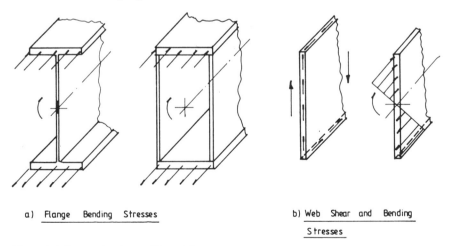

a) Flange Bending Stresses

b) Web Shear and Bending Stresses

Figure 6.4 Stresses in plate and box girders

Plate and box girders are composed of flat plate elements supported on one or both edges and loaded in plane by bending and shear. The way in which the girder acts is determined by the behaviour of the individual plates.

6.2.2 Elastic buckling of plates

The components of the plate and box girder under stress can be represented by the four plates loaded as shown in Figure 6.5. The way in which the plates buckle and the critical buckling stress depend on the edge conditions, dimensions and loading. The buckled plate patterns are also shown in the figure.

In all cases the critical buckling stress can be expressed by the equation: [12,17]

$$p_{cr} = K.\frac{\pi^2 E}{12(1 - v^2)}\left(\frac{t}{b}\right)^2$$

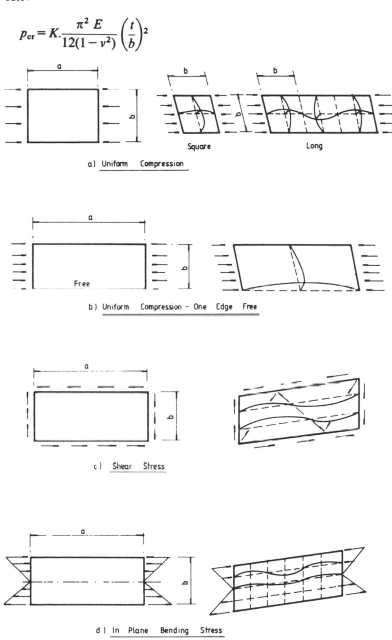

a) Uniform   Compression

b) Uniform   Compression – One   Edge   Free

c) Shear   Stress

d) In Plane   Bending   Stress

All edges simply supported except as noted in (b).

Figure 6.5 Elastic buckling of plates

K = buckling coefficient that depends on the ratio of plate length to width a/b (the edge conditions and loading case)

E = Young's modulus

v = Poisson's ratio

t = plate thickness

Some values of K for the four plates are shown in Figure 6.6. Note that the plate length a shown is also the stiffener spacing on a plate girder.

Plate and Load	$\dfrac{\text{Length}}{\text{Width}} = \dfrac{a}{b}$	Buckling Coefficient, k	Limiting Value of b/t for P_{cr} = yield stress
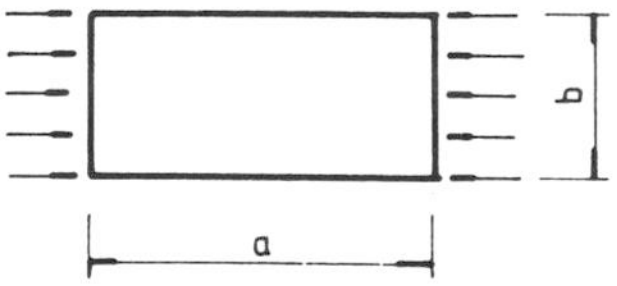	1·0 5·0	4·0 4·0	51·9 51·9
	1·0 ∞	1·425 0·425	30·9 16·9
	1·0 ∞	9·35 5·35	78·1 60·0
	1·0	25·6 minimum 23	131·3 124·4

All edges simply supported except as noted.

Figure 6.6 Buckling coefficients and limiting values of width/thickness ratios

The critical stress depends on the width/thickness ratio b/t. Limiting values of b/t, where the critical stress equals the yield stress, are also shown in Figure 6.6. These values are for Grade 43 steel for plate up to 16 mm thick, where the yield stress $p_y = 275$ n/mm^2. The values form the basis for the semi-compact section classification given in Table 7 in BS 5950.

The web plates of girders are subjected to combined stresses caused by direct bending stress and shear. An interaction formula is used to obtain critical stress combinations. Discussion of this topic is outside the scope of this book, where simplified design procedures given in the code are used. The reader should consult references 17 and 18 and Appendix H of BS 5950: Part 1.

6.2.3 Post-buckling strength of plates

(1) Plates in compression

The plate supported on two long edges shown in Figure 6.7(a) can support more load on the outer parts following buckling of the centre portion. The behaviour can be approximated by assuming that the load is carried by strips at the edge, as shown in Figure 6.7(b). The load is considered to be carried on an effective width of plate.[12] This principle is used in the design of thin plate members. An alternative method is adopted in the code, when the design strength of a slender cross-section is reduced in proportion to its slenderness. (See Section 3.6 of the code.)

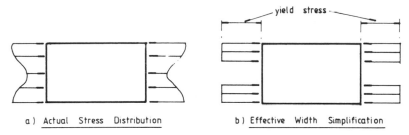

a) Actual Stress Distribution   b) Effective Width Simplification

Figure 6.7 Post-buckling strength: plate in compression

A plate supported on one long edge buckles more readily than the plate above and the strength gain is not as great. Stiffeners increase the load that can be carried (see Figure 6.6).

(2) Plates in edge bending

These plates can sustain load in excess of that causing buckling.[12] Longitudinal stiffeners in the compression region are very effective in increasing the load that can be carried. Such stiffeners are commonly provided on deep plate girders used in bridges (see Figure 6.5).

(3) Plates in shear

A strength gain is possible with plates in shear where tension field action is considered. This is a new inclusion in the code.

Thin unstiffened plates cannot carry much load after buckling. Referring to Figure 6.6, the critical buckling stress is increased if stiffeners are added. However, the stiffened plate can carry more load after buckling in the diagonal tension field, as shown in Figure 6.8. The flanges, stiffeners and tension field act like a truss. [13,19]

If the bending strength of the flanges is ignored, the tension field develops between the stiffeners, as shown in Figure 6.8(a). If the flange contribution is included the tension field spreads as shown in Figure 6.8(b). Failure in the girder panel occurs when plastic hinges form in the flanges and a yield zone in the web, as shown in Figure 6.8(c).[19]

Design formulae based on theoretical and experimental work have been developed to take tension field action into account. The design method in the code also includes the flange contribution. The resistance of the web is thus the

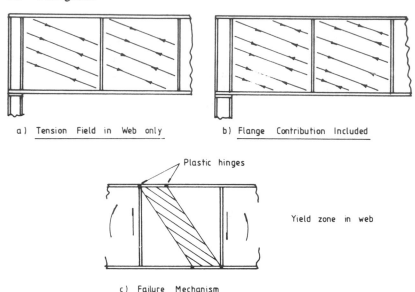

a) Tension Field in Web only   b) Flange Contribution Included

c) Failure Mechanism

Figure 6.8 Tension field action and failure mechanism

sum of the elastic buckling strength, the tension field and the flange strength.

Note that in the internal panels tension fields in adjacent panels support each other. In the end panels the end post must be designed as a vertical beam supported by the flanges to carry the tension field (see Figures 6.8(a) and (b)). Expressions have been derived for loads on the end post.[13]

6.3 Design to BS 5950: Part 1

6.3.1 Classification of girder cross sections

The classification of cross sections from Section 3.5 of Bs 5950: Part 1 was given in Section 5.3 (Beams). The limiting proportions for flanges and webs for built-up sections from Table 7 in the code are given in Figure 6.9. The limits for welded sections are lower than those for rolled sections because welded sections have more severe residual stresses and fabrication errors can also adversely affect behaviour. (The reader should refer to the code for treatment of slender cross sections.)

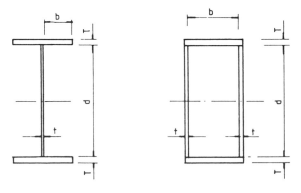

Type of Element		Class of Section		
		Plastic	Compact	Semi-compact
Outstand element of compression flange	$\dfrac{b}{T} \leqslant$	7·5 ε	8·5 ε	13 ε
Internal element of compression flange	$\dfrac{b}{T} \leqslant$	23 ε	25 ε	28 ε
Web with neutral axis at mid depth	$\dfrac{d}{T} \leqslant$	79 ε	98 ε	120 ε

$$\epsilon = (275/p_y)^{0.5}$$

Figure 6.9 Classification of girder cross sections

6.3.2 Moment capacity

The moment capacity for plate girders is given in Section 4.4 of BS 5950: Part 1. For sections where flanges meet the requirements for plastic compact or semi-compact section given in Figure 6.9 the moment capacity is given by:

(1) If the depth/thickness ratio d/t for the web is less than 63ε, the moment capacity is determined in the same way as for beams given in Section 5.4.2. The stress distribution is shown in Figure 6.10(a).
(2) If the depth/thickness ratio d/t for the web is greater than 63ε, the moment capacity should be calculated using one of the following methods:
    (a) Moment and axial load are assumed to be resisted by the flange only with each flange subjected to a uniform stress p_y. The web is designed to resist shear only;
    (b) Moment and axial load are assumed to be resisted by the whole section. The web is designed for combined shear and longitudinal stress;
    (c) Part of the load may be assumed to be resisted by method (a) and part by method (b).
    Only method (a) will be considered in this book.

The stress distribution in bending for this case is shown in Figure 6.10(b). The moment capacity for a girder with restrained compression flange is:

$$M_c = BT(d+T)p_y$$

Where B = flange
T = flange thickness
d = web depth

For cases where the compression flange is not restrained, lateral torsional buckling may occur. This is treated in the same way that was set out for beams in Section 5.5. The bending strength p_b for welded sections is taken from Table 12 in the code.

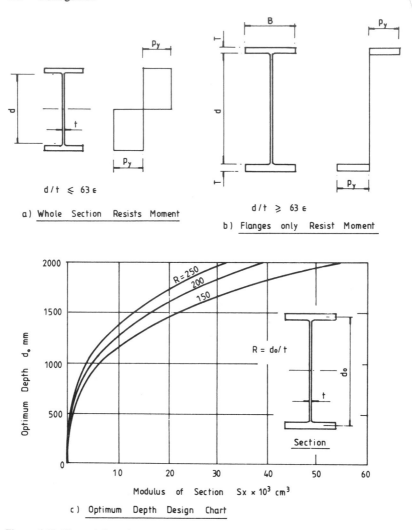

Figure 6.10 Plate girder stresses and optimum depth

6.3.3 Optimum depth

The optimum depth based on minimum area of cross section may be derived as follows. This treatment applies to a girder with restrained compression flange for a given ratio of web depth to thickness. Define terms:

d_0 = distance between centres of flanges

= d, clear depth of web approximately

R = ratio of web depth/thickness = d_0/t

A_f = area of flange

S = plastic modulus based on the flanges only

= $A_f d_0$

$A_f = S/d_0$

A_w = area of web = $d_o t = d_o^2/R$

A = total area = $2S/d_0 + d_0^2/R$

Differentiate with respect to d_0 and equate to zero to give

$$d_0 = (R\,S)^{1/3}$$

Curves drawn for depth d_0 against plastic modulus S for values of R of 150, 200 and 250 are shown in Figure 6.10(b). For the required value of

$$S = M/p_y$$

the optimum depth d_0 can be read from the chart for a given value of R, where

M = applied moment.

6.3.4 Shear capacity and web design

(1) Minimum thickness of web

This is given in Section 4.4.2 of BS 5950: Part 1. The following two conditions must be satisfied for webs with intermediate transverse stiffeners:

(1) Serviceability to prevent damage in handling:

Stiffener spacing $a > d$ $t \geqslant d/250$

Stiffener spacing $a < d$ $t \geqslant \dfrac{d}{250}\left(\dfrac{a}{d}\right)^{0.5}$

(2) To avoid the flange buckling into the web. This type of failure has been observed in girders with thin webs [13]

Stiffener spacing $a \leqslant 1.5d$ $t \geqslant \dfrac{d}{250}\left(\dfrac{p_{yf}}{455}\right)^{0.5}$

Where d = depth of web

t = thickness of web

p_{yf} = design strength of compression flange

(2) Design for strength

The design of thin webs is covered in Section 4.4.5 of BS 5950: Part 1 and applies to webs carrying shear stress only. This states that webs with intermediate stiffeners may be designed either without tension field action or with tension field action. Different panels between stiffeners may be designed by either method.

(3) Design without utilizing tension field action

The shear buckling resistance is given in Section 4.4.5.3 of BS 5950: Part 1 by

$$V_{cr} = q_{cr}dt$$

where q_{cr} = critical shear strength given in Tables 21(a) – (d) in the code.

Values of q_{cr} depend on the web depth to thickness ratio d/t and the stiffener spacing a given in terms of the ratio a/d. Note that for girders with no stiffeners a/d should be taken as infinity.

(4) Design utilizing tension field action

The shear buckling resistance of a stiffened panel is given in Section 4.4.5.4 by

$$V_b = q_b dt$$

where q_b = basic shear strength given in Tables 22(a) – (d) in the code.

The stress q_b includes the critical shear stress and an additional stress to take account of tensile membrane action in the web.

Note that the above expression can only be used to design internal and end panels meeting requirements set out in Sections 4.4.5.2 and 4.4.5.4.3, as appropriate, in the code.

The code states that if the flanges are not fully stressed the shear resistance may be increased to:

$$V_b = [q_b + q_f(K_f)^{0.5}]dt$$
$$\leqslant 0.6\, p_y dt$$

where q_f = flange-dependent shear strength factor taken from Tables 23(a)–(d).

$$K_f = \frac{M_{pf}}{4\, M_{pw}} \left(1 - \frac{f}{p_{yf}}\right)$$

where M_{pf} = plastic moment capacity of the flange
$$= p_{yf} B T^2 / 4$$

M_{pw} = plastic moment capacity of the web
$$= p_{yw} t d^2 / 4$$

f = mean longitudinal stress in the flange
$$= M/(d_0 B T)$$

p_{yf} = design strength of the flange

M = applied moment

The dimensions B, T, t, d and d_o are defined above and are shown in Figure 6.10.

6.3.5 Stiffener design

Two main types of stiffeners used in plate girders are:

(1) Intermediate stiffeners. These divide the web into panels and prevent it from buckling. They resist direct forces from tension field action if utilized and possible external loads.
(2) Load-bearing stiffeners. These are required at all points where substantial loads are applied and at supports to prevent local buckling and crushing of the web.

The stiffeners at the supports are termed 'end posts'. The design of the end posts where tension field action is taken into account is dealt with in Section 6.3.8. Note that other special-purpose stiffeners are defined in BS 5950: Part 1 in Section 4.5.1. Only the types mentioned above will be discussed in this book.

6.3.6 Intermediate transverse stiffeners

Transverse stiffeners may be placed on one or both sides of the web, as shown in Figure 6.11. Flats are the most common stiffener section used. The requirements and design procedure are set out in Section 4.4.6 of Bs 5950: Part 1. Only stiffeners not subjected to external loads are considered here. The code should be consulted for design of stiffeners subjected to external loads. The design process is:

(1) Spacing. This depends on:
    (a) Minimum web thickness (see Section 6.3.4(1));
    (b) Web shear capacity required. The closer the spacing, the greater the shear capacity (see Section 6.3.4(2)).
(2) Outstand. This is given in Section 4.5.1.2 of the code. The oustand should not exceed $19t_s\varepsilon$ (see Figure 6.11), where

$$t_s = \text{thickness of stiffener}$$
$$\varepsilon = (275/p_y)^{0.5}$$

When the outstand is between $13t_s\varepsilon$ and the $19t_s\varepsilon$ the design is to be based on a core section with an outstand of $13t_s\varepsilon$.

(3) Minimum stiffness

For the case where the stiffener spacing a is less than $\sqrt{2}d$ the moment of inertia I_s of the stiffener about the centre line of the web is to be such that:

$$I_s \geqslant 1.5\, d^3 t^3 / a^2$$

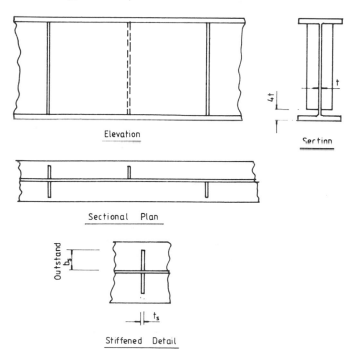

Elevation

Section

Sectional Plan

Outstand b_s

t_s

Stiffened Detail

Figure 6.11 Intermediate transverse stiffeners

where d = depth of web

t = minimum required web thickness for spacing a using tension field action

a = actual stiffener spacing

Note that additional stiffness is required where stiffeners are subject to lateral loads or to moments due to eccentricity of transverse loads relative to the web. No increase is needed where transverse loads are in line with the web.

(4) Buckling check

This check is only required for intermediate stiffeners in webs designed using tension field action. The stiffener should be checked for buckling for a force:

$$F_q = V - V_s \leqslant P_q$$

where V = maximum shear adjacent to the stiffener

V_s = shear buckling resistance of the web without using tension field action (see Section 6.3.4).

$= V_{cr} = q_{cr}dt$

P_q = buckling resistance of an intermediate stiffener (see Section 6.3.7)

(5) Connection of intermediate stiffeners to web

The connection between each plate and the web is to be designed for a shear of not less than:

$$t^2/(5b_s) \text{ kN/mm}$$

where t = web thickness (mm)

b_s = outstand of the stiffener (mm)

The code states that intermediate stiffeners may be cut off at about $4t$ above the tension flange. The stiffeners should extend to the compression flange but need not be connected to it (see Figure 6.11).

6.3.7 Load-bearing stiffeners

Load-bearing stiffeners are required to prevent local buckling and crushing of the web due to concentrated loads applied through the flange when the web itself cannot support the load. The capacity of the web alone in buckling and bearing was discussed in Sections 5.8.1 and 5.8.2, respectively.

The design procedure for these stiffeners is set out in Section 4.5 of BS 5950: Part 1. The process is as follows:

(1) Outstand. This is the same as set out for intermediate stiffeners in Section 6.3.6(2).

(2) Buckling resistance of stiffeners. This is set out in Section 4.5.1.5 of BS

5950: Part 1 (see Figure 6.12(a)). The stiffener is designed as a strut of area A_s at the centre of the girder where:

A_s = area of stiffener plus 20 times the web thickness on either side of the centre line of the stiffener.

$$= 2b_s t_s + 40t^2$$

where b_s = stiffener outstand

t_s = stiffener thickness

t = web thickness

The radius of gyration is taken about the centroidal axis of the strut area parallel to the web. The effective length L_E to be used in calculating the slenderness ratio of the stiffener acting as the strut is:

(a) Intermediate transverse stiffeners:

$$L_E = 0.7L$$

(b) Load-bearing stiffeners where the flange through which the load is applied is restrained against lateral movement is:

(i) Where the flange is restrained against rotation in the plane of the stiffener by other elements:

$$L_E = 0.7L$$

(ii) Where the flange is not so restrained:

$$L_E = 1.0L$$

where L = length of stiffener

Note that the code states that if no effective lateral restraint is provided the stiffener should be designed as part of the compression member applying the load.

The design strength from Table 6 is the minimum for the web or stiffener. It should be reduced by 20 N/mm² for welded construction. (See Clause 4.7.5 in the code.) The compressive strength p_c is taken from Table 27(c). The buckling resistance is:

Intermediate stiffener $P_q = p_c A_s \geqslant F_q$

Load bearing stiffener $P_x = p_c A_0 \geqslant F_x$

where F_q = intermediate stiffener force (see Section 6.3.6(4) above)

F_x = external load or reaction

If the load-bearing stiffener also acts as an intermediate stiffener the code states that it must be designed for the combined load $F_q + F_x$. This will only apply when the web is designed for tension field action.

(3) Bearing resistance

This is set out in Section 4.5.4.2 of BS 5950: Part 1. The area of stiffener A in contact with the flange is to be such that:

$$A > \frac{0.8F_x}{p_{ys}}$$

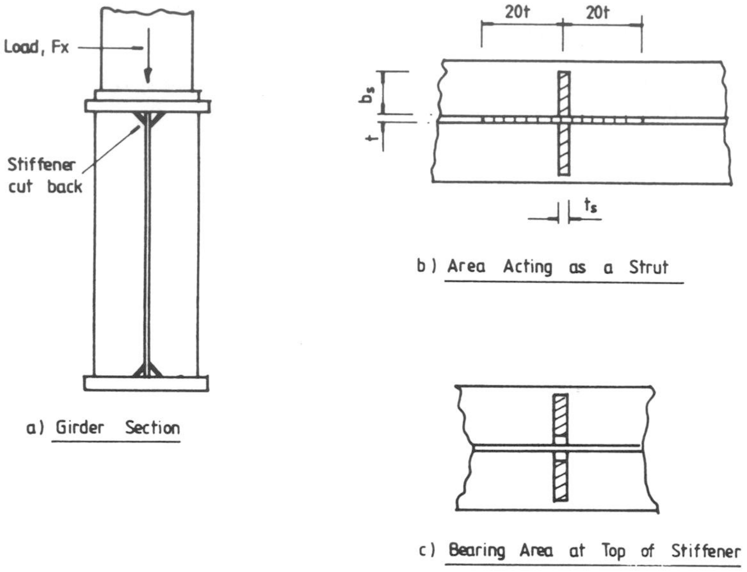

a) Girder Section

b) Area Acting as a Strut

c) Bearing Area at Top of Stiffener

Figure 6.12 Load-bearing stiffeners

where F_x = external load or reaction

p_{ys} = design strength of stiffener

The area A is shown in Figure 6.12(c). Note that the stiffener has been chamfered at the top to clear the web/flange weld.

(4) Web check between stiffeners

It may be necessary to check the compression edge of the web if loads are applied to it direct or through a flange between web stiffeners. A procedure to make this check is set out in Section 4.5.2.2 of BS 5950: Part 1. (The reader is referred to the code.)

6.3.8 End-post design

The end post of a plate girder may consist of a single end plate or double stiffeners, as shown on Figure 6.13. The design procedure is set out in Sections 4.45.4.2 and 4.4.5.4.4. of BS 5950: Part 1. This is summarized as follows:

(1) End panel not designed utilizing tension field action. Design the end post as a load-bearing stiffener as set out in Section 6.3.7.
(2) End panel and internal panels designed utilizing tension field action. In addition to carrying the reaction the end post must be designed as a beam spanning between the flanges. The two cases shown in the figure are discussed below.

References should be made to the code for the case where the interior panels are designed utilizing tension field action but the end panel is not.

(1) Single end plate (see Figure 6.13(a))

The single end plate acts as both bearing stiffener and end post. It must be connected by full-strength welds to the flanges. The design is made for:

(1) Compression due to the vertical reaction.
(2) Compression due to a moment $2/3\ M_{tf}$ from the tension field force. (M_{tf} is defined below.)

(2) Double stiffeners (see Figure 6.13(b)

The inner bearing stiffener carries the vertical reaction from the girder. It is checked for bearing at the end and for buckling at the centre (see Section 6.3.7).

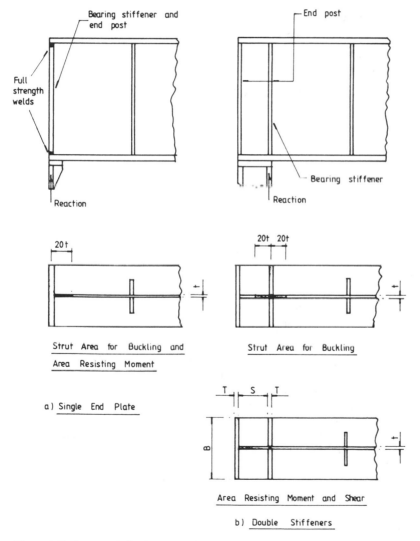

Figure 6.13 End-post design

The end posts is checked as a beam spanning between the flanges of the girder. It is designed to resist a shear force R_{tf} and a moment M_{tf} due to the tension field force.

The expressions to derive the shear and moment given in the code are:

Shear $R_{tf} = H_q/2$

Moment $M_{tf} = H_q d/10$

The force from the tension field

$$H_q = 0.75 \, dtp_y \left(1 - \frac{q_{cr}}{0.6p_y}\right)^{0.5}$$

f_v = applied shear stress

q_b = basic shear stress

q_{cr} = critical shear stress for the panel designed
     for tension field action

d = web depth

t = web thickness

If $f_v < q_b$, H_q may be reduced by the ratio

$$\left(\frac{f_v - q_{cr}}{q_b - q_{cr}}\right)$$

The shear capacity of the end post is:

$P_v = 0.6 \, p_y St$

S = length of web between stiffeners

t = web thickness

The shear capacity P_v must exceed the shear from the tension field R_{tf}.

The moment capacity of the end post at the centre of the girder, assuming that the flanges resist the whole moment, is:

$$M_{cx} = p_y \, BT \, (S + T)$$

where B = stiffener width

T = stiffener thickness

Note that proportions should be selected so that the plates selected are semi-compact as a minimum requirement. The moment capacity M_{cx} must exceed the moment due to tension field action M_{tf}. At the centre of the girder the bearing stiffener is subject to compression from the reaction and to tension from the moment due to tension field action. The separate checks only are made at this section.

A local capacity check for combined bearing and bending is made for the bearing stiffener at the bottom joint with the flange. The interaction expression is:

$$\frac{F_x}{A \, p_y} + \frac{M_{tf}}{M_{cx}} \leqslant 1$$

where F_x = reaction

A = area of bearing stiffener in contact with flange

M_{tr} = moment due to tension field action

M_{cx} = moment capacity of the end post based on
area A for the bearing stiffener

$= p_y A\,(S+T)$

The welds between the bearing stiffener and web must be designed to carry the reaction and the shear from the end-post beam action.

The application of the design procedure is given in the example in Section 6.6.

6.3.9. Flange to web welds

Fillet welds are used for the flange to web welds (see Figure 6.14). The welds are designed for the horizontal shear per weld:

$$= \frac{FAy}{2I_x}$$

where F = applied shear

A = area of flange

y = distance of the centroid of A from the centroid
of the girder

I_x = moment of inertia of the girder about the XX axis

The fillet weld can be intermittent or continuous, but continuous welds made by automatic welding are generally used.

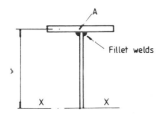

Figure 6.14 Flange-to-web weld

6.4 Design of a plate girder

A simply supported plate girder has a span of 12 m, and carries two concentrated loads on the top flange at the third points, consisting of 450 kN dead load and 300 kN imposed load. In addition, it carries a uniformly distributed dead load of 20 kN/m, which includes an allowance for self weight and an imposed load of 10 kN/m. The compression flange is fully restrained laterally. The girder is supported on a heavy stiffened bracket at each end. The material is Grade 43 steel. Design the girder without utilizing tension field action.

6.4.1 Loads, shears and moments

The factored loads are:

Concentrated loads $= (1.4 \times 450) + (1.6 \times 300) = 1110\,\text{kN}$
Distributed load $\quad = (1.4 \times 20) + (1.6 \times 10) \quad = 44\,\text{kN/m}$

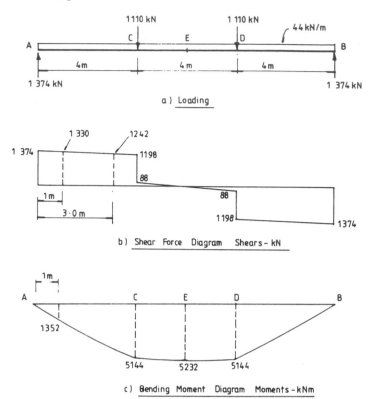

Figure 6.15 Load, shear and moment diagrams

The loads and reactions are shown in Figure 6.15(a) and the shear force diagram in Figure 6.15(b). The moments are:

$$M_c = (1374 \times 4) - (44 \times 4 \times 2) \qquad\qquad = 5144\,\text{kNm}$$
$$M_E = (1374 \times 6) - (1110 \times 2) - (44 \times 6 \times 3) = 5232\,\text{kNm}$$

The bending moment diagram is shown in Figure 6.15(c).

6.4.2 Girder section for moment

(1) Design for girder depth span/10

Take the overall depth of the girder as 1200 mm and assume that the flange plates are over 40 mm thick. Then the design strength from BS 5950: Part 1, Table 6, for plates is $p_y = 245\,\text{N/mm}^2$.

The flanges resist all the moment by a couple with lever arm of, say, 1140 mm, as shown in Figure 6.16(a). The flange area is:

$$= \frac{5232 \times 10^6}{1140 \times 245} \qquad = 18\,733\,\text{mm}^2$$

Make the flange plates 450 mm × 45 mm, giving an area of 20 250 mm². The

girder section with web plate 10 mm thick is shown in Figure 6.16(b).
The flange projection b is 220 mm and the ratio:

$b/T = 220/45 = 4.89$

Referring to Table 7 of the code, the ratio:

$$\frac{b}{T} = 4.89 \leqslant 7.5\varepsilon = 7.5\left(\frac{275}{245}\right)^{0.5} = 7.94$$

The flanges are plastic and the area of cross section is 51 600 mm².

(2) Design using the optimum depth chart

Redesign the girder using the optimum depth chart shown in Figure 6.10.
Assume that the flange plates are between 16 and 40 mm thick. Then the
design strength from Table 6 of the code is:

$p_y = 265 \text{ N/mm}^2$
Plastic modulus $S_x = 5232 \times 10^3/265 = 19.74 \times 10^3 \text{ cm}^3$
Using curve $d_0/t = 150$, the optimum depth $d_0 = 1450$ mm
Make the depth 1500:
$$\text{Flange area} = \frac{5232 \times 10^6}{1500 \times 265} = 13\,162 \text{ mm}^2$$

Provide flanges 500 × 30 mm giving an area of 15 000 mm². The girder section
with web plate 10 mm thick is shown in Figure 6.16(c). Note that the actual
d_0/t ratio is 144.

The flange projection b is 245 mm and the ratio $b/T = 245/30 = 8.17$. Referring to the limits in Table 7 in the code, the flanges are compact. The area of
cross-section is 44 400 mm². The saving in material compared with the first
design is 13.9 per cent.

The design will be based on a depth of 1200 mm because of headroom
restriction.

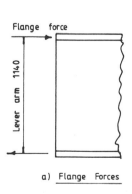

a) Flange Forces

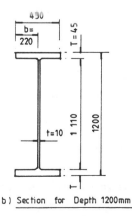

b) Section for Depth 1200mm

Figure 6.16 (contd. overleaf)

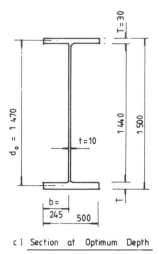

c) Section at Optimum Depth

Figure 6.16 Plate girder sections

6.4.3 Design of web

(1) Minimum thickness of web (Section 4.4.2 of BS 5950)

An arrangement for the stiffeners is set out in Figure 6.17. The design strength of the web $p_y = 275 \, \text{n/mm}^2$ from Table 6 of BS 5950: Part 1 for plate less than 16 mm thick. The minimum thickness is the greater of:

(1) Serviceability. Stiffener spacing $a >$ depth d in the centre of the girder. Web thickness $t \geqslant 1110/250 = 4.4 \, \text{mm}$.
(2) To prevent the flange buckling into the web:
    Stiffener spacing $a < 1.5$ depth d:

$$\text{Web thickness } t \geqslant \frac{1110}{250}\left(\frac{275}{455}\right)^{0.5} = 3.45 \, \text{mm}$$

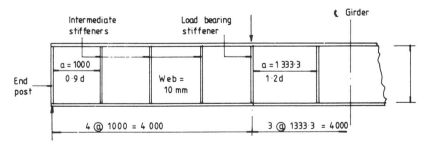

Figure 6.17 Stiffener arrangement

(2) Buckling resistance of web (Section 4.4.5.3 of Bs 5950)

Try a 10 mm thick web plate. The buckling resistance is checked for the maximum shear in the end panel:

Web depth/thickness ratio $d/t = 111$

Stiffener spacing/web depth ratio
$$a/d = 100/1110 = 0.9$$

From Table 21(b) in the code the critical shear strength:
$$q_{cr} = 136.8 \, \text{N/mm}^2$$

Shear buckling resistance:

$V_{cr} = 136.8 \times 10 \times 1110/10^3 \quad = 1518.4 \, \text{kN}$

Factored shear F_v $\qquad = 1374 \, \text{kN}$

The stiffener arrangement and web thickness are satisfactory.

6.4.4 Intermediate stiffener

(1) Trial size and outstand (Section 4.5.1.2 of BS 5950)

Try stiffeners composed of 2 No. 60 mm × 8 mm flats:

Design strength $p_y = 275 \, \text{n/mm}^2$ (Table 6)

Factor $\varepsilon = 1.0$

Outstand $60 < 13 \times 8 = 104 \, \text{mm}$

(2) Minimum stiffness (Section 4.4.6.4 of BS 5950)

The intermediate stiffener is shown in Figure 6.18. The moment of inertia about the centre of the web is:

$I_s = 8 \times 130^3/12 \qquad = 1.464 \times 10^6 \, \text{mm}^4$

$$> \frac{1.5 \times 1110^3 \times 8^3}{1000^2} \quad = 1.05 \times 10^6 \, \text{mm}^4$$

when the spacing $a = 1000 \, \text{mm}$
$$< \sqrt{2} \, (1100) = 1569.5 \, \text{mm}$$

Note that t, the minimum required web thickness for spacing $a = 1000 \, \text{mm}$

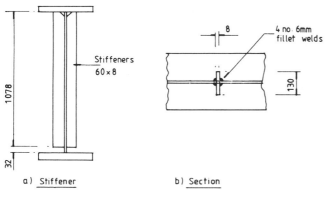

a) Stiffener   b) Section

Figure 6.18 Intermediate stiffener

using tension field action, is 8 mm. (See Section 6.5 below.) The stiffener is satisfactory with respect to stiffness. In a conservative design $t = 10$ mm.

$$I_s \geqslant 1.5 \times 1110 \times 10^3/1000^2 = 2.05 \times 10^6 \text{ mm}^4$$

Stiffeners 70 mm × 8 mm are then required.

(3) Connection to web, (Section 4.4.6.7 of BS 5950)

Shear between each flat and web
$$= 10^2/8 \times 60 \qquad = 0.208 \text{ kN/mm on two welds}$$
Use 6 mm fillet weld, strength 0.9 kN/mm.
Four continuous fillet welds are provided.

6.4.5 Load-bearing stiffener

(1) Trial size and oustand

Try stiffeners composed of 2 No. 150 mm × 15 mm plates as shown in Figure 6.19:
$$\text{Outstand } 150 < 13 \times 15 \qquad = 195 \text{ mm}$$
The stiffener is fully effective in resisting load.

(2) Bearing check (Section 4.5.4.2 of BS 5950)

The area in bearing at the top of the stiffener is shown in Figure 6.19(b). The stiffeners have been cut back 15 mm to clear the web to flange welds:

Design strength of stiffener $p_{ys} = 275$ N/mm²
$$A = 2 \times 15 \times 135 \qquad = 4050 \text{ mm}^2$$
$$> 0.8 \times 1110 \times 10^3/275 \qquad = 3229 \text{ m}^2 \text{ (satisfactory)}$$

(3) Buckling check

The stiffener area at the centre of the girder acting as a strut is shown in Figure 6.19(c). The stiffener properties are calculated from the dimensions shown:

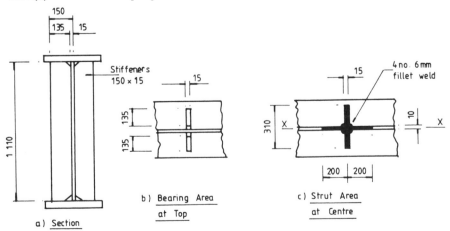

a) Section

b) Bearing Area at Top

c) Strut Area at Centre

Figure 6.19 Load-bearing stiffener

$$A = (2 \times 150 \times 15) + (400 \times 10) = 8500 \, \text{mm}^2$$
$$I_x = 15 \times 310^3/12 \qquad = 37.23 \times 10^6 \, \text{mm}^4$$
$$r_x = (37.23 \times 10^6/8500)^{0.5} \qquad = 66.1 \, \text{mm}$$

Assume that the flange is restrained against lateral movement and against rotation in the plane of the stiffeners:

Slenderness $\lambda = 0.7 \times 1110/66.1 = 11.8$

Design strength $= 275 - 20 = 255 \, \text{N/mm}^2$ (Table 6) (This is reduced for welded construction)

Compressive strength $p_c = 255 \, \text{N/mm}^2$ (Table 27(c))

Buckling resistance:

$$P_x = 255 \times 8500/10^3 \qquad = 2167 \, \text{kN}$$

The size selected is satisfactory.

(4) Connection to web

Shear between each flat and web:

$$= \frac{10^2}{8 \times 150} + \frac{1110}{2 \times 1110} = 0.583 \, \text{kN/mm on two welds}$$

Use 6-mm continuous fillet weld, strength 0.9 kN/mm. Four fillet welds are provided.

Note that the bearing area required controls the stiffener size.

6.4.6 End post

(1) Trial size and outstand

The trial size for the end post consisting of a single plate 450 mm × 15 mm is shown in Figure 6.20(a). The end post is designed as a load-bearing stiffener:

Outstand $= 220 \, \text{mm} > 13 \times 15 = 195 \, \text{mm}$
$$< 19 \times 15 = 285 \, \text{mm}$$

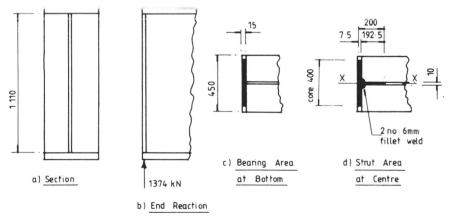

a) Section

b) End Reaction 1374 kN

c) Bearing Area at Bottom

d) Strut Area at Centre

Figure 6.20 End post

Base design on a stiffener core 400 mm × 15 mm

Design strength $= 275 - 20 = 255$ N/mm^2 (Table 6)

(2) Bearing check

The bearing area is shown in Figure 6.20(c):

$$A = 15 \times 400 \qquad\qquad = 6000 \text{ mm}^2$$
$$> 0.8 \times 1374 \times 10^3/255 = 4310.5 \text{ mm}^2 \text{ (Satisfactory)}$$

(3) Buckling check

The area at the centre line acting as a strut is shown in Figure 6.20(d):

$$A = (400 \times 15) + (192.5 \times 10) = 7925 \text{ mm}^2$$
$$I_x = 15 \times 400^3/12 \qquad\qquad = 80 \times 10^6 \text{mm}^4$$
$$r_x = (80.0 \times 10^6/7925)^{0.5} \qquad = 100.4$$
$$\lambda = 0.7 \times 1110/100.4 \qquad\quad = 7.7$$
$$p_c = 255 \text{ N/mm}^2 \text{ (Table 27(c))}$$
$$P_x = 255 \times 7925/10^3 \qquad\quad = 2020.8 \text{ kN}$$
$$\text{Load carried} \qquad\qquad\quad = 1374 \text{ kN}$$

The size is satisfactory.

(4) Connection to web

Shear between end plate and web:

$$= \frac{1374}{2 \times 1110} = 0.62 \text{ kN/mm per weld}$$

Provide 6-mm continuous fillet, weld strength 0.9 kN/mm. Two lengths of weld are provided.

6.4.7 Flange to web weld

See Figure 6.16(b) for the girder dimension:

$$I_x = (450 \times 1200^3 - 440 \times 1110^3)/12 = 14.65 \times 10^9 \text{ mm}^4$$

Horizontal shear per weld (see Section 6.3.9):

$$= \frac{1374 \times 450 \times 45 \times 577.5}{14.65 \times 10^9 \times 2} = 0.548 \text{ kN/mm}$$

Provide 6-mm continuous fillet, weld strength 0.9 kN/mm.

6.4.8 Design drawing

A design drawing of the girder is shown in Figure 6.21.

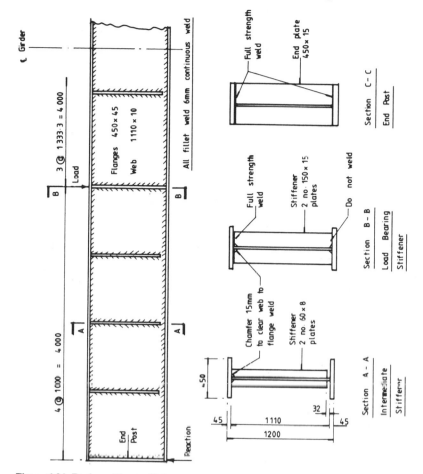

Figure 6.21 Design without utilizing tension field action

6.5 Design utilizing tension field action

Redesign the web, stiffeners and end post for the girder in Section 6.4 using tension field action.

6.5.1 Design of the web

Try an 8-mm thick web with the stiffeners spaced at 1000 mm in the end 4 m of the girder, as shown in Figure 6.22. The web design is set out in Section 4.4.5.4 of BS 5950:

$$d/t = 1110/8 \quad = 138.75$$
$$a/d = 1000/1110 = 0.9$$

The basic shear strength from Table 22(b) in the code:

$$q_b = 131 \text{ N/mm}^2$$

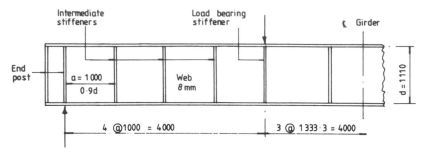

Figure 6.22 Stiffener arrangement

The shear buckling resistance of the stiffened panel is:

$$V_b = 131 \times 1110 \times 8/10^3 = 1163.3 \,\text{kN}$$

This is less than the applied shear of 1374 kN. The contribution to shear strength from the flanges is calculated.

The flange-dependent shear strength factor from Table 23(b) for $p_y = 275\,\text{N}/\text{mm}^2$ is:

$$q_f = 312\,\text{N/mm}^2$$

The plastic moment capacity of the flange where the design strength of the flange $p_{yf} = 245\,\text{N/mm}^2$:

$$M_{pf} = \frac{245 \times 450 \times 45^2}{4 \times 10^6}\quad 55.8\,\text{kNm}$$

The plastic moment capacity of the web where the design strength of the web $p_{yw} = 275\,\text{N/mm}^2$:

$$M_{pw} = \frac{275 \times 8 \times 1110^2}{4 \times 10^6} = 677.6\,\text{kNm}$$

The maximum moment in the end panel is 1352 kNm (see Figure 6.15(c)). The mean longitudinal stress in the flange due to moment:

$$f = \frac{1352 \times 10^6}{1155 \times 45 \times 450} = 57.8\,\text{N/mm}^2$$

$$K_f = \frac{55.8}{4 \times 667.6}\left(1 - \frac{57.8}{245}\right) = 0.0159$$

Total shear resistance:

$$V_b = [131 + 312\,(0.0159)^{0.5}]\,1110 \times 8/10^3 = 1512.6\,\text{kN}$$
$$\not> 0.6 \times 275 \times 8 \times 1110/10^3 = 1465.2\,\text{kN}$$

This exceeds the applied shear of 1374 kN.

Check the web in the panel 3.0 to 4 m from the support:

$$\text{Shear} = 1242\,\text{kN} > V_b = 1163.3\,\text{kN}$$

$$f = \frac{5144 \times 10^6}{1155 \times 45 \times 450} = 219.93\,\text{N/mm}^2$$

$$K_f = \frac{55.8}{4 \times 677.6} \left(1 - \frac{219.93}{245}\right) = 0.002107$$

Total shear resistance:

$$V_b = [131 + 312 \, (0.002107)^{0.5}] \, 1110 \times 8/10^3$$
$$= 1290.4 \, \text{kN}$$

The girder is satisfactory for the stiffener arrangement assumed.

6.5.2 Intermediate stiffener

(1) Minimum stiffness

Try stiffeners composed of two No. 80 mm × 8 mm flats (see Section 6.4.4 above). The outstand is satisfactory. The stiffener is shown in Figure 6.23:

$$I_s = 8 \times 168^3/12 = 3.161 \times 10^6 \, \text{mm}^4$$
$$> 1.05 \times 10^6 \, \text{mm}^4$$

(2) Buckling check

Maximum shear adjacent to the stiffener at 1000 mm from support (see Figure 6.15(b)):

$$V = 1330 \, \text{kN}$$

Shear buckling resistance of web without tension field action. From Table 21(b) for $d/t = 138.75$, $a/d = 0.9$:

$$q_{cr} = 102.7 \, \text{N/mm}^2$$
$$V_{cr} = 102.7 \times 1110 \times 8/10^3 = 911.9 \, \text{kN}$$

Stiffener force $F_q = 1330 - 911.9 = 418.1 \, \text{kN}$. The stiffener properties are:

$$A = (160 \times 8) + (320 \times 8) \qquad\qquad = 3840 \, \text{mm}^2$$
$$r_x = (3.161 \times 10^6/3840)^{0.5} \qquad\qquad = 28.69 \, \text{mm}$$
$$\lambda = 0.7 \times 1110/28.69 \qquad\qquad = 27.1$$
$$p_v = 241.6 \, \text{N/mm}^2 \text{ from Table 27(c) for } p_y = 255 \, \text{N/mm}^2$$

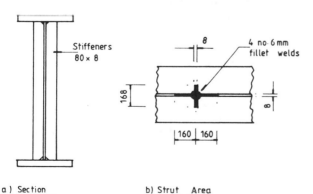

a) Section   b) Strut   Area

Figure 6.23 Intermediate stiffener

Buckling resistance

$$P_q = 241.6 \times 3840/10^2 \qquad\qquad = 927.8\,\text{kN} > F_q$$

The stiffener is satisfactory.
(Note that these stiffeners are to extend from flange to flange.)

(3) Connection to web

Provide 6-mm continuous fillet weld.

6.5.3 Load-bearing stiffener

Try stiffeners composed of 2 No. 150 mm × 20 mm plates as shown in Figure
6.24. The stiffener outstand will be satisfactory and the bearing area will also
be adequate. (Refer to Section 6.4.5.)

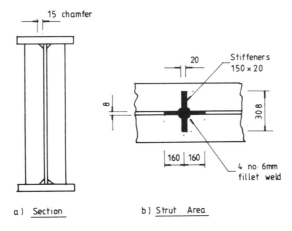

a) Section   b) Strut Area

Figure 6.24 Load-bearing stiffener

The buckling check is carried out first:

$$A = (2 + 150 \times 20) + (320 \times 8) \qquad = 8560\,\text{mm}^2$$
$$I_x = 20 \times 308^3/12 \qquad = 48.69 \times 10^6\,\text{mm}^4$$
$$r_x = (48.69 \times 10^6/8520)^{0.5} \qquad = 75.6$$
$$\lambda = 0.7 \times 1110/75.6 \qquad = 10.28$$
$$p_y = 265\,\text{N/mm}^2\ (\text{Table 6}) - 20 = 245\,\text{N/mm}$$
$$p_c = 245\,\text{N/mm}^2\ (\text{Table 27(c)})$$
$$P_x = 245 \times 8560/10^3 \qquad = 2097.2\,\text{kN}$$

Combined load = 1110 + 1198 − 911.9 = 1396.1 kN

The size selected is satisfactory. Provide 6-mm continuous fillet weld
between stiffeners and web.

6.5.4 End post

The design will be made using double stiffeners. The trial arrangement is
shown in Figure 6.25.

(1) Bearing check

The reaction is carried on the inner stiffener. The stiffener ends are chamfered to clear the web to flange welds and the bearing area is shown in Figure 6.25(c).

$$p_{ys} = 265 \text{ N/mm}^2 \text{ (Table 6)} - 20 = 245 \text{ N/mm2}$$
$$A = 2 \times 20 \times 206 = 8240 \text{ mm}^2$$
$$> 0.84 \times 1374 \times 10^3/245 = 4486.5 \text{ mm}^2$$

Note that oustand $= 221 \text{ mm} < 13\left(\dfrac{275}{265}\right)^{0.5} 20 = 264.9$

The full area of the stiffener is effective and the stiffener is satisfactory for bearing.

(2) Buckling check

The area at the centre line of the bearing stiffener acting as a strut is shown in Figure 6.25(d):

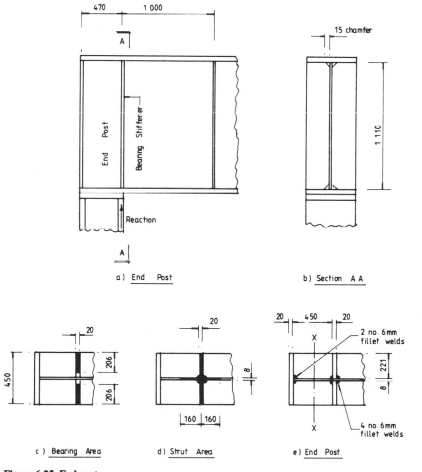

a) End Post

b) Section A A

c) Bearing Area

d) Strut Area

e) End Post

Figure 6.25 End post

$A = (2 \times 20 \times 221) + (2 \times 160 \times 8) = 11\,400\,\text{mm}^2$

$I_x = 20 \times 450^3/12 \qquad\qquad\qquad = 151.87 \times 10^6\,\text{mm}^4$

$r_x = (151.87 \times 10^6/11\,400)^{0.5} \qquad = 115.4\,\text{mm}$

$\lambda = 0.7 \times 1110/115.4 \qquad\qquad = 6.73$

$p_c = 245\,\text{N/mm}^2$ (Table 27(c))

$P_x = 245 \times 11400/10^3 \qquad\qquad = 2793\,\text{kN}$

This exceeds the reaction 1374 kN.

(3) Shear and moment from the tension field

Refer to Section 4.4.5.4.3 of BS 5950: Part 1. The critical shear strength for the end panel for:

$d/t = 138.75$ and $a/d = 0.9$ from Table 21(b)

$q_{cr} = 102.7\,\text{N/mm}^2$

Basic shear strength (see Section 6.5.1):

$q_b = 131\,\text{N/mm}^2$

Applied shear stress:

$f_v = 1374 \times 10^3/(8 \times 1110) = 154.73\,\text{N/mm}^2$

Shear stress f_v is greater than the basic shear strength q_b, so tension field force H_q cannot be reduced. This force is:

$$H_q = \frac{0.75 \times 1110 \times 8 \times 275}{10^3}\left(1 - \frac{102.7}{0.6 \times 275}\right)^{0.5}$$

$$= 1125.3\,\text{kN}$$

Shear from the tension field force:

$R_{tf} = 1125.3/2 \qquad\qquad = 562.6\,\text{kN}$

Moment in the end post:

$$M_{tf} = \frac{1125.3 \times 1110}{10 \times 10^3} \qquad = 124.9\,\text{kNm}$$

(4) Shear capacity of end post

The end post is shown in Figure 6.25(e). The web 450 mm × 8 mm resists shear (see Table 7 of code for limiting proportions for webs):

$d/t = 450/8 = 56.25 < 79\varepsilon < 79$

The section is plastic.

Shear capacity $= 0.6 \times 275 \times 450 \times 8/10^3 = 594\,\text{kN}$

The end post is satisfactory with respect to shear.

(5) Moment capacity

Referring to Figure 6.25(e), the flange proportions are:

$b/T = 221/20 \qquad = 11.05$

Design strength $p_y = 265 \, \text{N/mm}^2$ (Table 6 in the code).

Note that p_y is not reduced in moment calculations. From Table 7, $b/T < 13\varepsilon$ = 13.2. The flanges are semi-compact.

Moment capacity check at the centre of the girder:

$M_{cx} = 265 \times 450 \times 20 \times 470/10^6 = 1120.9 \, \text{kNm}$

This exceeds the moment from tension field action of 124.9 kNm.

(6) Combined bearing and bending

A local capacity check for combined bearing and bending at the bottom joint of the bearing stiffener with the flange gives:

$$\frac{1374 \times 10^6}{8240 \times 245} + \frac{124.9 \times 10^6}{8240 \times 470 \times 265} = 0.803 \; < 1.0$$

The end post is satisfactory.

(7) Weld sizes

The four fillet welds shown in Figure 6.25(e) to connect the bearing stiffener to the web are designed first. The welds must support the reaction and the beam shear from the end post:

End post $I_x = (450 \times 490^3 - 442 \times 450^3)/12$

$\qquad\qquad = 1055 \times 10^6 \, \text{mm}^4$

$$\text{Weld force} = \frac{1374}{4 \times 1110} + \frac{562.6 \times 2 \times 221 \times 20 \times 234}{4 \times 1055 \times 10^6}$$

$\qquad\qquad = 0.31 + 0.28$

$\qquad\qquad = 0.59 \, \text{kN/mm}$

Provide 6 mm continuous fillet weld; strength 0.9 kN/mm.

This size of weld will be satisfactory for the welds between the end plate and web.

6.5.5. Design drawing

A drawing of the girder designed utilizing tension field action is shown in Figure 6.26.

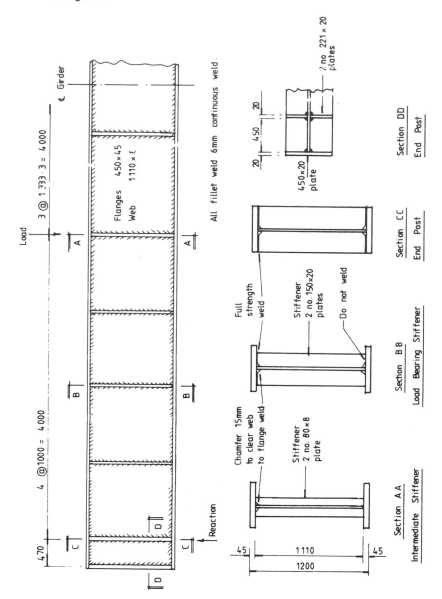

Figure 6.26 Design utilizing tension field action

Problems

6.1 A welded plate girder fabricated from Grade 43 steel is proportioned as shown in Figure 6.27. It spans 15.0 m between centres of brackets and supports a 254 × 254 UC 107 column at mid-span. The loading is shown in the figure. The compression flange is effectively restrained over the span and intermediate stiffeners are provided at 1.875 m centres between supports and the centre load.

Assuming that the plate girder and fire-protection casing weigh 20 kN/m, carry out the following work:

(1) Check the adequacy of the section with respect to bending, shear and deflection.
(2) Design a suitable load-bearing stiffener for the supports and concentrated load positions.
(3) Determine the weld size required at the point of maximum shear.

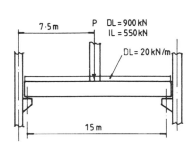 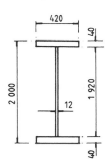

Figure 6.27

6.2 A welded plate girder of Grade 43 steel carries two concentrated loads transmitted from 254 × 254 UB 107 columns at the third points. The columns rest on the top flange and the loads are each 400 kN dead load and 250 kN imposed load, respectively. The plate girder is 12 m span and is simply supported at its ends. The compression flange has adequate lateral restraint at the points of concentrated loads and at the supports. Assume that the weight of the girder is 4 kN/m and that the girder is supported on brackets at each end.

(1) Assuming that the depth limit is 1400 mm for the plate girder, design the girder section.
(2) Design the load-bearing and intermediate stiffeners.
(3) Design the web-to-flange weld.
(4) Sketch the arrangement and details of the plate girder.

6.3 The framing plans for a four-storey building are shown in Figure 6.28. The front elevation is to have a plate girder at first-floor level to carry wall and floors and give clear access between columns B and C.

The plate girder is simply supported with a shear connection between the girder end plates and the column flanges. Columns B and C are 305 × 305 UC 158. The loading from floor, roof and wall is as follows:

Dead loads:

Front wall between B and C
(includes glazing and columns) $= 0.7 \, kN/m^2$

Floors of r.c. slab
(including screed, finish, ceilings, etc.) $= 6.0 \, kN/m^2$

Roof of r.c. slab
(including screed, finish, ceilings, etc.) $= 4.0 \, kN/m^2$

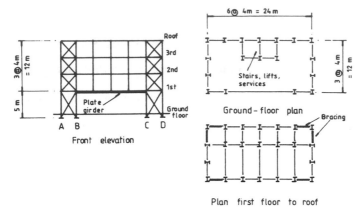

Figure 6.28 Framing plans for a three-storey building

Imposed loads:

Roof $= 1.5\,\text{kN/m}^2$

Floors $= 2.5\,\text{kN/m}^3$

(1) Calculate the loads on the girder.
(2) Design the plate girder and show all design information on a sketch.

7

Tension members

7.1 Uses, types and design considerations

7.1.1 Uses and types

A tension member transmits a direct axial pull between two points in a structural frame. A rope supporting a load or cables in a suspension bridge are obvious examples. In building frames tension members occur as:

(1) Tension chords and internal ties in trusses;
(2) Tension bracing members;
(3) Hangers supporting floor beams.

Examples of these members are shown in Figure 7.1
  The main sections used for tension members are:

(1) Open sections such as angles, channels, tees, joists, universal beams and columns;
(2) Closed sections. Circular, square and rectangular hollow sections;

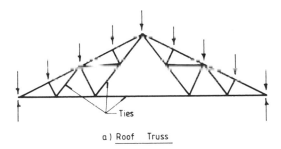

a) Roof   Truss

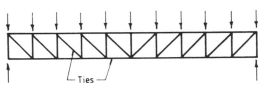

b) Lattice   Girder

Figure 7.1 (contd. overleaf)

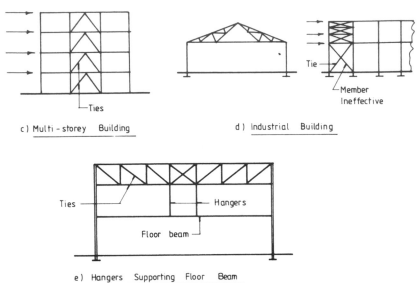

c) Multi-storey   Building

d) Industrial   Building

e) Hangers   Supporting   Floor   Beam

Figure 7.1 Tension members in buildings

(3) Compound and built-up sections. Double angles and double channels are
common compound sections used in trusses. Built-up sections are used in
bridge trusses.

Round bars, flats and cables can also be used for tension members where there
is no reversal of load. These elements as well as single angles are used in cross
bracing, where the tension diagonal only is effective in carrying a load, as
shown in Figure 7.1(d). Common tension member sections are shown in
Figure 7.2.

7.1.2 Design considerations

Theoretically the tension member is the most efficient structural element, but
its efficiency may be seriously affected by the following factors:

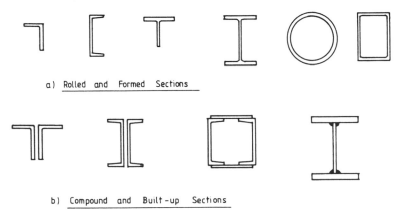

a) Rolled   and   Formed   Sections

b) Compound   and   Built-up   Sections

Figure 7.2 Tension member sections

(1) The end connections. For example, bolt holes reduce the member section.
(2) The member may be subject to reversal of load, in which case it is liable to buckle because a tension member is more slender than a compression member.
(3) Many tension members must also resist moment as well as axial load. The moment is due to eccentricity in the end connections or to lateral load on the member.

7.2 End connections

Some common end connections for tension members are shown in Figures 7.3(a) and (b). Comments on the various types are:

(1) Bolt or threaded bar. The strength is determined by the tensile area at the threads.
(2) Single angle connected through one leg. The outstanding leg is not fully effective, and if bolts are used the connected leg is also weakened by the bolt hole.

Full-strength joints can be made by welding. Examples occur in lattice girders made from hollow sections. However, for ease of erection most site joints are bolted, and welding is normally confined to shop joints.

Site splices are needed to connect together large trusses that have been fabricated in sections for convenience in transport. Shop splices are needed in long members or where the member section changes. Examples of bolted and welded splices in tension members are shown in Figures 7.2(c) and (d).

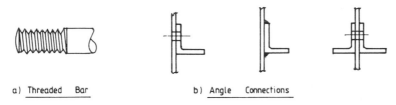

a) Threaded   Bar   b) Angle   Connections

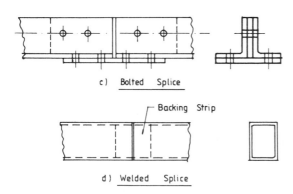

c)   Bolted   Splice

Backing   Strip

d) Welded   Splice

Figure 7.3 End connections and splices

7.3 Structural behaviour of tension members

7.3.1 Direct tension

The tension member behaves in the same way as a tensile test specimen. In the elastic region:

Tensile stress $f_t = P/A$

Elongation $\delta = PL/AE$

where $P =$ load on the member

$A =$ area of cross section

$L =$ length

7.3.2 Tension and moment: elastic analysis

(1) Moment about one axis

Consider the I section shown in Figure 7.4(b), which has two axes of symmetry. If the section is subjected to an axial tension P and moment M_x about the XX axis, the stresses are:

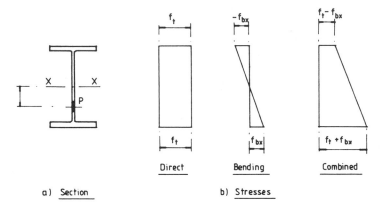

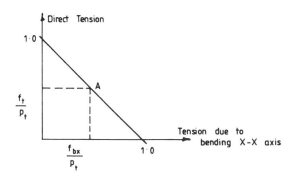

Figure 7.4 Elastic analysis: tension and moment about one axis

Direct tensile stress f_t $\qquad = P/A$
Tensile bending stress f_{bx} $\quad = M_x/Z_x$
Maximum tensile stress $f_{max} = f_t + f_{bx}$

where Z_x = elastic modulus for the XX axis.

The stress diagrams are shown in Figure 7.4(b). Define the allowable stresses:

p_t —direct tension

p_{bt}—tension due to bending

Then the interaction expression

$$\frac{f_t}{p_t} + \frac{f_{bx}}{p_{bt}} \leqslant 1$$

gives permissible combinations of stresses. This is shown graphically in Figure 7.4(c). A section with one axis of symmetry may be treated similarly.

(2) Moment about two axes

If the section is subjected to axial tension P and moments M_x and M_y about the XX and YY axes, respectively, the individual stresses and maximum stress are:

Direct tensile stress $\qquad\qquad f_t \;\; = P/A$
Tensile bending stress XX axis $f_{bx} \;\; = M_x/Z_x$
Tensile bending stress YY axis $f_{by} \;\; = M_y/Z_y$
Maximum stress $\qquad\qquad\quad f_{max} = f_t + f_{bx} + f_{by}$
Z_y = elastic modulus for the YY axis.

These stresses are shown in Figures 7.5(b)–(d).

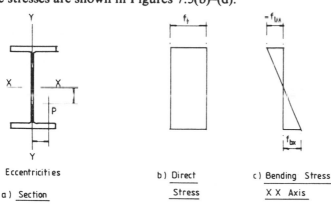

Eccentricities

a) Section

b) Direct
Stress

c) Bending Stress
X X Axis

d) Bending Stress
Y Y Axis

Maximum stress

$f_{max} = f_t + f_{bx} + f_{by}$

Figure 7.5 (contd. overleaf)

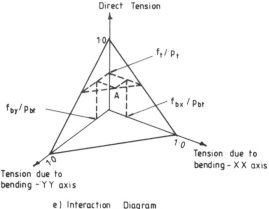

e) Interaction   Diagram

Figure 7.5 Elastic analysis: tension and moments about two axes

The interaction expression to give permissible combinations of stresses is:

$$\frac{f_t}{p_t}+\frac{f_{bx}}{p_{bt}}+\frac{f_{by}}{p_{bt}}\leqslant 1$$

This may be represented graphically by the plane in Figure 7.5(e).

Sections with one axis of symmetry or with no axis of symmetry which are free to bend about the principal axes can be treated similarly.

7.3.3 Tension and moment: plastic analysis

(1) Moment about one axis

For a section with two axes of symmetry (as shown in Figure 7.6(a)) the moment is resisted by two equal areas extending inwards from the extreme fibres. The central core resists the axial tension. The stress distribution is shown in Figure 7.6(b) for the case where the tension area lies in the web. At higher loads the area needed to support tension spreads to the flanges, as shown in Figure 7.6(c).

For design strength p_y the maximum tension the section can support is:

$$P_t = p_y A$$

Compression from Moment

Direct Tension

Tension from Moment

Section

Axial Tension

Moment

Resultant Stress

a) Tension Area in Web

b) Stress Distribution

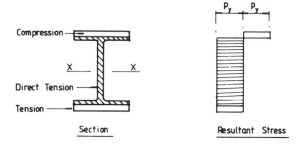

Section

Resultant Stress

c) Tension Area Spread to Flanges

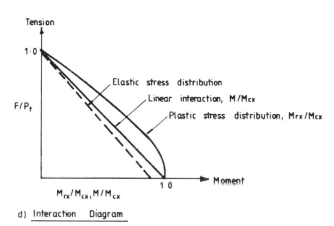

d) Interaction   Diagram

Figure 7.6 Plastic analysis: tension and moment about one axis

If moment only is applied, the section can resist:

> Plastic moment $M_{cx} = p_y S_x$ and an
> Elastic moment $M_{EX} = p_y Z_x$

where S_x = plastic modulus

> Z_x = elastic modulus

For values F of tension less than P_t if the tension area is in the web (as shown in Figure 7.6(a)) the length a of web supporting F is:

> $a = F/p_y t$

where t = web thickness.

The reduced moment capacity in the presence of axial load is:

$$M_{rx} = (S_x - \frac{ta^2}{4}) p_y$$

A more complicated formula is needed for the case where the tension area enters the flanges, as shown in Figure 7.6(b). The curve of F/P_t against M_{rx}/M_{cx} for an I section bent about the XX axis is convex (see Figure 7.6(d)), but a conservative design results if the straight line joining the end points is adopted. This gives the linear interaction expression:

$$\frac{F}{P_t} + \frac{M_x}{M_{cx}} = 1$$

where M_x = applied moment

The elastic curve is also shown. The strength gain due to plasticity is the area between the two curves.

In calculating the reduced moment capacity it is convenient to use a reduced plastic modulus. This was given for the case above by:

$$S_{rx} = S_x - ta^2/4$$

If the average stress on the whole section of area A:

$$f = F/A$$

then the formula for reduced plastic modulus can be written after substituting for a as:
$$S_{rx} = S_x - n^2 A^2/4t$$

where $n = f/p_y$

The expression is more complicated when the tension area spreads into the flanges.

These are the formulae given in *Steelwork Design*, Guide to BS 5980: Part 1: 1985, Volume 1, to calculate the reduced plastic modulus. The change value of n is given to indicate when the tension area enters the flanges. Note that the reduced plastic modulus is not required if the linear interaction expression is adopted.

The analysis for sections with one axis of symmetry is more complicated.[14]

(2) Moment about two axes

Solutions can be found for sections subject to axial tension and moments about both axes at full plasticity. I sections with two axes of symmetry have been found to give a convex failure surface, as shown in Figure 7.7. This interaction surface is constructed in terms of:

$$F/P_t, \quad M_{rx}/M_{cx}, \quad M_{ry}/M_{cy}$$

where F = axial tension

P_t = tension capacity

M_{cx} = moment capacity for the XX axis in the absence of axial load

M_{rx} = reduced moment capacity for the XX axis in the presence of axial load

M_{cy} = moment capacity for the YY axis in the absence of axial load

M_{ry} = reduced moment capacity for the YY axis in the presence of axial load.

In practice, M_{cy} is restricted with some sections. Any point A on the failure surface gives the permissible combination of axial load and moments the section can support.

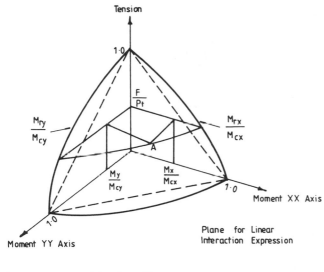

Figure 7.7 Plastic analysis: tension and moment about two axes

A plane may be drawn through the terminal points on the failure surface. This can be used to give a simplified and conservative linear interaction expression:

$$\frac{F}{P_t}+\frac{M_x}{M_{cx}}+\frac{M_y}{M_{cy}}=1$$

where M_x = applied moment about the XX axis

M_y = applied moment about the YY axis

7.4 Design of tension members

7.4.1 Axially loaded tension members

The tension capacity is given in Section 4.6.1 of BS 5950: Part 1. This is:

$P_t = A_e p_y$

A_e = effective area of the section defined in Sections 3.3.3 and 4.6.2–4.6.4 of the code.

From Section 3.3.3, the effective area of each element of a member is given by:

$A_e = K_e \times$ net area where holes occur

   $\leqslant$ gross area

$K_e = 1.2$ for Grade 43 and 1.1. for Grade 50 steel

(Net area = gross area less holes.)

Tests show that holes do not reduce the capacity of a member in tension provided that the ratio of net area to gross area is greater than the ratio of yield strength to ultimate strength.

7.4.2 Angle tension members

(1) Single angles connected through one leg

These may be designed in accordance with Section 4.6.3 of the code as axially loaded members with an effective area (see Figure 7.3(b)):

$$A_e = a_1 + \left(\frac{3\,a_1}{3\,a_1 + a_2}\right) a_2$$

a_1 = net sectional area of the connected leg
a_2 = sectional area of the unconnected leg

(2) Double angles connected to one side of a gusset

$$A_e = a_1 + \left(\frac{5a_1}{5a_1 + a_2}\right) a_2$$

(3) Double angles connected to both sides of a gusset

If these members are connected together as specified in the code they can be designed as axially loaded members using the net area specified in Section 3.3.2 of the code. This is the gross area minus the deduction for holes.

7.4.3 Tension members with moments

The code states in Sections 4.6.2 and 4.8.1 that moments from eccentric end connections and other causes must be taken into account in design. Single angles, double angles and T sections carrying direct tension only may be designed as axially loaded members, as set out in Section 4.6.3 of the code.

Design of tension members with moments is covered in Section 4.8.2 of the code. This states that the member should be checked for capacity at points of greatest moment using the simplified interaction expression:

$$\frac{F}{A_e p_y} + \frac{M_x}{M_{cx}} + \frac{M_y}{M_{cy}} \leqslant 1$$

where F = applied axial load
$\quad\quad A_e$ = effective area
$\quad\quad M_x$ = applied moment about the XX axis
$\quad\quad M_{cx}$ = moment capacity about the XX axis in the absence of axial load
$\quad\quad M_y$ = applied moment about the YY axis
$\quad\quad M_{cy}$ = moment capacity about the YY axis in the absence of axial load.

The interaction expression was discussed in Section 7.3.3(2) above. (See Section 5.4.2 for calculation of M_{cx} and M_{cy}.) For bending about one axis the terms for the other axis are deleted.

An alternative expression given in the code takes account of convexity of the failure surface. This leads to greater economy in the design of plastic and compact sections.

7.5 Design examples

7.5.1 Angle connected through one leg

Design a single angle to carry a dead load of 80 kN and an imposed load of 35 kN.

(1) Bolted connection

Factored load $= (1.4 \times 80) + (1.6 \times 35) = 168$ kN.

Try $80 \times 60 \times 7$ angle connected through the long leg, as shown in Figure 7.8(a). The bolt hole is 22 mm diameter for 20 mm diameter bolts.

Design strength from Table 6 in the code $p_y = 275$ N/mm^2:

$a_1 =$ net area of connected leg

$\quad = (76.5 - 22)\ 7 = 381.5$ mm^2

$a_2 =$ area of unconnected leg

$\quad = 56.5 \times 7 \qquad = 395.5$ mm^2

Effective area

$$A_e = 381.5 + \left(\frac{3 \times 381.5}{3 \times 381.5 + 395.5} \right) 395.5$$

$$\quad = 675.5 \text{ mm}^2$$

Tension capacity:

$$P_t = 675.5 \times 275/10^3 = 185.7 \text{ kN}$$

The angle is satisfactory.

Note that the connection would require either 3 No. Grade 8.8 or 3 No. friction-grip 20 mm diameter bolts to support the load.

(2) Welded connection

Try $75 \times 50 \times 6$ L connected through the long leg (see Figure 7.8(b)):

$a_1 = 72 \times 6 \qquad\qquad\qquad\qquad = 432$ mm^2

$a_2 = 47 \times 6 \qquad\qquad\qquad\qquad = 282$ mm^2

$A_e = 432 + \left(\dfrac{3 \times 432}{3 \times 432 + 282} \right) 282 \qquad = 663.6$ mm^2

$P_t = 663.6 \times 275/10^3 \qquad\qquad = 182.5$ kN

The angle is satisfactory.

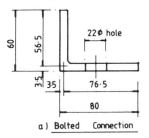

a) Bolted Connection

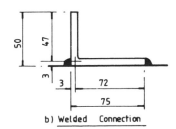

b) Welded Connection

Figure 7.8 Single angle connected through one leg

7.5.2 Hanger supporting floor beams

A high-strength Grade 55 steel hanger consisting of a 203 × 203 UC 46 carries the factored loads from beams framing into it and from the floor below, as shown in Figure 7.9(a). Check the hanger at the main floor beam connection.

The design strength from Table 6 of BS 5950: Part 1 for sections less than 16 mm thick is:

$$p_y = 450 \, \text{N/mmm}^2$$

The net section is shown in Figure 7.9(b). For Grade 55 steel the effective section is equal to the net section. The factor K_e from Section 3.3.3 of the code is 1.0. The connection plates are not considered.

Check the limiting proportions of the flanges using Table 7a in the code:

$$\varepsilon = (275/450)^{0.5} \qquad = 0.781$$
$$b/t = 101.6/11 = 9.23 < 15 \, \varepsilon = 11.72$$

Values of b and T are shown in Figure 7.9(b).

The section is semi-compact. The moment capacity is calculated using the elastic properties. This can be calculated using first principles, and the properties are:

Location of the centroidal axis is shown
Effective area = 53.1 cm^2
Minimum value of elastic modulus Z $\quad$ = 365 cm^3

The moment capacity for the major axis:

$$M_{cx} = 363 \times 450/10^3 \qquad\qquad = 163.35 \, \text{kNm}$$

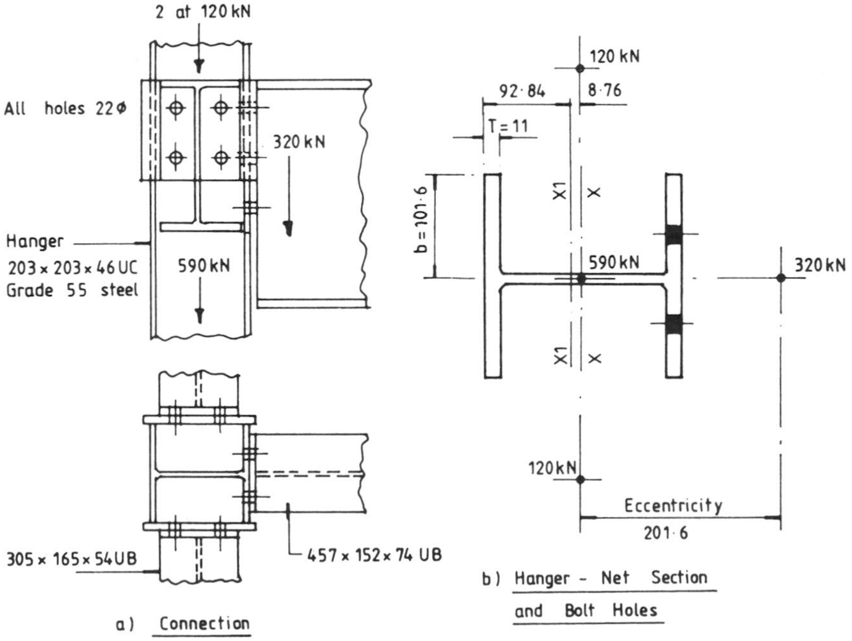

a)   Connection

b)   Hanger - Net Section
and   Bolt Holes

Figure 7.9 High-strength hanger

The applied axial load:

$$F = (2 \times 120) + 590 + 320 \qquad\qquad = 1150 \text{ kNm}$$

The applied moment about the X_1–X_1 axis:

$$M_x = (320 + 0.21) + (2 \times 120 + 590)0.0088 = 74.5 \text{ kNm}$$

Substitute into the interaction expression:

$$\frac{F}{A_e p_y} + \frac{M_x}{M_{cx}} = \frac{1150 \times 10}{53.1 \times 450} + \frac{74.5}{163.35} \qquad = 0.94 \; < 1$$

The hanger is satisfactory.

Problems

7.1 A tie member in a roof truss is subjected to an ultimate tension of 1000 kN. Design this member using Grade 43 steel and an equal angle section.

7.2 A tension member in Grade 43 steel consists of 2 No. $150 \times 100 \times 8$ mm unequal angles placed back to back. At the connection, two rows of 2 No. 22 mm diameter holes are drilled through the longer legs of the angles. Determine the ultimate tensile load that can be carried by the member.

7.3 A tension member from a heavy truss is subjected to an ultimate axial load and bending moment of 2000 kN and 500 kNm, respectively. Design a suitable universal beam section in Grade 43 steel. Assume that the gross section will resist the load and moment.

7.4 A tie member in a certain steel structure is subjected to tension and biaxial bending. The ultimate tensile load was found to be 3000 kN while the ultimate moments about the major and minor axes were 160 kNm and 90 kNm, respectively. Check whether a 305×305 UC 158 in Grade 43 steel is adequate. Assume that the gross section resists the loads and moments.

8

Compression members

8.1 Types and uses

8.1.1 Types of compression members

Compression members are one of the basic structural elements, and are described by the terms 'columns', 'stanchions' or 'struts', all of which primarily resist axial load.

Columns are vertical members supporting floors, roofs and cranes in buildings. Though internal columns in buildings are essentially axially loaded and are designed as such, most columns are subjected to axial load and moment. The term 'strut' is often used to describe other compression members such as those in trusses, lattice girders or bracing. Some types of compression members are shown in Figure 8.1. Building columns will be discussed in this chapter and trusses and lattice girders are dealt with in Chapter 9.

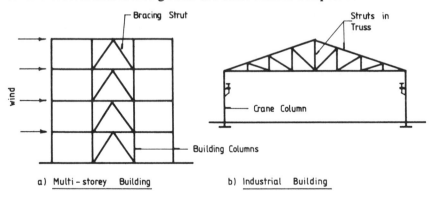

Figure 8.1 Types of compression members

8.1.2 Compression member sections

Compression members must resist buckling, so they tend to be stocky with square sections. The tube is the ideal shape, as will be shown below. These are in contrast to the slender and more compact tension members and deep beam sections.

Rolled, compound and built-up sections are used for columns. Universal columns are used in buildings where axial load predominates, and universal beams are often used to resist heavy moments that occur in columns in industrial buildings. Single angles, double angles, tees, channels and structural hollow sections are the common sections used for struts in trusses, lattice girders and bracing. Compression member sections are shown in Figure 8.2.

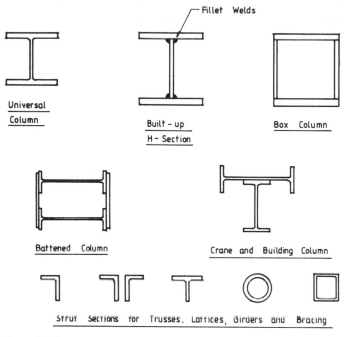

Figure 8.2 Compression member sections

8.1.3 Construction details

Construction details for columns in buildings are:

(1) Beam-to-column connections;
(2) Column cap connections;
(3) Column splices;
(4) Column bases.

(1) Beam-to-column cap connections

Typical beam-to-column connections and column cap connections are shown in Figures 8.3(a) and (b), respectively.

(2) Column splices

Splices in compression members are discussed in Section 6.1.7.2 of BS 5950: Part 1. The code states that where the members are not prepared for full contact in bearing, the splice should be designed to transmit all the moments and forces to which the member is subjected. Where the members are prepared for full contact the splice should provide continuity of stiffness about both axes and resist any tension caused by bending.

In multi-storey buildings splices are usually located just above floor level. If butted directly together the ends are usually machined for bearing. Fully bolted splices and combined bolted and welded splices are used. If the axial load is high and the moment does not cause tension the splice holds the columns' lengths in position. Where high moments have to be resisted, high-strength or friction-grip bolts or a full-strength welded splice may be required. Some typical column splices are shown in Figure 8.3(c).

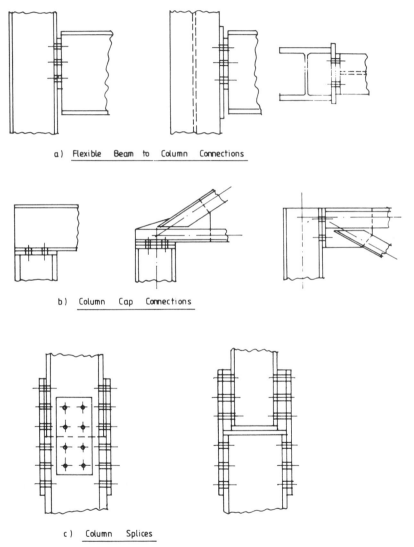

a) Flexible   Beam to   Column   Connections

b)   Column   Cap   Connections

c)   Column   Splices

Figure 8.3 Column construction details

(3) Column bases

Column bases are discussed in Section 8.10.

8.2 Loads on compression members

Axial loading on columns in buildings is due to loads from roofs, floors and walls transmitted to the column through beams and to self weight (see Figure 8.4(a)). Floor beam reactions are eccentric to the column axis, as shown, and if the beam arrangement or loading is asymmetrical, moments are transmitted to the column. Wind loads on multi-storey buildings designed to the simple design method are usually taken to be applied at floor levels and to be resisted by the bracing, and so do not cause moments.

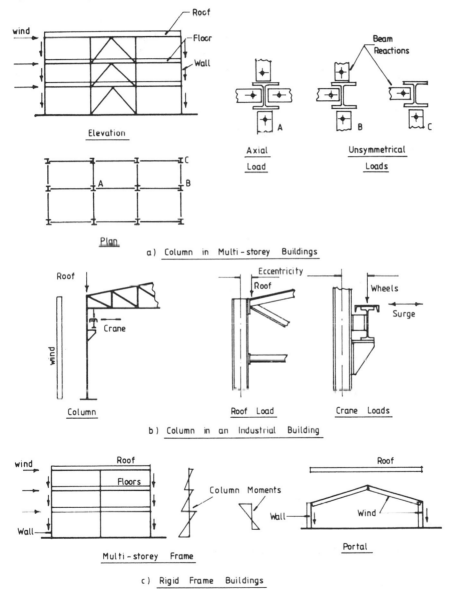

Figure 8.4 Loads and moments on compression members

In industrial buildings loads from cranes and wind cause moments in columns, as shown in Figure 8.4(b). In this case the wind is applied as a distributed load to the column through the sheeting rails.

In rigid frame construction moments are transmitted through the joints from beams to column, as shown in Figure 8.4(c). Rigid frame design is outside the scope of this book.

8.3 Classification of cross sections

The same classification that was set out for beams in Section 5.3 is used for compression members. That is, to prevent local buckling, limiting proportions for flanges and webs in axial compression are given in Table 7 BS 5950: Part 1. The proportions for rolled and welded column sections are shown in Figure 8.5.

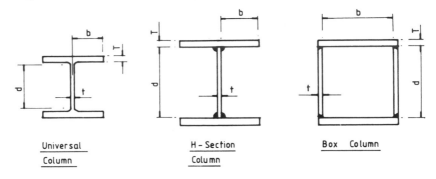

Limiting   Proportions

Element	Section Type		Plastic Section	Compact Section	Semi – Compact Section
Outstand element of compression flange	Rolled b/T	$\leqslant$	8·5 ϵ	9·5ϵ	15 ϵ
	Welded b/T	$\leqslant$	7·5 ϵ	8·5ϵ	13 ϵ
Internal element of compression flange	Welded b/T	$\leqslant$	23 ϵ	25 ϵ	28 ϵ
Web subject to compression throughout	Rolled d/T	$\leqslant$			28 ϵ
	Welded d/T	$\leqslant$			39 ϵ

$$\epsilon = (275 / Py)^{0.5}$$

All elements in compression due to axial load.

Figure 8.5 Limiting proportions for rolled and welded column sections

8.4 Axially loaded compression members

8.4.1 General behaviour

Compression members may be classified by length. A short column, post or pedestal fails by crushing or squashing, as shown in Figure 8.6(a). The squash load P_y in terms of the design strength is:

$$P_y = p_y A$$

where A = area of cross section

A long or slender column fails by buckling, as shown in Figure 8.6(b). The failure load is less than the squash load and depends on the degree of slenderness. Most practical column fail by buckling. For example, a universal column under axial load fails in flexural buckling about the weaker YY axis (see Figure 8.6(c)).

The strength of a column depends on its resistance to buckling. Thus the column of tubular section shown in Figure 8.6(d) will carry a much higher load than the bar of the same cross-sectional area.

This is easily demonstrated with a sheet of A4 paper. Open or flat, the paper cannot be stood on edge to carry its own weight; but rolled into a tube it will carry a considerable load. The tubular section is the optimum column section having equal resistance to buckling in all directions.

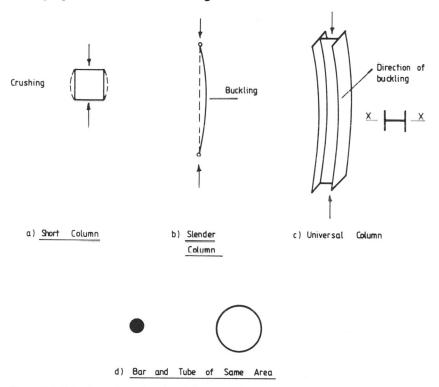

Figure 8.6 **Behaviour of members in axial compression**

8.4.2 Basic strut theory

(1) Euler load

Consider a pin-ended straight column. The critical value of axial load P is found by equating disturbing and restoring moments when the strut has been

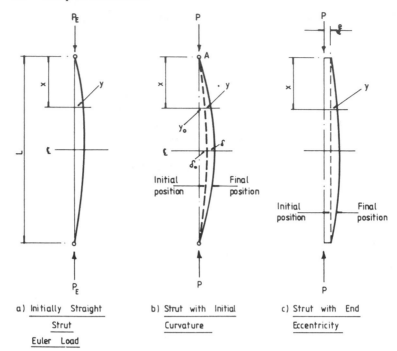

a) Initially Straight
Strut

Euler Load

b) Strut with Initial
Curvature

c) Strut with End
Eccentricity

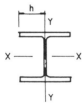

d) Column   Section   **Figure 8.7 Load cases for struts**

given a small deflection y, as shown in Figure 8.7(a). The equilibrium equation is:

$$EI_y d^2y/dx^2 = -Py$$

This is solved to give the Euler or lowest critical load.[6]

$$P_E = \pi^2 EI_y/L^2$$

In terms of stress the equation is:

$$p_E = \frac{\pi^2 E}{(L/r_y)^2} = \frac{\pi^2 E}{\lambda^2}$$

where I_y = moment of inertia about the minor axis YY

     L = length of the strut

     P = axial load

     r_y = radius of gyration for the minor axis YY

        $= (I_y/A)^{0.5}$

$p_E = P_E/A = $ Euler critical stress

λ = slenderness ratio = L/r_y

The slenderness λ is the only variable affecting the critical stress. At the critical load the strut is in neutral equilibrium. The central deflection is not defined and may be of unlimited extent. The curve of Euler stress against slenderness for a universal column section is shown in Figure 8.9.

(2) Strut with initial curvature

In practice, columns are generally not straight, and the effect of out of straightness on strength is studied in this section. Consider a strut with an initial curvature bent in a half sine wave, as shown in Figure 8.7(b). If the initial deflection at x from A is y_o and the strut deflects y further under load P, the equilibrium equation is:

$$EI d^2 y/dx^2 = P(y + y_o)$$

where deflection $y = \sin(\pi x/L)$

If δ_o is the initial deflection at the centre and δ the additional deflection caused by P, then it can be shown by solving the equilibrium equation that:

$$\delta = \frac{\delta_o}{(P_E/P) - 1}$$

The maximum stress at the centre of the strut is given by:

$$p_{max} = \frac{P}{A} + \frac{P(\delta_o + \delta)h}{I_y}$$

where h is shown in Figure 8.7(d).

Put $p_{max} = p_y$—design strength

p_c = P/A—average stress

p_E = P_E/A—Euler stress

I_y = $A r_y^2$ = moment of inertia about the YY axis

A = area of cross section

r_y = radius of gyration for the YY axis

h = half the flange breath

The equation for maximum stress can be written:

$$p_y = p_c + p_c \left\{ 1 + \frac{1}{(p_E/p_c) - 1} \right\} \frac{\delta_o h}{r_y^2}$$

Put $\eta = \delta_o h/r_y^2$ and rearrange to give:

$$(p_E - p_c)(p_y - p_c) = \eta p_E p_c$$

The value of p_c, the limiting strength at which the maximum stress equals the design strength, can be found by solving this equation and η is the Perry factor. This is redefined in terms of slenderness. (See Section 8.4.3 (2) below. The design strength curve is also discussed in that section.)

(3) Eccentrically loaded strut

Most struts are eccentrically loaded, and the effect of this on strut strength is examined here. A strut with end eccentricities e is shown in Figure 8.7(c). If y is deflection from the initially straight strut the equilibrium equation is:

$$EI\, d^2y/dx^2 = -P\,(e+y)$$

This can be solved to give the secant formula for limiting stress.[6]

Theoretical studies and tests show that the behaviour of a strut with end eccentricity is similar to that of one with initial curvature. Thus the two cases can be combined with the Perry factor, taking account of both imperfections.

8.4.3 Practical strut behaviour and design strength

(1) Residual stresses

As noted above, in general, practical struts are not straight and the load is not applied concentrically. In addition, rolled and welded strut sections have residual stresses which are locked in when the section cools.

A typical pattern of residual stress for a hot-rolled H section is shown in Figure 8.8. If the section is subjected to a uniform load the presence of these stresses causes yielding to occur first at the ends of the flanges. This reduces the

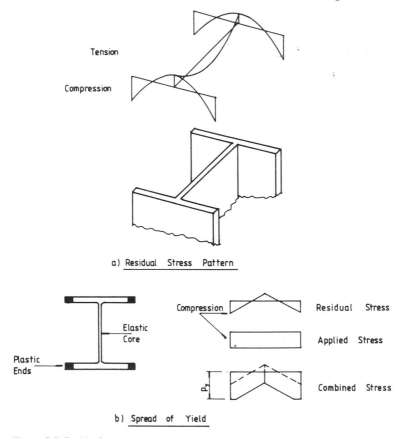

a) Residual   Stress   Pattern

b) Spread   of   Yield

Figure 8.8 Residual stresses

flexural rigidity of the section, which is now based on the elastic core, as shown in Figure 8.8(b). The effect on buckling about the YY axis is more severe than for the XX axis. Theoretical studies and tests show that the effect of residual stresses can be taken into account by adjusting the Perry factor η

(2) Column tests and design strengths

An extensive column-testing programme has been carried out, and this has shown that different design curves are required for:

(1) Different column sections;
(2) The same section buckling about different axes;
(3) Sections with different thicknesses of metal.

For example, H sections have high residual compressive stresses at the ends of the flanges, and these affect the column strength if buckling takes place about the minor axis.

The total effect of the imperfections discussed above (initial curvature, end eccentricity and residual stresses on strength) are combined into the Perry constant η. This is adjusted to make the equation for limiting stress p_c a lower bound to the test results.

The constant η is defined by:

$$\eta = 0.001 \, a \, (\lambda - \lambda_o)$$
$$\lambda = 0.2 \, (\pi^2 E / p_y)^{0.5}$$

The value λ_o gives the limit to the plateau over which the design strength p_y controls the strut load.

The Robertson constant a is assigned different values to give the different design curves. For H sections buckling about the minor axis a has the value 5.5 to give design curve (c) (Table 27(c)).

A strut table selection is given in Table 25 in BS 4950: Part 1. For example, for rolled and welded H sections with metal thicknesses up to 40 mm the following design curves are used:

(1) Buckling about the major axis XX—curve (b) (Table 27(b)).
(2) Buckling about the minor axis YY—curve (c) (Table 27(c)).

The compressive strength is given by the smaller root of the equation that was derived above for a strut with initial curvature. This is:

$$(p_E - p_c) \, (p_y - p_c) = \eta p_E p_c$$
$$p_c = \frac{p_E p_y}{(\phi + \phi^2 - p_E p_y)^{0.5}}$$
$$\phi = [p_y + (\eta + 1) \, p_E]/2$$

The curves for Euler stress p_E and limiting stress or compressive strength p_c for a rolled H section column buckling about the minor axis are shown in Figure 8.9. It can be noted that short struts fail at the design strength while slender ones approach the Euler critical stress. For intermediate struts, the compressive strength is a lower bound to the test results, as noted above. Compressive strengths for struts for curves a, b, c and d are given in Tables 27(a)–(d) in BS 5950: Part 1.

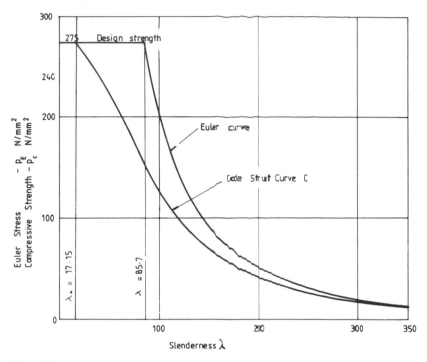

Figure 8.9 Strut strength curves

8.4.4 Effective lengths

(1) Theoretical considerations

The actual length of a compression member on any plane is the distance between effective positional or directional restraints in that plane. A positional restraint should be connected to a bracing system which should be capable of resisting 1% of the axial force in the restrained member. See Clause 4.7.1 of BS 5950.

The actual column is replaced by an equivalent pin-ended column of the same strength that has an effective length:

$$L_E = KL$$

where L = actual length and K = effective length ratio

where K is to be determined from the end conditions.

An alternative method is to determine the distance between points of contraflexure in the deflected strut. These points may lie within the strut length or they may be imaginary points on the extended elastic curve. The distance so defined is the effective length.

The theoretical effective lengths for standard cases are shown in Figure 8.10.[6] Note that for the cantilever and sway case the point of contraflexure is outside the strut length.

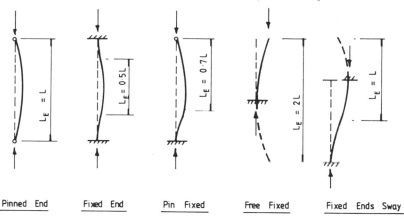

Pinned End   Fixed End   Pin Fixed   Free Fixed   Fixed Ends Sway

Figure 8.10 Figure effective lengths

(2) Code definitions and rules

The effective length is defined in Section 1.2.14 of BS 5950: Part 1 as the length between points of effective restraint of a member multiplied by a factor to take account of the end conditions and loading.

Effective lengths for compression members are set out in Section 4.7.2 of the code. This states that for members other than angles, channels and T sections the effective length should be determined from the actual length and conditions of restraint in the relevant plane. The code specifies:

(1) That restraining members which carry more than 90 per cent of their moment capacity after reduction for axial load shall be taken as incapable of providing directional restraint.
(2) Table 24 is used for standard conditions of restraint.
(3) Appendix D1 is used for stanchions in single-storey buildings of simple construction (see Section 8.6).
(4) Appendix E is used for members forming part of a frame with rigid joints.

The normal effective lengths L_E are given in Table 24 of the code. Some values from this table for various end conditions where L is the actual length are:

(1) Effectively held in position at both ends
    Restrained in direction at both ends $L_E = 0.7\ L$
    Partially restrained in direction
      at both ends $L_E = 0.85\ L$
    Not restrained in direction at either end $L_E =\quad L$
(2) One end effectively held in position and restrained in direction.
    Other end not held in position
    Partially restrained in direction $L_E = 1.5\ L$
    Not restrained in direction $L_E = 2.0\ L$

The reader should consult the table in the code for other cases.

Note the case for the fixed end strut, where the effective length is given as $0.7\ L$, is to allow for practical ends where true fixity is rarely achieved. The theoretical value shown in Figure 8.10 is $0.5\ L$.

8.4.5 Slenderness

The slenderness λ is defined in Section 4.7.3 of the code as:

$$\lambda = \frac{\text{Effective length}}{\text{Radius of gyration about relevant axis}} = \frac{L_E}{r}$$

The code states that, for members resisting loads other than wind load, λ must not exceed 180. Wind load cases are dealt with in Chapter 9 of this book.

8.4.6 Compression resistance

The compression resistance of a strut is defined in Section 4.7.4 of BS 5950: Part 1 as:

(1) Plastic, compact or semi-compact sections: $P_c = A_g p_c$

(2) Slender sections: $P_c = A_g p_{cs}$

   where A_g = gross sectional area defined in Section 3.3.1 of the code.
   p_c = compressive strength from Section 4.7.5 and Tables 27(a)–(d) of the code.
   p_{cs} = compressive strength for slender sections as defined in Sections 3.6 and 4.7.5 of the code.

8.4.7 Column design

Column design is indirect, and the process is as follows (the tables referred to are in the code):

(1) The steel grade and section is selected.
(2) The design strength p_y is taken from Table 6.
(3) The effective length L_E is estimated using Table 24 for the appropriate end conditions.
(4) The slenderness λ is calculated for the relevant axis.

Table 8.1 Compression Resistance of Grade 43 steel U.C. sections

Serial size in mm	Mass per metre in Kg	2	2.5	3	3.5	4	5	6	8	10
254 × 254	167	5230	4990	4730	4460	4180	3590	3010	2080	3580
universal	132	4160	3960	3750	3530	3300	2820	2360	1620	1140
column	107	3360	3200	3020	2840	2650	2260	1880	1280	909
	89	2790	2650	2510	2360	2200	1860	1550	1060	742
	73	2350	2230	2110	1970	1830	1550	1270	860	602
203 × 203	86	2570	2400	2220	2030	1830	1460	1150	740	–
universal	71	2130	1980	1830	1670	1510	1200	943	605	–
column	60	1820	1700	1560	1410	1270	995	778	494	–
	52	1600	1480	1360	1230	1100	865	676	1429	–
	46	1410	1310	1200	1080	968	757	590	374	–
152 × 152	37	1030	910	787	671	568	411	306	–	–
universal	30	825	727	627	533	450	325	241	–	–
column	23	632	552	472	397	334	239	177	–	–

Compression resistances in kilonewtons for effective lengths in metres

(5) The strut curve is selected from Table 25.
(6) The compressive strength is read from the appropriate part of Tables 27(a)–(d).
(7) The compression resistance P_c is calculated (see Section 8.4.6 above).

For a safe design P_c should just exceed the applied load, and successive trials are needed to obtain an economical design. Load tables can be formed to give the compression resistance for various sections for different values of effective length. Table 8.1 gives compression resistances for some universal column sections. Column sizes may be selected from tables in the *Guide* to BS 5950: Part 1: 1985, Volume 1, Section Properties, Member Capacities, Constrado. A computer program for column design is given in Chapter 11 of this book.

8.4.8 Example: universal column

A part plan of an office floor and the elevation of internal column stack *A* are shown in Figures 8.11(a) and (b). The roof and floor loads are as follows:

Roof   Dead load   (total) $= 5$ kN/m²
          Imposed load $= 1.5$ kN/m²
Floors Dead load   (total) $= 7$ kN/m²
          Imposed load $= 3$ kN/m²

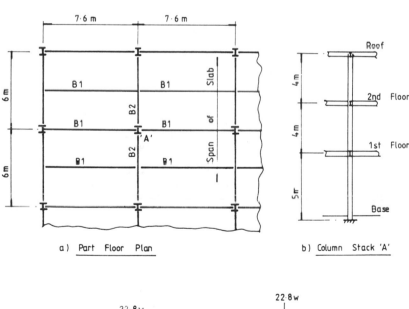

a) Part Floor Plan

b) Column Stack 'A'

c) Beam Loads

Figure 8.11 Column design example

Design column *A* for axial load only. The self weight of the column, including fire protection, may be taken as 1 kN/m. The roof and floor steel have the same layout. Use Grade 43 steel.

When calculating the loads on the column lengths, the imposed loads may be reduced in accordance with Table 2 of BS 6399: Part 1. This is permitted, because it is unlikely that all floors will be fully loaded simultaneously. Values from the table are:

Number of floors carried by member	*Reduction in imposed load (%)*
1	0
2	10
3	20

The roof is regarded as a floor for reckoning purposes.

The slabs for the floor and roof are precast one-way spanning slabs. The dead and imposed loads are calculated separately.

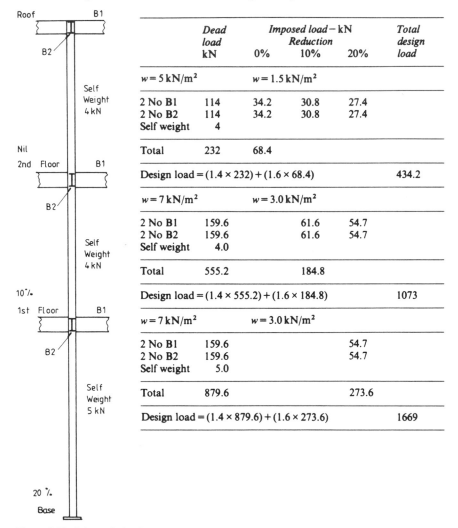

	Dead load kN	*Imposed load – kN* *Reduction*			*Total design load*
		0%	10%	20%	
$w = 5\,\text{kN/m}^2$		$w = 1.5\,\text{kN/m}^2$			
2 No B1	114	34.2	30.8	27.4	
2 No B2	114	34.2	30.8	27.4	
Self weight	4				
Total	232	68.4			
Design load = $(1.4 \times 232) + (1.6 \times 68.4)$					434.2
$w = 7\,\text{kN/m}^2$		$w = 3.0\,\text{kN/m}^2$			
2 No B1	159.6		61.6	54.7	
2 No B2	159.6		61.6	54.7	
Self weight	4.0				
Total	555.2		184.8		
Design load = $(1.4 \times 555.2) + (1.6 \times 184.8)$					1073
$w = 7\,\text{kN/m}^2$		$w = 3.0\,\text{kN/m}^2$			
2 No B1	159.6			54.7	
2 No B2	159.6			54.7	
Self weight	5.0				
Total	879.6			273.6	
Design load = $(1.4 \times 879.6) + (1.6 \times 273.6)$					1669

Figure 8.12 Column design loads

(1) Loading

Four floor beams are supported at column A. These are designated as B1 and B2 in Figure 8.11(a). The reactions from these beams in terms of a uniformly distributed load are shown in Figure 8.11(c):

Load on beam $B1 = 7.6 \times 3 \times 10 = 22.8 \, w \, kN$

where w is the uniformly distributed load. The dead and imposed loads must be calculated separately in order to introduce the different load factors. The self weight of beam B2 is included in the reaction from beam B1.

The design loading on the column can be set out as shown in Figure 8.12. The design loads are required just above the first floor, the second floor and the base.

(2) Column design

(1) Top length = Roof to second floor

Design load $= 434.2 \, kN$

Try 152×152 UC 30

$A = 38.2 \, cm^2$; $r_y = 3.82 \, cm$

Design strength $p_y = 275 \, N/mm^2$ (Table 6)

where section thickness is less than 16 mm.

If the beam connections are the shear type discussed in Section 5.8.3, where end rotation is permitted, the effective length, from Table 24:

$L_E = 0.85 \times 4000 = 3400 \, mm$

Slenderness $\lambda = 3400/38.2 = 89$

For a rolled H section thickness less than 40 mm buckling about the minor YY axis, use Table 27(c):

Compressive strength $p_c = 144 \, N/mm^2$

Compressive resistance $P_c = 144 \times 38.2/10 = 550.1 \, kN$

The column splice and floor beam connections at second-floor level are shown in Figure 8.13(a). The net section at the splice is shown in Figure 8.13(b) with 4 No. 22 mm diameter holes. The section is satisfactory.

(2) Intermediate length—first floor to second floor. Design load $= 1073 \, kN$.

Try 203×203 UC 46.

$A = 58.8 \, cm^2$; $r_y = 5.11 \, cm$

$p_y = 275 \, N/mm^2$

$\lambda = 3400/51.1 = 66.5$

$p_c = 188 \, N/mm^2$

$P_c = 188 \times 58.8/10 = 1105.4 \, kN$

The section is satisfactory.

(3) Bottom length—base to first floor. Design load $= 1669 \, kN$.

Try 254×254 UC 73.

$A = 92.9 \, cm^2$; $r_y = 6.46 \, cm$

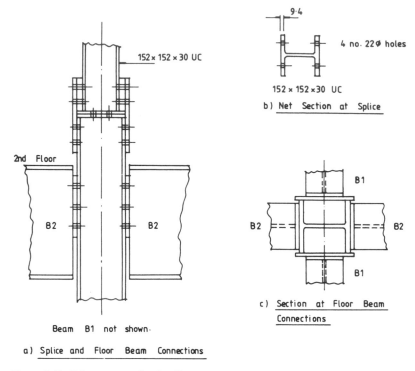

152 × 152 × 30 UC

9·4

4 no. 22∅ holes

152 × 152 ×30 UC

b) Net Section at Splice

2nd Floor

B2   B2

Beam B1 not shown·

a) Splice and Floor Beam Connections

B1

B2   B2

B1

c) Section at Floor Beam
Connections

Figure 8.13 Column connection details

The flange is 14.2 mm thick. The design strength from Table 6 in the code is

$p_y = 275 \, \text{N/mm}^2$

The beam connections do not restrain the column in direction at the first-floor level. The base can be considered fixed. The effective length is taken as:

$L_E = 0.85 \times 5000$ $\quad = 4250 \, \text{mm}$

$\lambda \;\; = 4250/64.6$ $\quad = 65.8$

$p_c \;= 189.4 \, \text{N/mm}^2$ (Table 27(c))

$P_c = 92.9 \times 189.4/10$ $\quad = 1759.5 \, \text{kN}$

The section selected is satisfactory. The same sections could have been selected from Table 8.1.

8.4.9 Built-up column: design

The two main types of columns built up from steel plates are the H and box sections shown in Figure 8.2. The classification for cross sections is given in Figure 8.5.

For plastic, compact or semi-compact cross sections the local compression capacity is based on the gross section. The code states in Section 4.7.5 that the design strength p_y for sections fabricated by welding is to be the value from Table 6 reduced by 20 N/mm². This takes account of the severe residual stresses and possible distortion due to welding.

Slender cross sections are dealt with in Section 3.6 of the code. The capacity of these sections is limited by local buckling and the design strengths are to be taken as follows:

(1) Flanges. The design strength for a slender flange is to be reduced by the factor given in Table 8 of the code. The factors for welded sections are:

$$\text{Outstand element } \frac{10}{(b/T\varepsilon)-3}$$

$$\text{Internal element } \frac{21}{(b/T\varepsilon)-7}$$

(2) Webs. The code states in Section 3.6.3 that the design strength p_y for webs subjected to moments and axial loads is to be such that the limiting proportions for a semi-compact section are met. This gives:

$$28\,(275/p_y)^{0.5} = d/t$$
$$\text{or } p_y = 215600/(d/t)^2$$

where d/t = width/thickness ratio for the web

The code specifies in Section 4.7.5 that the compressive strength p_c depends on:

(1) Slenderness λ based on the gross section;
(2) Design strength p_y reduced for welded fabrication and for slenderness, if necessary, as set out above.
(3) Relevant strut curve selected from Table 25.

The value of p_c may be taken from Tables 27(b), (c), (d) or calculated from the formula in appendix C.

8.4.10 Example: built-up column

Determine the compression resistance of the column section shown in Figure 8.14. The effective length of the column is 8 m and the steel is Grade 43.

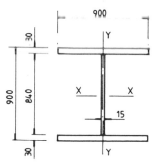

Figure 8.14 Built-up H column

(1) Flanges

The design strength from Table 6 for plate 30 mm thick $p_y = 265\,\text{N/mm}^2$. Reducing by 20 N/mm² for a welded section gives:

$p_y = 245\,\text{N/mm}^2$

$\varepsilon = (275/245)^{0.5} = 1.059$

Flange outstand $b = 442.5 = 14.75\ T$
$$> 13\varepsilon T\ = 313.77\ T$$

Referring to Table 7, the flange is slender. The design strength for the slender outstand is to be reduced by the factor from Table 8:

$$\text{Stress reduction factor} = \frac{10}{b/(T\varepsilon) - 3} = \frac{10}{[442.5/(30 \times 1.059)] - 3}$$
$$= 0.915$$

Reduced design strength $p_y = 245 \times 0.915 = 224.2\ \text{N/mm}^2$

(2) Web

This is an internal element in axial compression.
$$p_y = 275 - 20 = 255\ \text{N/mm}^2$$
$$\varepsilon = (275/255)^{0.5} = 1.038$$

Stress reduction factor $= 21/[(840/15 \times 1.038) - 7] = 0.42$
Reduced design strength $p_y = 255 \times 0.42 = 107.1\ \text{N/mm}^2$

(3) Properties of the gross section and slenderness

Area $= (2 \times 30 \times 900) + (84 \times 15)$		$= 6.66 \times 10^4\ \text{mm}^2$
I_y	$= (60 \times 900^3/12) + (\text{neglect web}) = 3.645 \times 10^2\ \text{mm}^4$	
r_y	$= [3.645 \times 10^9/6.66 \times 10^4]^{0.5}$	$= 233.9\ \text{mm}$
λ	$= 8000/233.9$	$= 34.2$

(4) Compressive strength p_c for the flanges

From Table 25 the strut table to be used is 27(c) for a welded H section with plate up to 40 mm thick. The compressive strength p_c is calculated from the formula in Appendix C:

$$\text{Euler strength } p_E = \frac{\pi^2 \times 205 \times 10^3}{34.2^2} = 1729.8\ \text{N/mm}^2$$

Robertson constant $\quad a = 5.5$ (curve (c))

$$\text{Limiting slenderness } \lambda_o\quad = 0.2 \left(\frac{\pi^2 \times 205 \times 10^3}{224.2} \right)^{0.5} = 19.0$$

Perry factor $\eta = 0.001 \times 5.5\ (34.2 - 19.0) = 0.0836$

$$\phi = \frac{224.2 + (0.0836 + 1)1729.8}{2} = 1049.3$$

Compressive strength:

$$p_c = \frac{1729.8 \times 224.2}{1049.3 + (1049.3^2 - 1729.8 \times 224.2)^{0.5}}$$
$$= 204.8\ \text{N/mm}^2$$

(5) Compressive strength p_c for the web

The calculation is similar to that for flanges in (4) above.

$$p_c = 98.4 \, \text{N/mm}^2$$

(6) Compressive resistance of the column

$$P_c = \frac{2 \times 30 \times 900 \times 204.8}{10^3} + \frac{15 \times 840 \times 98.4}{10^3}$$

$$= 11059.2 + 1239.8 = 12299 \, \text{kN}$$

8.4.11 Cased columns: design

(1) General requirements

Solid concrete casing acts as fire protection for steel columns and the casing assists in carrying the load and preventing the column from buckling about the weak axis. Regulations governing design are set out in Section 4.14 of BS 5950: Part 1.

The column must meet the following general requirements:

(1) The steel section is either a single-rolled or fabricated I or H section with equal flanges. Channels and compound sections can also be used. (Refer to the code for requirements.)
(2) The steel section is not to exceed 1000 mm × 500 mm. The dimension 1000 mm is in the direction of the web.
(3) Primary structural connections should be made to the steel section.
(4) The steel section is unpainted and free from dirt, grease, rust, scale, etc.
(5) The steel section is encased in concrete of at least Grade 20, to BS 8110.
(6) The cover on the steel is to be not less than 50 mm. The corners may be chamfered.
(7) The concrete extends the full length of the member and is thoroughly compacted.
(8) The casing is reinforced with bars not less than 5 mm diameter at a maximum spacing of 200 mm to form a cage of closed links and longitudinal bars. The reinforcement is to pass through the centre of the cover, as shown in Figure 8.15(a).
(9) The effective length is not to exceed $40 \, b_c$, $100 \, b_c^2/d_c$ or $250 \, r$, whichever is the least, where

b_c = minimum width of solid casing

d_c = minimum depth of solid casing

r = minimum radius of gyration of the steel section.

(2) Compression resistance

The design basis set out in Section 4.14.3 of the code is as follows:

(1) The radius of gyration about the YY axis shown in Figure 8.15, r_y should be taken as $0.2 \, b_c$ but not more than $0.2 \, (B + 150)$, where B = overall width of the steel flange. The radius of gyration for the XX axis r_x should be taken as that of the steel section.

(2) The compression resistance P_c is

$$P_c = \left(A_g + 0.45 \frac{f_{cu}}{p_y} A_c \right) p_c$$

$\not>$ than the short strut capacity P_{cs}

$$P_{cs} = \left(A_g + 0.25 \frac{f_{cu}}{p_y} A_c \right) p_y$$

where A_c = gross sectional area of the concrete. Casing in excess of 75 mm from the steel section is neglected. Finish is neglected

A_g = gross area of the steel section

f_{cu} = characteristic strength of the concrete at 28 days. This is not to exceed 40 N/mm²

p_c = compressive strength of the steel section determined using r_x and r_y, in the determination of which $p_y \leqslant 355$ N/mm²

p_y = design strength of the steel.

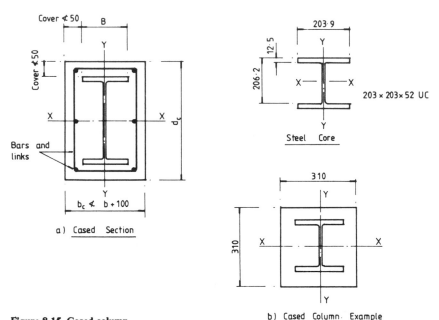

Figure 8.15 Cased column

b) Cased Column. Example

8.4.12 Example: cased column

An internal column in a multi-storey building has an actual length of 4.2 m centre to centre of floor beams. The steel section is a 203 × 203 UC 52. Calculate the compression resistance of the column if it is cased in accordance with Section 4.14 of BS 5950: Part 1. The steel is Grade 43 and the concrete Grade 25. The steel core and cased section are shown in Figure 8.15(b). The casing has been made 310 mm square.

The properties of the steel section are:

$A = 66.4 \text{ cm}^3$, $r_x = 8.9$ cm, $r_y = 5.16$ cm

For the cased section:

$$r_y = 0.2 \times 310 \qquad = 62\,\text{mm}$$
$$\not> 0.2\,(203.9 + 150) = 70.78\,\text{mm}$$

Because the column is cased throughout, the effective length is taken from Table 24 as 0.7 of the actual length:

Effective length $L_E = 0.7 \times 4200 \qquad\qquad = 2940\,\text{mm}$

The effective length L_E is not to exceed:

$40\,b_c$	$= 40 \times 310$	$= 12400\,\text{mm}$
$100\,b_c^2/d_c$	$= 100 \times 310$	$= 3100\,\text{mm}$
$250\,r$	$= 250 \times 51.6$	$= 12\,900\,\text{mm}$
Slenderness $\lambda = 2940/62$		$= 47.4$

The design strength from Table 6, $p_y = 275\,\text{N/mm}^2$. For buckling about YY, select curve (c) from Table 25. Compressive strength from Table 27(c):

$$p_c = 225.2\,\text{N/mm}^2$$

The gross sectional area of the concrete:

$$A_c = 310 \times 310 = 96\,100\,\text{mm}^2$$

Compressive resistance of the cased section:

$$P_c = \left(66.4 \times 10^2 + \frac{0.45 \times 25 \times 96\,100}{275} \right) \frac{225.2}{10^3}$$
$$= 1495.3 + 885.3$$
$$= 2380.6\,\text{kN}$$

This is not to exceed the short strut capacity:

$$P_{cs} = \left(66.4 \times 10^2 + \frac{0.25 \times 25 \times 96100}{275} \right) \frac{275}{10^3}$$
$$= 1826 + 600.6$$
$$= 2426.6\,\text{kN}$$

The compression resistance is 2380.6 kN.

8.5 Beam columns

8.5.1 General behaviour

(1) Behaviour classification

As already stated at the beginning of this chapter, most columns are subjected to bending moment in addition to axial load. These members, termed 'beam-columns', represent the general load case of an element in a structural frame. The beam and axially loaded column are limiting cases.

Consider a plastic or compact H section column as shown in Figure 8.16(a). The behaviour depends on the column length, how the moments are applied

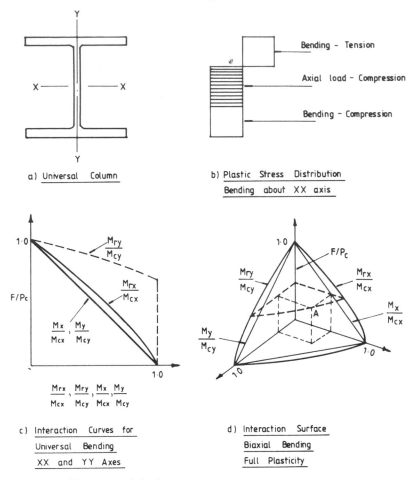

a) Universal Column

b) Plastic Stress Distribution
   Bending about XX axis

c) Interaction Curves for
   Universal Bending
   XX and YY Axes

d) Interaction Surface
   Biaxial Bending
   Full Plasticity

Figure 8.16 Short-column behaviour

and the lateral support, if any, provided. The behaviour can be classified into the following five cases:

Case 1 A short column subjected to axial load and uniaxial bending about either axis or biaxial bending. Failure generally occurs when the plastic capacity of the section is reached. Note limitations set in (2) below.

Case 2 A slender column subjected to axial load and uniaxial bending about the major axis XX. If the column is supported laterally against buckling about the minor axis YY out of the plane of bending, the column fails by buckling about the XX axis. This is not a common case (see Figure 8.17(a)). At low axial loads or if the column is not very slender a plastic hinge forms at the end or point of maximum moment

Case 3 A slender column subjected to axial load and uniaxial bending about the minor axis YY. The column does not require lateral support and there is no buckling out of the plane of bending. The column fails by buckling about the YY axis. At very low axial loads it will reach the bending capacity for YY axis (see Figure 8.17(b)).

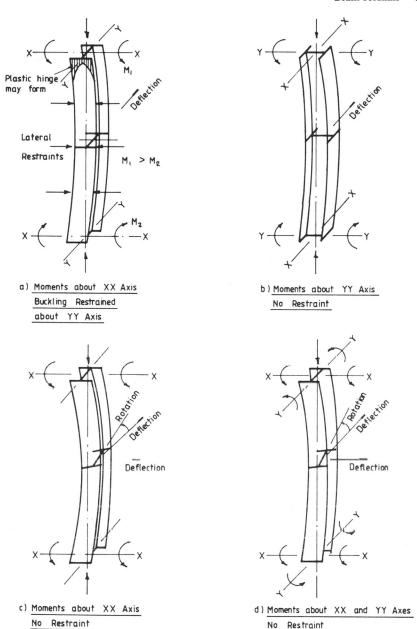

a) Moments about XX Axis
   Buckling Restrained
   about YY Axis

b) Moments about YY Axis
   No Restraint

c) Moments about XX Axis
   No Restraint

d) Moments about XX and YY Axes
   No Restraint

Figure 8.17 Slender columns subjected to axial load and moment

Case 4 A slender column subjected to axial load and uniaxial axial bending about the major axis XX. This time the column has no lateral support. The column fails due to a combination of column buckling about the YY axis and lateral torsional buckling where the column section twists as well as deflecting in the XX and YY planes (see Figure 8.17(c)).

Case 5 A slender column subject to axial load and biaxial bending. The column has no lateral support. The failure is the same as in Case 4 above but minor axis buckling will usually have the greatest effect. This is the general loading case (see Figure 8.17(d)).

Some of these cases are discussed in more detail below.

(2) Short-column failure

The behaviour of short columns subjected to axial load and moment is the same as for tension members subjected to identical loads. This was discussed in Section 7.3.3

The plastic stress distribution for uniaxial bending is shown in Figure 8.16(b). The moment capacity for plastic or compact sections in the absence of axial load is given by:

$$M_c = S p_y$$
$$\leqslant 1.2 \, Z p_y \text{ (see Section 4.2.5 of BS 5950: Part 1)}$$

where S = plastic modulus for the relevant axis
Z = elastic modulus for the relevent axis.

The interaction curves for axial load and bending about the two principal axes separately are shown in Figure 8.18(a). Note the effect of the limitation of bending capacity for the YY axis.

These curves are in terms of F/P_c against M_{rx}/M_{cx} and M_{ry}/M_{cy}, where:

F = applied axial load

P_c = $p_y \, A$, the squash load

M_{rx} = reduced moment capacity about the XX axis in the presence of axial load

M_{cx} = moment capacity about the XX axis in the absence of axial load

M_{ry} = reduced moment capacity about the YY axis in the presence of axial load

M_{cy} = moment capacity about the YY axis in the absence of axial load.

Values for M_{rx} and M_{ry} are calculated using equations for reduced plastic modulus given in the *Guide* to BS 5950: Part 1: 1985, Volume 1, Section Properties, Member Capacities, Constrado.

Linear interaction expressions can be adopted. These are:

$$F/P_c + M_x/M_{cx} = 1$$
and $F/P_c + M_y/M_{cy} = 1$

where M_x = applied moment about the XX axis
M_y = applied moment about the YY axis.

This simplification gives a conservative design.

Plastic and compact H sections subjected to axial load and biaxial bending are found to give a convex failure surface, as shown in Figure 8.18(a). At any point A on the surface the combination of axial load and moments about the XX and YY axes M_{rx} and M_{ry}, respectively, that the section can support can be read off.

A plane drawn through the terminal points of the surface gives a linear interaction expression:

$$\frac{F}{P_c} + \frac{M_x}{M_{cx}} + \frac{M_y}{M_{cy}} = 1$$

This results in a conservative design.

(3) Failure of slender columns

With slender columns, buckling effects must be taken into account. These are minor axis buckling from axial load and lateral torsional buckling from moments applied about the major axis. The effect of moment gradient must also be considered.

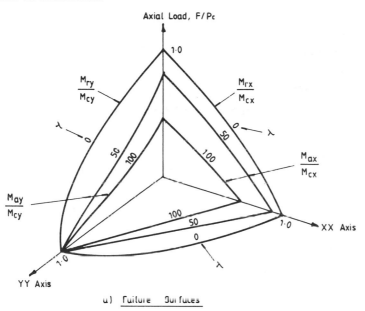

a) Failure Surfaces

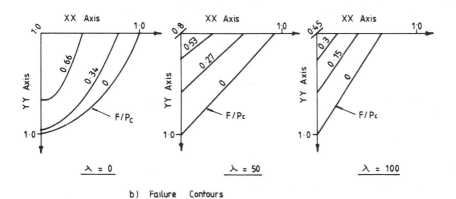

b) Failure Contours

Figure 8.18 Failure surfaces for slender beam: columns

All the imperfections, initial curvature, eccentricity of application of load and residual stresses affect the behaviour. The end conditions have to be taken into account in estimating the effective length.

Theoretical solutions have been derived and compared with test results. Failure surfaces for H-section columns plotted from the more exact approach given in the code are shown in Figure 8.18(a) for various values of slenderness. Failure contours are shown in Figure 8.18(b). These represent lower bounds to exact behaviour.

The failure surfaces are presented in the following terms:

Slenderness $\lambda = 0$ $- F/P_c$; M_{rx}/M_{cx}; M_{ry}/M_{cy}
$$\lambda = 50, 100 - F/P_c; \; M_{ax}/M_{cx}; \; M_{ay}/M_{cy}$$

Some of the terms were defined in Section 8.5.1(2) above. New terms used are:

M_{ax} = maximum buckling moment about the XX axis in the presence of axial load

M_{ay} = maximum buckling moment about the YY axis in the presence of axial load.

The following points are to be noted.

(1) M_{cy}, the moment capacity about the YY axis, is not subjected to the restriction $1.2 p_y Z_y$.
(2) At zero axial load, slenderness does not affect the bending strength of an H section about the YY axis.
(3) At high values of slenderness the buckling resistance moment M_b about the XX axis controls the moment capacity for bending about that axis.
(4) As the slenderness increases, the failure curves in the F/P_c, YY-axis plane change from convex to concave, showing the increasing dominance of minor axis buckling.

For design purposes the results are presented in the form of an interaction expression, and this is discussed in the next section.

8.5.2 Code design procedure

The code design procedure for compression members with moments is set out in Section 4.8.3 of BS 5950: Part 1. This requires the following two checks to be carried out:

(1) Local capacity check; and
(2) Overall buckling check.

In each case two procedures are given. These are a simplified approach and a more exact one. Only the simplified approach will be used in design examples in this book.

(1) Local capacity check

The member should be checked at the point of greatest bending moment and axial load. This is usually at the end, but it could be within the column height if lateral loads are also applied. The capacity is controlled by yielding or local buckling. (Local buckling was considered in Section 8.3.)

The interaction relationship for semi-compact and slender cross sections and the simplified approach for compact cross sections given in Section 4.8.3.2 of the code is:

$$\frac{F}{A_g p_y} + \frac{M_x}{M_{cx}} + \frac{M_y}{M_{cy}} \leqslant 1$$

where F = applied axial load

A_g = gross cross-sectional area

M_x = applied moment about the major axis XX

M_{cx} = moment capacity about the major axis XX in the absence of axial load

M_y = applied moment about the minor axis YY

M_{cy} = moment capacity about the minor axis YY in the absence of axial load.

A more rigorous interaction relationship for plastic and compact sections is given in the code. This is based on the convex failure surface discussed above and gives greater economy in design.

(2) Overall buckling check

The simplified interaction relationship to be satisfied is given in Clause 4.8.3.3.1 of the code. This is:

$$\frac{F}{A_g p_c} + \frac{m M_x}{M_b} + \frac{m M_y}{p_y Z_y} \leqslant 1$$

where m = equivalent uniform moment factor from Table 18 in the code

M_b = buckling resistance moment capacity about the major axis XX

Z_y = elastic modulus of section for the minor axis YY

The value for M_b is determined using the methods set out in Section 5.5 of this book (dealing with lateral torsional buckling of beams). A more exact approach is also given in the code. This uses the convex failure surfaces discussed above.

8.5.3 Example of beam column design

A braced column 4.5 m long is subjected to the factored end loads and moments about the XX axis, as shown in Figure 8.19(a). The column is held in position but only partially restrained in direction at the ends. Check that a 203 × 203 UC 52 in Grade 43 steel is adequate.

(1) Column-section classification

Design strength from Table 6 $p_y = 275\,\text{N/mm}^2$

Factor $\varepsilon = (275/p_y)^{0.5}$ $= 1.0$

See Figure 8.19(b)

Flange $b/T = 101.95/12.5$ $= 8.156 < 8.5$

Web $d/t = 160.8/8.0$ $= 20.1$ < 39

Referring to Table 7, the flanges are plastic and the web semi-compact.

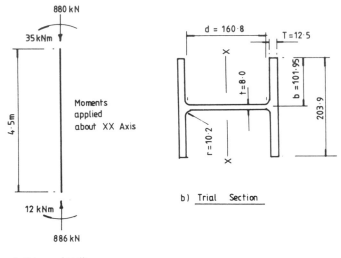

a) Column Length
and Loads   **Figure 8.19 Beam column design example**

(2) Local capacity check

Section properties for 203 × 203 UC 52 are:

$A = 66.4 \, \text{cm}^2$; $Z_x = 510 \, \text{cm}^3$; $r_y = 516 \, \text{cm}$
$x = 15.8$; $u = 0.848$ $S_x = 568 \, \text{cm}^3$

Moment capacity about the XX axis:

$M_{cx} = 275 \times 568/19^3$ $= 156.2 \, \text{kNm}$
         $< 1.2 \times 275 \times 510/10^3 = 168.4 \, \text{kNm}$

Interaction expression:

$$\frac{880 \times 10}{66.4 \times 275} + \frac{35}{156.2} = 0.48 + 0.22 = 0.7 < 1$$

The section is satisfactory with respect to local capacity.

(3) Overall buckling check

The effective length from Table 24:

$L_E = 0.85 \times 4500$ $= 3825$
Slenderness $\lambda = 3825/51.6 = 74.1$

From Table 25 select Table 27(c) for buckling about the YY axis. From Table 27(c), compressive strength $p_c = 172.8 \, \text{N/mm}^2$.

Referring to Table 9, the support conditions for the beam column are that it is laterally restrained and restrained against torsion but partially free to rotate in plan:

Effective length $L_E = 0.85 \times 4500 = 3825$ mm

Slenderness $\quad\quad \lambda \quad\quad\quad\quad\quad\quad\quad = 74.1$

The ratio of end moments:

$\beta = 12/35 = 0.342$

From Table 18 the equivalent uniform moment factor $m = 0.697$. The factor $n = 1.0$. The equivalent slenderness:

$\lambda_{LT} = uv\lambda^{\bullet}$

$u \quad = 0.848$—buckling parameter for H section

$N \quad = 0.5$—for uniform section with equal flanges

$x \quad = 15.8$—torsional index

$\lambda/x = 74.1/15.8 = 4.69$

$v \quad = 0.832$—slenderness factor from Table 14

$\lambda_{LT} = 0.848 \times 0.832 \times 74.1 = 52.2$

From Table 11 the bending strength:

$p_b = 232.7$ N/mm^2

Buckling resistance moment:

$M_b = 232.7 \times 568/10^3 = 132.1$ kNm

Interaction expression:

$$\frac{880 \times 10}{66.4 \times 172.8} + \frac{0.697 \times 35}{132.1} = 0.77 + 0.18 = 0.95 < 1.0$$

The section is also satisfactory with respect to overall buckling.

8.6 Eccentrically loaded columns in buildings

8.6.1 Eccentricities from connections

The eccentricities to be used in column design in simple construction for beam and truss reactions are given in Clause 4.7.6 of BS 5950: Part 1. These are as follows:

(1) For a beam supported on a cap plate the load should be taken as acting at the face of the column or edge of the packing.
(2) For a roof truss on a cap plate the eccentricity may be neglected provided that simple connections are used.
(3) In all other cases the load should be taken as acting at a distance from the face of the column equal to 100 mm or at the centre of the stiff bearing, whichever gives the greater eccentricity.

The eccentricities for the various connections are shown in Figure 8.20.

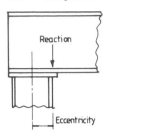

Reaction

Eccentricity

a) Beam to Column Connection

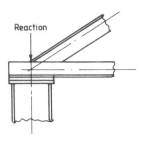

Reaction

b) Truss to Column Connection

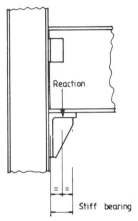

Reaction

Stiff bearing

c) Beam Supported on

Bracket

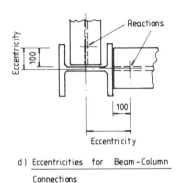

Eccentricity

100

Reactions

100

Eccentricity

d) Eccentricities for Beam–Column

Connections

Figure 8.20 Eccentricities for end reactions

8.6.2 Moments in columns of simple construction

The design of columns is set out in Section 4.7.7 of the code. The moments are calculated using eccentricities given in Section 8.6.1 above. For multi-storey columns effectively continuous at splices, the net moment applied at any one level may be divided between lengths above and below in proportion to the stiffness $I/1$ of each length. When the ratio of stiffness does not exceed 1.5 the moments may be divided equally. These moments have no effect at levels above or below that at which they are applied.

The following interaction equation should be satisfied for the overall buckling check:

$$\frac{F_c}{A_g p_c} + \frac{M_x}{M_{bs}} + \frac{M_y}{p_y Z_y} \leqslant 1$$

where M_{bs} = buckling resistance moment for a simple column calculated
using an equivalent slenderness

$\lambda_{LT} = 0.5L/r_y$
I = moment of inertia of the column about the relevant axis

L = distance between levels at which both axes are restrained

r_y = radius of gyration about the minor axis

F_c = compressive force in the column

Other terms are defined in Section 8.5.2.

8.6.3 Example: Corner column in a building

The part plan of the floor and roof steel for an office building is shown in Figure 8.21(a) and an elevation of the corner column is shown in Figure 8.21(b). The roof and floor loadng is as follows:

Roof —total dead load $= 5\,kN/m^2$
     imposed load $= 1.5\,kN/m^2$
Floors—total dead load $= 7\,kN/m^2$
     imposed load $= 3\,kN/m^2$

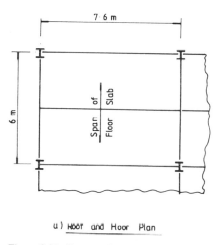 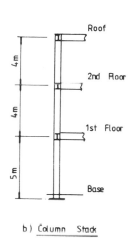

a) Roof and Floor Plan   b) Column Stack

Figure 8.21 Corner-column design example

The self weight of the column, including fire protection, is $1.5\,kN/m$. The external beams carry the following loads due to brick walls and concrete casing (they include self weight):

Roof beams—parapet and casing $= 2\,kN/m$
Floor beams—walls and casing $= 6\,kN/m$

The reinforced concrete slabs for the roof and floors are one-way slabs spanning in the direction shown in the figure.

Design the corner column of the building using Grade 43 steel.

In accordance with Table 2 of BS 6399: Part 1, the imposed loads may be reduced as follows:

One floor carried by member   —no reduction

Two floors carried by member —10% reduction

Three floors carried by member—20% reduction

The roof is counted as a floor. Note that the reduction is only taken into account in the axial load on the column. The full imposed load at that section is taken in calculating the moments due to eccentric beam reactions.

(1) Loading and reactions to floor beams

Mark numbers for the floor beams are shown in Figure 8.22(a). The end reactions are calculated below:

Roof

B1	Dead load	$=(5 \times 3.8 \times 1.5)+(2 \times 3.8)$	$= 36.1 \, \text{kN}$
	Imposed load	$= 1.5 \times 3.8 \times 1.5$	$= 8.55 \, \text{kN}$
B2	Dead load	$= 5 \times 3.8 \times 3$	$= 57.0 \, \text{kN}$
	Imposed load	$= 1.5 \times 3.8 \times 3$	$= 17.1 \, \text{kN}$
B3	Dead load	$= (0.5 \times 57.0)+(2 \times 3)$	$= 34.5 \, \text{kN}$
	Imposed load	$= 0.5 \times 17.1$	$= 8.55 \, \text{kN}$

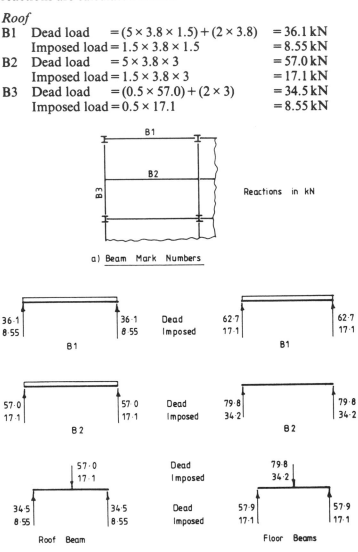

a) Beam Mark Numbers

b) Roof and Floor Beam Reactions

Figure 8.22 Floor-beam reactions

Floors

B1   Dead load $= (7 \times 3.8 \times 1.5) + (6 \times 3.8)$ $= 62.7\,\text{kN}$
     Imposed load $= 3 \times 3.8 \times 1.5$ $= 17.1\,\text{kN}$
B2   Dead load $= 7 \times 3.8 \times 3$ $= 79.8\,\text{kN}$
     Imposed load $= 3 \times 3.8 \times 3$ $= 34.2\,\text{kN}$
B3   Dead load $= (0.5 \times 79.8) + (6 \times 3)$ $= 57.9\,\text{kN}$
     Imposed load $= 0.5 \times 34.2$ $= 17.1\,\text{kN}$

The roof and floor beam reactions are shown in Figure 8.22(b).

(2) Loads and moments at roof and floor levels

The loading at the roof, second floor, first floor and base is calculated from values shown in Figure 8.22(b). The values for imposed load are calculated separately, so that reductions permitted can be made and the appropriate load factors for dead and imposed load introduced to give the design loads and moments.

The moments due to the eccentricities of the roof and floor beam reactions are based on the following assumed sizes for the column lengths:

Roof to second floor
—152×152 UC where the inertia I is proportional to 1.0
Second floor to first floor
—203×203 UC where the inertia I is proportional to 2.5
First floor to base
—203×203 UC where the inertia I is proportional to 2.5

Further, it will be assumed initially that the moments at the floor levels can be divided between the upper and lower column lengths in proportion to the stiffnesses which are based on the inertia ratios given above. The actual values are not required.

The division of moments is made as follows:

(1) Joint at second floor level
    Upper column length—stiffness $I/L = 1/4$ $= 0.25$
    Lower column length—stiffness $I/L = 2.5/4$ $= 0.625$
    If M is the moment due to the eccentric floor beam reaction then the moment in the upper column length is:
    $M_\text{u} = [0.25/(0.25 + 0.625)]\, M = 0.286\, M$
    Moment in the lower column length is:
    $M_\text{L} = (1 - 0.286)\, M$ $= 0.714\, M$
(2) Joint at first level
    It will be assumed that the same column section will be used for the two lower lengths. Hence the moments of inertia are the same and the stiffnesses are inversely proportional to the column lengths.
    Upper column length—stiffness $= 1/4 = 0.25$
    Lower column length—stiffness $= 1/5 = 0.20$
    The stiffness of the upper column length does not exceed 1.5 times the stiffness of the lower length. Thus the moments may be divided equally between the upper and lower lengths.

The eccentricities of the beam reactions and the column loads and moments from dead and imposed loads are shown in Figure 8.23.

Column Stack	Column Sections	Position	Dead Load	Imposed Load	Reduced Imp.Load	Dead Mx	Imposed Mx	Dead My	Imposed My
Roof × Nil I ÷ 1·0	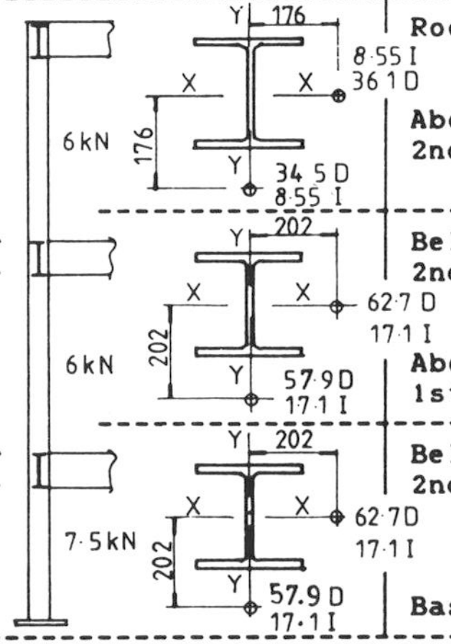 Y 176 8·55 I 36 I D X — X 6kN 34·5 D 8·55 I Y 176	Roof	70.6	17.1	17.1	6.07	1.51	6.35	1.51
		Above 2nd Floor	76.6	17.1	17.1	3.34	0.99	3.62	0.99
2nd Floor × 10% I ÷ 2·5	202 62·7 D 17·1 I X — X 6kN 57·9 D 17·1 I Y 202	Below 2nd Floor	197.2	51.3	46.2	8.35	2.47	9.01	2.47
		Above 1st Floor	203.2	51.3	46.2	5.84	1.73	6.33	1.73
1st Floor × 20% I ÷ 2·5	202 62·7 D 17·1 I X — X 7·5kN 57·9 D 17·1 I Y 202	Below 2nd Floor	323.8	85.5	68.4	5.84	1.73	6.33	1.73
		Base	331.3		68.4	–	–	–	–

× 10% Permitted values for reduction in imposed loads

Loads are in kN; Moments in kNm

Figure 8.23 Loads and moments from dead and imposed loads

(3) Column design

Roof to second floor

Referring to Figure 8.23, the design load and moments at roof level are:

Axial load F $=(1.4 \times 70.6)+(1.6 \times 17.1)=126.2\,\text{kN}$

Moment $M_x=(1.4 \times 6.07)+(1.6 \times 1.51)=10.92\,\text{kNm}$

$M_y=(1.4 \times 6.35)+(1.6 \times 1.51)=11.32\,\text{kNm}$

Try 152×152 UC 30, the properties of which are:

$A = 38.2\,\text{cm}^2$; $r_y=3.82\,\text{cm}$; $Z_x = 221.2\,\text{cm}^3$

$Z_y = 73.06\,\text{cm}^3$; $S_x = 247.1\,\text{cm}^3$; $S_y = 111.2\,\text{cm}^3$

The roof beam connections and column section dimensions are shown in Figure 8.24(a).

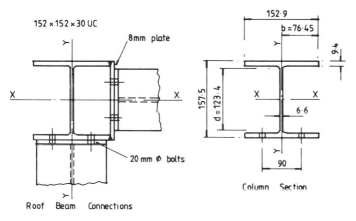

a) Column – Roof to Second Floor

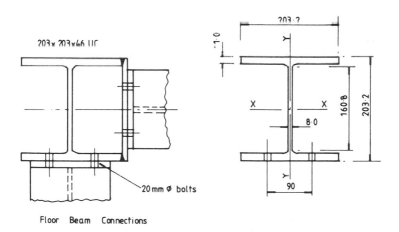

b) Column – Second Floor to Base

Figure 8.24 Column connections and section dimensions

The design strength from Table 6, $p_y = 275\,\text{N/mm}^2$
Flange $b/T = 76.45/9.4 = 8.13 < 8.5$—Plastic
Web $d/T = 123.4/6.6 = 18.7 < 39$—Semi-compact
The limiting proportions are from Table 7 of the code.

Local capacity check:

Moment capacities for the XX and YY axes are:

$M_{cx} = 247.1 \times 275/10^3$ $= 67.95\,\text{kNm}$
      $< 1.2 \times 221.2 \times 275/10^3$ $= 73.0\,\text{kNm}$
$M_{cy} = 111.2 \times 275/10^3$ $= 30.58\,\text{kNm}$
      $< 1.2 \times 275 \times 73.06/10^3$ $= 24.10\,\text{kNm}$

Interaction expression:

$$\frac{126.2 \times 10}{38.2 \times 275} + \frac{10.92}{67.95} + \frac{11.31}{24.10} = 0.75 < 1.0$$

The section is satisfactory.
Overall buckling check:
The column is effectively held in position and partially restrained in direction at both ends. From Table 24, the effective length is:

$L_E = 0.85 \times 4000 = 3400\,\text{mm}$
$\lambda = 3400/38.2 = 89$
From Table 27(c)
$p_c = 144\,\text{N/mm}^2$

The axial load at the centre of the column is

$$= 126.2 + (3 \times 1.4) = 130.4\,\text{kN}$$

The buckling resistance moment M_b is calculated using Section 4.7.7 of the code:

$\lambda_{LT} = 0.5 \times 4000/38.2 = 52.35$
$p_b = 232.2\,\text{N/mm}^2$ (Table 11)
$M_b = 232.2 \times 247.1/10^3 = 57.4\,\text{kNm}$

Interaction expression:

$$\frac{130.4 \times 10}{38.2 \times 144} + \frac{10.92}{57.4} + \frac{11.31 \times 10^3}{275 \times 73.06} = 0.98 < 1.0$$

The section is satisfactory.

Second floor to base

The same column section will be used from the second floor to the base. The lower column length between first floor and base will be designed.
    Referring to Figure 8.23, the design load and moments just below first floor level are:

$$F = (1.4 \times 323.8) + (1.6 \times 68.4) = 562.76\,\text{kN}$$

$$M_x = (1.4 \times 5.84) + (1.6 \times 1.73) = 10.94 \, \text{kNm}$$
$$M_y = (1.4 \times 6.33) + (1.6 \times 1.73) = 11.63 \, \text{kNm}$$

Try 203 × 203 UC 46, the properties of which are:

$$A = 58.8 \, \text{cm}^2; \, r_y = 5.11 \, \text{cm};$$
$$Z_y = 151.5 \, \text{cm}^3; \, S_x = 479.4 \, \text{cm}^3$$

Local capacity check:

The floor beam connections and column section dimensions are shown in Figure 8.24(b). The section is plastic and

$$p_y = 275 \, \text{N/mm}^2$$

The moment capacities are:

$$M_{cx} = 275 \times 497.4/10^3 \quad\quad = 136.8 \, \text{kNm}$$
$$M_{cy} = 1.2 \times 151.5 \times 275/10^3 = \quad 50.0 \, \text{kNm}$$

Interaction expression:

$$\frac{562.76 \times 10}{58.8 \times 275} + \frac{10.94}{136.8} + \frac{11.63}{50.0} = 0.66 \, < 1.0$$

Overall buckling check:

$$\lambda = 0.85 \times 5000/51 = 83.2$$
$$p_c = 155.2 \, \text{N/mm}^2 \, (\text{Table 27(c)})$$

Axial load at centre of column:

$$= 562.76 + (1.4 \times 3.75) = 568.01 \, \text{kN}$$
$$\lambda_{LT} = 0.5 \times 5000/51.1 = 48.9$$
$$P_b = 240.6 \, \text{N/mm}^2 - \text{Table 11}$$
$$M_b = 240.6 \times 497.4/10^3 = 119.6 \, \text{kNm}$$

Interaction expression:

$$\frac{568.01 \times 10}{58.8 \times 155.1} + \frac{10.94}{119.6} + \frac{11.63 \times 10^3}{275 \times 151.5} = 0.992 \, < 1.0$$

The section is satisfactory.

8.7 Cased columns subjected to axial load and moment

8.7.1 Code design requirements

The design of cased members subjected to axial load and moment is set out in Section 4.14.4 of BS 5950: Part 1. The member must satisfy two conditions.

(1) Capacity check

$$\frac{F_c}{P_{cs}} + \frac{M_x}{M_{cx}} + \frac{M_y}{M_{cy}} \leqslant 1$$

F_c = compressive force due to axial load

P_{cs} = compressive resistance of a cased strut with zero slenderness (see Section 8.4.11)

M_x = applied moment about the XX axis

M_y = applied moment about the YY axis

M_{cx} = moment capacity of the steel section about the XX axis

M_{cy} = moment capacity of the steel section about the YY axis

(2) Buckling resistance

$$\frac{F_c}{P_c} + \frac{mM_x}{M_b} + \frac{mM_y}{M_{cy}} < = 1$$

where P_c = compression resistance (see Section 8.4.11)

$\quad\quad m$ = equivalent uniform moment factor

$\quad\quad M_b$ = buckling resistance moment calculated using the radius of gyration r_y for a cased section.

8.7.2 Example

A column of length 7 m is subjected to the factored loads and moments as shown in Figure 8.25. Design the column using Grade 43 steel and Grade 30 concrete.

Try 203 × 203 UC 60, the properties of which are:

$A = 75.8\,\text{cm}^2,\quad S_x = 652\,\text{cm}^3,$

$r_x = 8.96\,\text{cm},\quad r_y = 5.19\,\text{cm},$

$u = 0.847,\quad\quad x = 14.1$

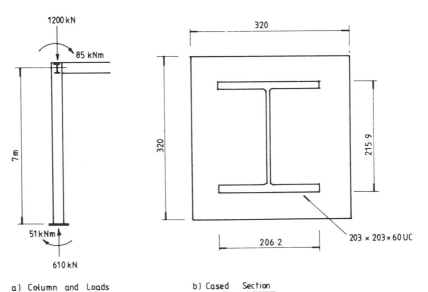

a) Column and Loads   b) Cased   Section

Figure 8.25 Cased column: design example

The section is plastic and design strength $p_y = 275 \, \text{N/mm}^2$. The cased section $320 \, \text{mm} \times 320 \, \text{mm}$ is shown in Figure 8.25(b).

(1) Capacity check

The terms for the interaction expression in Section 8.7.1(1) above are calculated:

$$P_{cs} = \left(75.8 + \frac{0.25 \times 30 \times 32^2}{275} \right) \frac{275}{10} = 2852 \, \text{kN}$$

$$M_{cx} = 652 \times 275/10^3 = 179.3 \, \text{kNm}$$

Interaction expression:

$$\frac{1200}{2852} + \frac{85}{179.3} = 0.42 + 0.474 = 0.894 \; < 1.0$$

This is satisfactory.

(2) Buckling resistance

For the cased section $r_y = 0.2 \times 320 = 64 \, \text{mm}$. The strut is taken to be held in position and partially restrained in direction at the ends:

$L_E = 7000 \times 0.85 = 5950 \, \text{mm}$ (Table 24)

$\lambda = 5950/64 = 93.0$

$p_c = 137 \, \text{N/mm}^2$ (Table 27(c))

$$P_c = \left(75.8 + \frac{0.45 \times 30 \times 32^2}{275} \right) \frac{137}{10} = 1727 \, \text{kN}$$

From Table 18, for $\beta = -51/85 = -0.6$:

$m = 0.43$

$\lambda = 93.0$ (same as above)

$\lambda/x = 93/14.1 = 6.59$

$v = 0.746$ (Table 14)

$\lambda_{LT} = 0.847 \times 0.746 \times 93 = 58.7$

$p_b = 216.3 \, \text{N/mm}^2$ (Table 11)

$M_b = 216.3 \times 652/10^3 = 141 \, \text{kNm}$

Interaction expression:

$$\frac{1200}{1727} + \frac{0.43 \times 85}{141} = 0.694 + 0.259 = 0.953 \; < 1.0$$

The section is satisfactory.

8.8 Side column for a single-storey industrial building

8.8.1 Arrangement and loading

The cross section and side elevation of a single-storey industrial building are shown in Figures 8.26(a) and (b). The columns are assumed to be fixed at the base and pinned at the top, and act as partially propped cantilevers in resisting lateral loads. The top of the column is held in the longitudinal direction by the eaves member and bracing, as shown on the side elevation.

The loading on the column is due to:

(1) Dead and imposed load from the roof and dead load from the walls and column; and
(2) Wind loading on roof and walls.

The load on the roof consists of:

(1) Dead load due to sheeting, insulation board, purlins and weight of truss and bracing. This is approximately 0.3–0.5 kN/m² on the slope length of the roof; and
(2) Imposed load due to snow, erection and maintenance loads. This is given in BS 6399: Part 1 as 0.75 kN/m² on plan area.

The loading on the walls is due to sheeting, insulation board, sheeting rails and the weight of the column and bracing. The weight is approximately the same as for the roof.

The wind load depends on the location and dimensions of the building. The method of calculating the wind load is taken from CP3: Chapter V: Part 2. This is shown in the following example

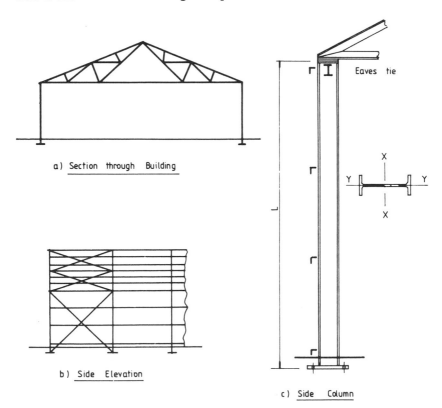

a) Section through Building

b) Side Elevation

c) Side Column

Figure 8.26 Side column in a single-storey industrial building

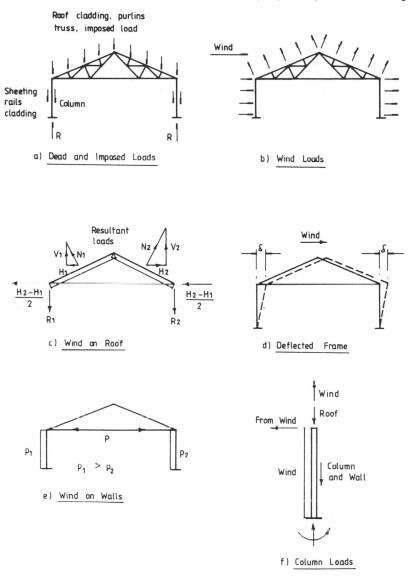

Figure 8.27 **Loads on side column of an industrial building**

The breakdown and diagrams for the calculation of the loading and moments on the column are shown in Figure 8.27, and the following comments are made on these figures.

(1) The dead and imposed loads give an axial reaction R at the base of the column (see Figure 8.27(a)).

(2) The wind on the roof and walls is shown in Figure 8.27(b). There may be a pressure or suction on the windward slope, depending on the angle of slope. The reactions from wind on the roof only are shown in Figure 8.27(c). The uplift results in vertical reactions R_1 and R_2. The net horiz-

ontal reaction is assumed to be divided equally between the two columns. This is $0.5(H_2 - H_1)$, where H_1 and H_2 are the horizontal components of the wind loads on the roof slopes.

(3) The wind on the walls causes the frame to deflect, as shown in Figure 8.27(d). The top of each column moves by the same amount S. The wind p_1 and p_2 on each wall, taken as uniformly distributed, will have different values, and this results in a force P in the bottom chord of the truss, as shown in Figure 8.27(e). The value of P may be found by equating deflections at the top of each column. For the case where p_1 is greater than p_2 there is a compression P in the bottom chord:

$$\frac{p_1 L^4}{8EI} - \frac{PL^3}{3EI} = \frac{p_2 L^4}{8EI} + \frac{PL^3}{3EI}$$

This gives $P = 3L(p_1 - p_2)/16$

where $I =$ moment of inertia of the column about the XX axis
        (same for each column)
      $E =$ Young's modulus
      $L =$ column height.

(4) The resultant loading on the column is shown in Figure 8.27(f), where the horizontal point load at the top is:

$$H = P + (H_2 - H_1)/2$$

The column moments are due entirely to wind load.

8.8.2 Column design procedure

(1) Section classification

Universal beams are often used for these columns where the axial load is small, but the moment due to wind is large. Referring to Figure 8.28(a), the classification is checked as follows:

(1) Flanges are checked using Table 7 of the code where limits for b/T are given, where

$b =$ flange outstand as shown in the figure

$T =$ flange thickness

(2) Webs are in combined axial and flexural compression. The classification can be checked using Table 7 and Section 3.5.4 of the code. For example, from Table 7 for webs generally a plastic section has the limit:

$$d/t \leqslant \frac{79\,\varepsilon}{0.4 + 0.6\alpha}$$

where $d =$ clear depth of web
        $t =$ thickness of web
        $\alpha = 2y_c/d$
where $y_c =$ distance from the plastic neutral axis to the edge of the web connected to the compression flange. This depends on the axial load.

(2) Effective length for axial compression

Effective lengths for cantilever columns connected by roof trusses are given in Appendix D of BS 5950: Part 1. The tops must be held in position longitudinally by eaves members connected to a braced bay.

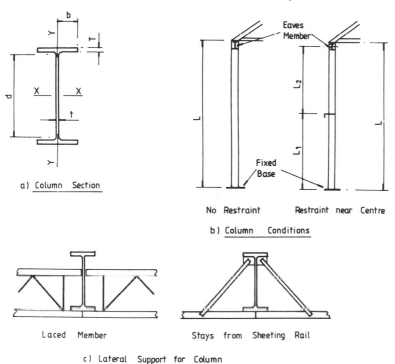

a) Column   Section

No   Restraint   Restraint   near   Centre

b)   Column   Conditions

Laced   Member   Stays   from   Sheeting   Rail

c)   Lateral   Support   for   Column

Figure 8.28 Side column design features

Two cases are shown in Figure 8.28(b):

(1) Column with no restraints:

XX axis $L_E = 1.5 L$

YY axis $L_E = 0.85$ L.
If the base is not effectively fixed about the YY axis:
$L_E = 1.0L$

(2) Column with restraints:

The restraint provides lateral support against buckling about the weak axis:

XX axis $L_E = 1.5 L$

YY axis $L_E = 0.85 L_1$ or L_2, whichever is the greater.

The restraint is often provided by a laced member or stays from a sheeting rail, as shown in Figure 8.28(c).

(3) Effective length for calculating the buckling resistance moment

The effective length of compression flange is estimated using Sections 4.3.5,

4.3.6 and Tables 9 and 10 of BS 5950: Part 1, and the effective lengths for the two cases shown in Figure 8.28(b) are:

(1) Column with no restraints (Table 10):
The column is fixed at the base and restrained laterally and torsionally at the top. For normal loading $L_E = 0.5\,L$. Note that the code specifies in this case that the uniform moment factor m and slenderness-correction factor n are to be taken as 1.0.
(2) Column with restraints:
This is to be treated as a beam and the effective length taken from Table 9:

$L_E = 0.85\,L_1$ or $1.0\,L_2$ in the case shown.

(4) Column design

The column moment is due to wind and controls the design. The load combination is then dead plus imposed plus wind load. The load factor from Table 2 of the code is $\gamma_f = 1.2$.

The following two checks are required in design:

(1) Local capacity check at base; and
(2) Overall buckling check.

The design procedure is shown in the example that follows.

(5) Deflection at the column cap

The deflection at the column cap must not exceed the limit given in Table 5 of the code for a single-storey building. The limit is height/300.

8.8.3 Example: design of a side column

A section through a single-storey building is shown in Figure 8.29. The frames are at 5 m centres and the length of the building is 30 m. The columns are pinned at the top and fixed at the base. The loading is as follows:

Roof: dead load—measured on slope
        sheeting, insulation board, purlins and truss   = 0.45 kN/m²
        imposed load—measured on plan   = 0.75 kN/m²
Walls: sheeting, insulation board, sheeting rails   = 0.35 kN/m²
Column: estimate   = 3.0 kN

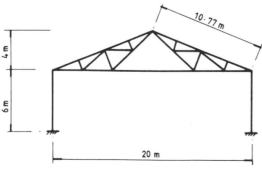

Figure 8.29 Section through building

Wind load. This is taken from CP3, Chapter V: Part 2. The wind loading is set out below, where all necessary data regarding the building are given.

Determine the loads and moments on the side column and design the member using Grade 43 steel. Note that the column is taken as not being supported laterally between the top and base.

(1) Column loads and moments

Dead and imposed load

Roof—dead load	$= 10 \times 5 \times 0.45 \times 10.77/10 = 24.23\,\text{kN}$	
—imposed load	$= 10 \times 5 \times 0.75$	$= 37.5\,\text{kN}$
Walls	$= 6 \times 5 \times 0.35$	$= 10.5\,\text{kN}$
Column		$= 3.0\,\text{kN}$
Total load at base		$= 75.23\,\text{kN}$

Wind load

Location—north-east England

Basic wind speed $V = 45\,\text{m/s}$
Topography factor $S_1 = 1.0$

Ground roughness 3. The location is on the outskirts of a city with obstructions up to 10 m in height.
Building size—Class B. The greatest horizontal or vertical dimension does not exceed 50 m.

Height to the top of the roof	$H = 10\,\text{m}$	
Height to the top of the walls	$H = 6\,\text{m}$	
From Table 3 Roof	$S_2 = 0.74$	
Walls	$S_2 = 0.668$	
Statistical factor	$S_3 = 1.0$	
Design wind speed		
Roof $V_s = 0.74 \times 45$	$= 33.3\,\text{m/s}$	
Walls $V_s = 0.668 \times 45$	$= 30.06\,\text{m/s}$	

Dynamic pressure

Roof $q = 0.613 \times 33.3^2/10^3$ $\qquad = 0.68\,\text{kN/m}^2$
Walls $q = 0.613 \times 30.06^2/10^3$ $\qquad = 0.554\,\text{kN/m}^2$

The external pressure coefficients C_{pe} for the roof and walls taken from Table 8 and 7, respectively, are shown in Figure 8.30(a). The values are for the wind angle $\alpha = 0$ degrees to cause moments in the columns and for a roof slope of 22 degrees.
The internal pressure coefficients C_{pi} for the case where there is only a negligible probability of a dominant opening occurring during a severe storm are taken from Appendix E of the code. The critical value for design is given by the more onerous of the values $+0.2$ or -0.3.

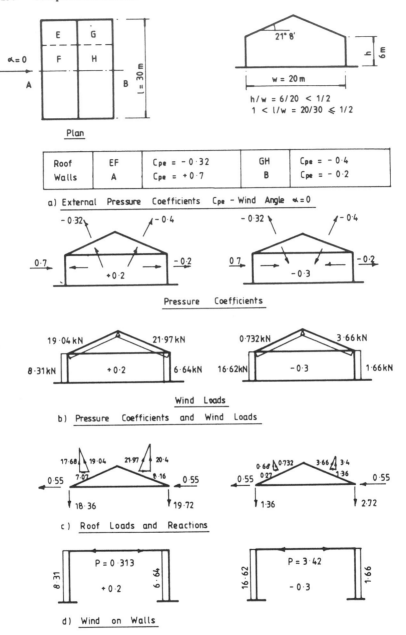

Figure 8.30 Wind-pressure coefficients and loads

The pressure coefficients for the building are shown in Figure 8.30(b). The wind load normal to the walls and roof slope is given by:

$W = 5\, qL\, (C_{pe} - C_{pi})$
L = height of wall or length of roof slope
q = dynamic pressure for walls or roof slopes.

The wind loads are also shown in Figure 8.30(b).

The resultant normal loads on the roof and the horizontal and vertical resolved parts are shown in Figure 8.30(c). The horizontal reaction is divided equally between each support and the vertical reactions are found by taking moments about supports. The reactions at the top of the columns for the two wind-load cases are shown in the figure.

The wind loading on the walls requires the analysis set out above in Section 8.8.1, where the column tops deflect by an equal amount and a force P is transmitted through the bottom chord of the truss. For the internal pressure case (see Figure 8.30(d)):

$$P = 3(8.31 - 6.64)/16 = 0.313 \, \text{kN}$$

The loads and moments on the columns are summarized in Figure 8.31.

Wind Case	Internal Pressure		Internal Suction	
Column	Windward	Leeward	Windward	Leeward
Dead	↓24.23	↓24.23	↓24.23	↓24.23
Imposed	↓39.5	↓37.5	↓37.5	↓37.5
Wind	↑18.36	↑19.72	↑1.36	↑2.72
	.237	.863	2.87	3.97
Wind Wall Column	8.31 13.5	6.64 13.5	16.6 13.5	13.5 1.66
Dead	↑37.73	↑37.73	↑37.73	↑37.73
Imposed	↑37.54	↑37.5	↑37.5	↑37.5
Wind	↓18.36	↓19.72	↓1.36	↓2.72
Wind Moment	26.35	25.1	32.64	18.84

Loads are in kN.   Moments are in kNm

Figure 8.31 Summary of loads and moments

Notional horizontal loads

To ensure stability, the structure is checked for a notional horizontal load in accordance with Clause 2.4.2.3 of BS 5950: Part 1. The notional force from the roof loads is taken as the greater of:

1 per cent of the factored dead loads $= 0.01 \times 1.4 \times 24.23 = 0.34 \, \text{kN}$

or 0.5 per cent of the factored dead load plus vertical imposed load

$$= 0.005 \, [(1.4 \times 24.23) + (1.6 \times 37.5)] = 0.5 \text{kN}$$

This load is applied at the top of each column and is taken to act simultaneously with 1.4 times the dead and 1.3 times the imposed vertical loads.

The design load at the base is

$$P = (1.4 \times 37.73) + (1.3 \times 37.5) = 101.57 \, kN$$

The moment is:

$$M = 0.5 \times 6 = 3.0 \, kNm$$

The design conditions for this case are less severe than those for the combination dead + imposed + wind loads.

(2) Column design

The maximum design condition is for the wind-load case of internal suction for the windward column. The load combination is dead plus imposed plus wind loads.

Local capacity check (see Section 8.52)

Design load $= 1.2(37.73 + 37.5 - 1.36)$ $= 88.64 \, kN$
Design moment $= 1.2 \times 32.64$ $= 39.17 \, kNm$

Try 406 × 140 UB 39, the properties of which are:

$A = 49.4 \, cm^2$; $S_x = 721 \, cm^3$; $Z_x = 627 \, cm^3$
$r_x = 15.88 \, cm$; $r_y = 2.89 \, cm$; $x = 47.4$
$I_x = 12452 \, cm^4$

Check the section classification using Table 7. The section dimensions are shown in Figure 8.32:

Design strength $p_y = 275 \, N/mm^2$ (Table 6)
Factor $\varepsilon = 1.0$
Flanges $b/T = 70.9/8.6 = 8.2 < 8.5$ (Plastic)
Web. This is in combined axial and flexural compression.

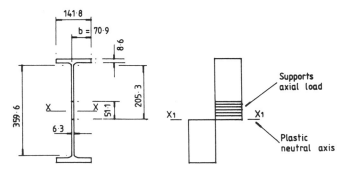

Figure 8.32 Column section

Design axial load = 88.64 kN

Length of web supporting the load at the design strength:

$$= \frac{88.64 \times 10^3}{275 \times 6.3} = 51.1 \, mm$$

The plastic neutral axis is 25.55 mm from the centre, as shown in Figure 8.32:

$y_c = (359.6/2) + 25.55 = 205.3$ mm

$x = 2 \times 205.3/359.6 = 1.14$

$\dfrac{d}{t} = \dfrac{359.6}{6.3} = 57.1 < \dfrac{79}{0.4 + (0.6 \times 1.14)} = 72$ (Plastic)

The moment capacity about the XX axis:

$M_{cx} = 275 \times 721/10^3 \qquad = 198.22$ kNm

$\qquad < 1.2 \times 275 \times 627/10^3 = 206.9$ kNm

Interaction expression:

$\dfrac{88.64 \times 10}{49.4 \times 275} + \dfrac{39.17}{198.22} = 0.07 + 0.2 = 0.27 < 1.0$

The section is satisfactory.

(3) Overall buckling check

Compressive strength

$\lambda_x = 1.5 \times 6000/158.5 \qquad = 56.78$

$\lambda_y = 0.85 \times 6000/28.9 \qquad = 176.47 < 180$

$p_c = 52.1$ Nmm^2 (Table 27(c))

Buckling resistance moment

$L_E = 0.5 \times 6000 = 3000$ (Table 10)
When this table is used, $m = n = 1$
$\lambda = 3000/28.9 = 103.8$

Use the conservative approach in Section 4.3.7.7 of the code:

$p_b = 146.8$ N/mm^2 from Table 19(b) for
$\lambda = 103.8$ and $x = 47.4$
$M_b = 146.8 \times 721/10^3 = 105.8$ kNm

Interaction expression

$\dfrac{88.64 = 10}{49.4 \times 52.1} + \dfrac{39.17}{105.8} = 0.35 + 0.37 = 0.71 < 1.0$

The section is satisfactory. The slenderness exceeds 180 for the next lightest section.

(4) Deflection at column cap

For the internal suction case:

$\delta = \dfrac{16.62 \times 10^3 \times 6000^3}{8 \times 205 \times 10^3 \times 12452 \times 10^4} - \dfrac{2.87 \times 10^3 \times 6000^3}{3 \times 205 \times 10^3 \times 12\,452 \times 10^4}$

$= 17.57 - 8.09 = 9.48$ mm

$\delta/\text{Height} = 9.48/6000 = 1/632 < 1/300$ (Table 5)

The column is satisfactory with respect to deflection.

8.9 Crane columns

8.9.1 Types

Three common types of crane columns used in single-storey industrial build-ings are shown in Figure 8.33. These are:

(1) A column of uniform section carrying the crane beam on a bracket.
(2) A laced crane column.
(3) A compound column fabricated from two universal beams or built up from plate.

Only the design of a uniform column used for light cranes will be discussed here. Types (2) and (3) are used for heavy cranes.

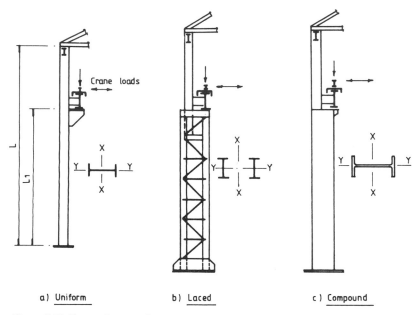

a) Uniform   b) Laced   c) Compound

Figure 8.33 Types of crane columns

8.9.2 Loading

A building frame carrying a crane is shown in Figure 8.34(a). The hook load is placed as far as possible to the left to give the maximum load on the column. The building, crane and wind loads are shown in the figure in (b), (c) and (d), respectively.

8.9.3 Frame action and analysis

In order to determine the values of moments in the columns the frame as a whole must be considered. Consider the frame shown in Figure 8.34(a), where the columns are of uniform section pinned at the top and fixed at the base. The separate load cases are discussed.

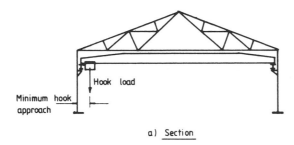

a) Section

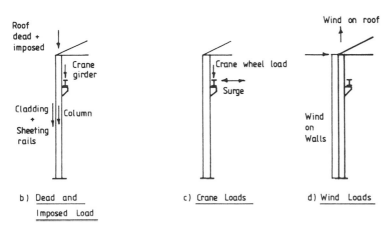

b) Dead and Imposed Load

c) Crane Loads

d) Wind Loads

Figure 8.34 Loads on crane columns

(1) Dead and imposed loads

The dead and imposed loads from the roof and walls are taken as acting axially on the column. The dead load from the crane girder causes moments as well as axial load in the column. (See the crane wheel load case below.)

(2) Vertical crane wheel loads

The vertical crane wheel loads cause moments as well as axial load in each column. The moments applied to each column are unequal, so the frame sways (as shown in Figure 8.35(a)) and a force P is transmitted through the bottom chord.

Consider the column ABC on Figure 8.35(a). The deflection at the top is calculated for the moment from the crane wheel loads M_1 and force P separately using the moment area method.[16] The separate moment diagrams are shown in Figure 8.35(b).

The deflection due to M_1 is:

$$\delta = M_1 L_1 (L - L_1/2)/EI$$

where L = column height
L_1 = height to the crane rail
EI = column rigidity

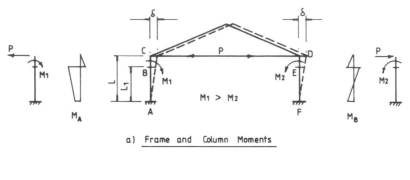

a) Frame and Column Moments

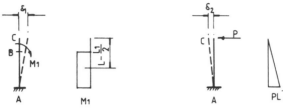

b) Moment and Load causing Deflection in Column ABC

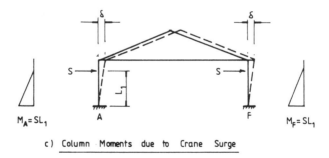

c) Column Moments due to Crane Surge

Figure 8.35 Vertical and horizontal crane-wheel loads and moments

The deflection due to the load P at the column top is

$$\delta_2 = PL^3/3EI$$

The frame deflection is

$$\delta = \delta_1 - \delta_2$$

Equating deflections at the top of each column gives:

$$\frac{M_1 L_1}{EI}\left(L - \frac{L_1}{2}\right) - \frac{PL^3}{3EI} = \frac{M_2 L_1}{EI}\left(L - \frac{L_1}{2}\right) + \frac{PL^3}{3EI}$$

where $P = 3L_1\,(L - L_1/2)\,(M_1 + M_2)/2L^3$

The moments in the column can now be calculated.

If the self weight of the crane girder applies a moment M to each column then the force in the bottom chord is:

$$P_1 = \frac{L_1 M}{L^3}(L - L_1/2)$$

(3) Crane surge

In Figure 8.35(c) the crane surge load S is the same each side and each column acts as a free cantilever. The loads and moments for this case are shown in the figure.

(4) Wind loads

Wind loads on this type of frame were treated in Section 8.8.1.

(5) Load combinations

The separate load combinations and load factors γ_f to be used in design are given in Table 2 of BS 5950: Part 1. The load cases and load factors are:

(1) 1.4 Dead + 1.6 Imposed + 1.6 Vertical Crane Load
(2) 1.4 Dead + 1.6 Imposed + 1.6 Horizontal Crane Load
(3) 1.4 Dead + 1.6 Imposed + 1.4 (Vertical and Horizontal Crane Loads)
(4) 1.2 (Dead + Imposed + Wind + Vertical and Horizontal Crane Loads)

It may not be necessary to examine all cases. Note that in case (2) there is no impact allowance on the vertical crane wheel loads.

8.9.4 Design procedure

(1) Effective lengths for axial compression

The effective lengths for axial compression for a uniform column carrying the crane girder on a bracket are given in Appendix D of BS 5950: Part 1.
    In Figure 8.33(a) the effective lengths are:

XX axis $L_E = 1.5 L$
YY axis $L_E = 0.85 L_1$

The code specifies that the crane girder must be held in position longitudinally by bracing in the braced bays. If the base is not fixed in the YY direction, $L_E = 1.0 L_1$.

(2) Effective length for calculating the buckling resistance moment

The reader should refer to Sections 4.3.5, 4.3.6 and Table 9 of BS 5950: Part 1. The crane girder forms an intermediate restraint to the cantilever column. Section 4.3.6 states in this case that the member is to be treated as a beam between restraints and Table 9 is to be used to determine the effective length L_E. A value of $L_E = 0.85 L_1$ may be used for this case.

(3) Column design

The column is checked for local capacity at the base and overall buckling.

(4) Deflection

The deflection limitation for columns in single-storey buildings applies. In Section 2.5.1 and Table 5 of BS 5950: Part 1 the limit for the column top is Height/300. However, the code also states that in the case of crane surge and wind only the greater effect of either need be considered in any load combination.

8.9.5 Example: design of a crane column

(1) Building frame and loading

The single-storey building frame shown in Figure 8.36(a) carries a 50 kN electric overhead crane. The frames are at 5 m centres and the length of the building is 30 m. The static crane wheel loads are shown in Figure 8.36(b). The crane beams are simply supported, spanning 5 m between columns, and the weight of a beam is approximately 4 kN. The arrangement of the column and crane beam with the end clearance and eccentricity are shown in Figure 8.36(c).

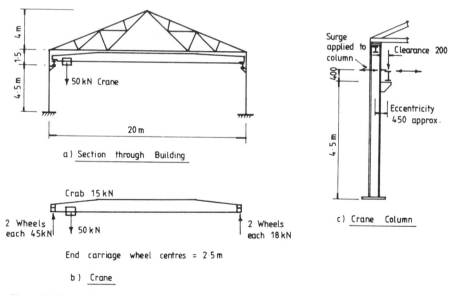

a) Section through Building

b) Crane

c) Crane Column

Figure 8.36 Building frame with crane

Dead and imposed loads

The roof and wall loads are the same as for the building in Section 8.8.3. The loads are:

Dead loads—Roof	= 24.23 kN
Walls	= 10.5 kN
Crane column + bracket	= 4.0 kN
Crane beam	= 4.0 kN
Dead load at column base	= 42.73 kN

Dead load at crane girder level   = 27.86 kN
Imposed load—Roof   = 37.5 kN

The eccentric dead load of the crane beams cause small moments in the columns.

In Figure 8.37(a) the applied moment to each column:

$$M = 4 \times 0.45 = 1.8 \, \text{kNm}$$

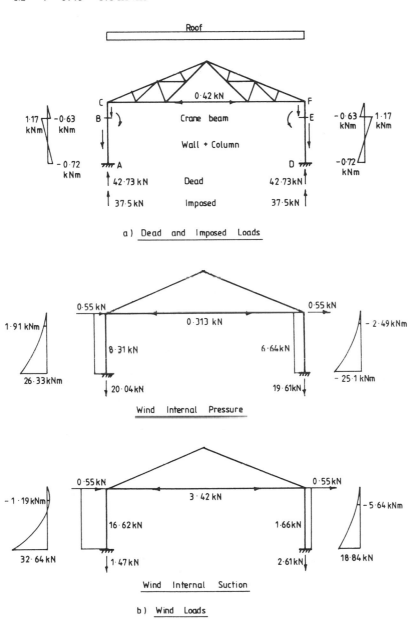

a) Dead and Imposed Loads

Wind Internal Pressure

Wind Internal Suction

b) Wind Loads

Figure 8.37 Crane column loads and moments

Dead load at crane girder level $= 27.86\,\text{kN}$
Imposed load—Roof $= 37.5\,\text{kN}$

The eccentric dead load of the crane beams cause small moments in the columns.

In Figure 8.37(a) the applied moment to each column:

$$M = 4 \times 0.45 = 1.8\,\text{kNm}$$

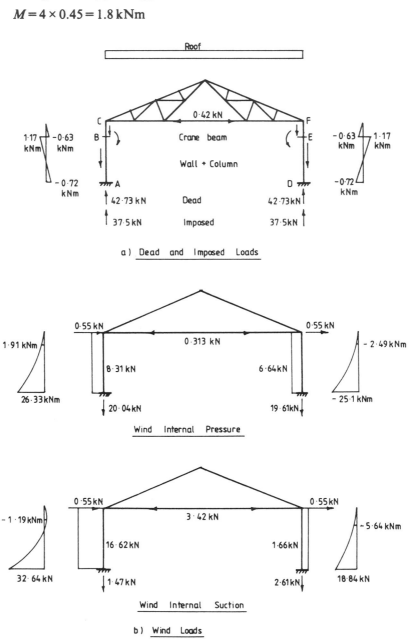

a) Dead and Imposed Loads

b) Wind Loads

Figure 8.37 Crane column loads and moments

Wind loads

The wind loads are the same as for the building in Section 8.8.3 and wind load and column moments are shown in Figure 8.37(b).

Vertical crane wheel loads

The crane wheel loads, including impact, are:

Maximum wheel loads $= 45 + 25\% = 56.25\,\text{kN}$
Light side-wheel loads $= 18 + 25\% = 22.5\,\text{kN}$

To determine the maximum column reaction the wheel loads are placed equidistant about the column, as shown in Figure 8.38(a). The column reaction and moment for the maximum wheel loads are:

$R_1 = 2 \times 56.25 \times 3.75/5 \qquad = 84.375\,\text{kN}$
$M_1 = 84.3 \times 0.45 \qquad\qquad = 37.87\ \text{kNm}$

For the light side-wheel loads:

$R_2 = 2 \times 22.5 \times 3.75/5 \qquad = 33.75\ \text{kN}$
$M_2 = 33.75 \times 0.45 \qquad\qquad = 15.19\ \text{kN}$

The load in the bottom chord is (see above):

$$P_2 = \frac{3 \times 4.5}{2 \times 6^3}\left(6 - \frac{4.5}{2}\right)(37.87 + 15.19) = 6.22\,\text{kN}$$

The moments for column ABC are:

$M_{BC} = -6.22 \times 1.5 \qquad\qquad = -\ \ 9.33\,\text{kNm}$
$M_{BA} = 37.87 - 9.33 \qquad\qquad = \quad 28.54\,\text{kNm}$
$M_A = 37.87 - (6.22 \times 6) \qquad = \quad\ \ 0.55\,\text{kNm}$

These moments and the moments for column DEF are shown in Figure 8.38(b).

Crane surge loads

The horizontal surge load per wheel

$\qquad = 0.1\ (50 + 15)/4 \qquad\quad = 1.63\,\text{kN}$

The column reaction from surge loads

$R_3 = 2 \times 1.63 \times 3.75/5 \qquad = 2.45\,\text{kN}$

The moments at the column base are

$M_A = 2.45 \times 4.9 \qquad\qquad 12.0\,\text{kNm}$

The loads and moments are shown in Figure 8.38(c).

(2) Design load combinations

Consider column ABC with wind internal suction case, maximum crane wheel loads and crane surge. The design loads and moments for three load combinations are:

(1) Dead + Imposed + Vertical Crane Loads
     Base
     $P = (1.4 \times 42.73) + (1.6 \times 37.5) + (1.6 \times 84.38)$ $= 254.83\,\text{kN}$
     $M = (1.4 \times 0.72) + (1.6 \times 0.55)$ $= -0.13\,\text{kNm}$
     Crane girder
     $P = (1.4 \times 27.86) + (1.6 \times 37.5) + (1.6 \times 84.38)$ $= 234.01\,\text{kN}$
     $M = (1.4 \times 1.17) + (1.6 \times 28.54)$ $= 47.3\,\text{kNm}$
(2) Dead + Imposed + Vertical and Horizontal Crane Loads
     Base
     $P = (1.4 \times 42.73) + (1.6 \times 37.5) + (1.4 \times 84.38)$ $= 237.95\,\text{kN}$
     $M = -(1.4 \times 0.72) + 1.4(0.55 - 12)$ $= -17.04\,\text{kNm}$
     Crane girder
     $P = (1.4 \times 27.86) + (1.6 \times 37.5) + (1.4 \times 84.38)$ $= 217.14\,\text{kN}$
     $M = (1.4 \times 1.17) \times (28.54 + 0.98)$ $= 42.97\,\text{kNm}$
(3) Dead + Imposed + Wind Internal Suction + Vertical and Horizontal
     Crane Loads
     Base
     $P = 1.2[42.73 + 37.5 - 1.47 + 84.38]$ $= 195.77\,\text{kN}$
     $M = 1.2[-0.72 + 32.64 + 0.55 + 12.0]$ $= 53.36\,\text{kNm}$
     Crane girder
     $M = 1.2[1.17 - 1.19 + 28.54 + 0.98]$ $= 35.4\,\text{kNm}$

Note that design conditions arising from notional horizontal loads specified in Clause 2.4.2.3 of BS 5950: Part 1 are not as severe as those in condition (3) above.

(3) Column design

Try 406 × 140 UB 46, the properties of which are:

$A = 59.0\,\text{cm}^2$; $S_x = 888.4\,\text{cm}^3$, $Z_x = 777.8\,\text{cm}^3$
$r_x = 16.29\,\text{cm}$, $r_y = 3.02\,\text{cm}$, $x\quad = 38.8$; $I_x = 15\,647\,\text{cm}^4$

Local capacity check at base

The reader should refer to the example in Section 8.8.3. The section can be shown to be plastic: design strength $p_y = 275\,\text{N/mm}^2$:

$M_{CX} = 275 \times 888.4/10^3$ $= 244.3\,\text{kNm}$

Interaction expression

Case (1) $\dfrac{254.83 \times 10}{59.0 \times 275}$ + Moment negligible $= 0.157\ < 1.0$

Case (2) $\dfrac{237.95 \times 10}{590.275} + \dfrac{17.04}{244.3}$ $= 0.217\ < 1.0$

Case (3) $\dfrac{195.77 \times 10}{59.0 \times 275} + \dfrac{53.36}{244.3}$ $= 0.339\ < 1.0$

The section is satisfactory. Case (3) is the most severe load combination.

Overall buckling check

Compressive strength:

$\lambda_x = 1.5 \times 6000/162.9$ .... $= 55.25$

$\lambda_y = 0.85 \times 4500/30.2$ .... $= 112.6$

$p_c = 106.4 \, \text{N/mm}^2$ (Table 27(c))

Bucking resistance moment:

$L_E = 0.85 \times 4500$ .... $= 3825 \, \text{mm}$

$\lambda = 3825/30.2$ .... $= 112.6$

$x = 38.8$

$n = 1.0$ (conservative estimate)

$p_b = 138.2 \, \text{N/mm}^2$ (Table 19)

$M_b = 128.2 \times 888.4/10^3$ .... $= 122.7 \, \text{kNm}$

Interaction expression:

$$\frac{195.77 \times 10}{59.0 \times 106.4} + \frac{53.36}{122.7} \qquad = 0.74 \; < 1.0$$

The column is satisfactory.

Deflection at column cap

The reader should refer to Figures 8.37 and 8.38:

Deflection due to crane surge δ_s
$EI \, \delta_s = 12 \times 10^6 \times 4900 \times 4367/2 = 1.284 \times 10^{14}$
Deflection due to wind δ_w
$EI \, \delta_w = 16\,620 \times 6000^3/8 - 2.87 \times 10^3 \times 6000^3/3 = 2.421 \times 10^{14}$
Add deflection from crane wheel loads to that caused by wind load:
$EI \, \delta = 2.421 \times 10^{14} + 37.87 \times 10^6 \times 4500 \times 3750 - 6220 \times 6000 \times 10^3/3$
$= 4.334 \times 10^{14}$

$$\delta = \frac{4.334 \times 10^{14}}{205 \times 10^3 \times 15647 \times 10^4} = 13.51 \, \text{mm}$$

$\delta/\text{height} = 13.51/6000$ .... $= 1/444 \; < 1/300$

The deflection controls the column size.

8.10 Column bases

8.10.1 Types and loads

Column bases transmit axial load, horizontal load and moment from the steel column to the concrete foundation. The main function of the base is to distribute the loads safely to the weaker material.

The main types of bases used are shown in Figure 8.39. These are:

(1) Slab base;
(2) Gusseted base; and
(3) Pocket base.

With respect to slab and gusseted bases, depending on the values of axial load and moment, there may be compression over the whole base or compression over part of the base and tension in the holding-down bolts. Bases subjected to moments about the major axis only are considered here. Horizontal loads are resisted by shear in the weld between column and base plates, shear in the holding-down bolts and friction and bond between the base and the concrete. The horizontal loads are generally small.

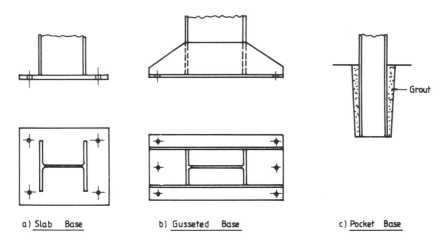

a) Slab Base   b) Gusseted Base   c) Pocket Base

Figure 8.39 Column bases

8.10.2 Design strengths

(1) Base plates

The design strength of the plate p_{yp} is given in Section 4.13.2.2 of BS 5950: Part 1. This is to be taken from Table 6 but is not to exceed 270 N/mm^2.

(2) Holding-down bolts

The strengths of bolts are given in Table 32 of the code (see Sections 4.2.2 and 4.2.3). The tensile stress area should be used in the design check for bolts in tension.

(3) Concrete

The column base is set on steel packing plates and grouted in. Mortar cube strengths vary from 15 to 25 N/mm^2. The bearing strength is given in Section 4.13.1 of the code as $0.4f_{cu}$, where f_{cu} is the cube strength at 28 days. For design of pocket bases the compressive strength of the structural concrete is taken from BS 8110: Part 1.

8.10.3 Axially loaded slab base

(1) Code requirements and theory

This type of base is used extensively with thick steel slabs being required for heavily loaded columns. The slab base is free from pockets where corrosion may start and maintenance is simpler than with gusseted bases.

The design of slab bases with concentric loads is given in Section 4.13.2.2 of BS 5950: Part 1. This states that where the rectangular plate is loaded by an I, H, channel, box or rectangular hollow section its minimum thickness should be:

$$t = \left(\frac{2.5}{p_{yp}} w \left(a^2 - 0.3 b^2 \right) \right)^{0.5}$$

but not less than the flange thickness of the column supported, where

a = greater projection of the plate beyond the column
b = lesser projection of the plate beyond the column
w = pressure on the underside of the base assuming uniform distribution
p_{yp} = design strength of the plate.

The above equation takes into account plate bending in two directions. The moment in the direction of the greater projection is reduced by the co-existent moment at right angles. Poisson's ratio 0.3 is introduced to allow for this effect.[6]

Consider an element at A and the two cantilever strips 1 mm wide shown in Figure 8.40. The moments at A are:

$M_X = w a^2 / 2$
$M_Y = w b^2 / 2$

The projection a is greater than b, so the net moment is:

$M_X = 1/2 w a^2 - 1/2 (0.3) w b^2$
   $= w/2 \left(a^2 - 0.3 b^2 \right)$
   $= 1.2 \, p_{yp} t^2 / 6$

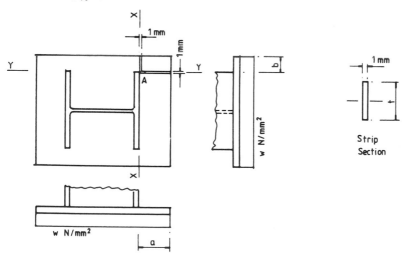

Figure 8.40 Column base plate moments

where the elastic modulus for the cantilever strip is $t^2/6$. The moment capacity for the base plate is given in Section 4.13.7.3 of the code as $1.2 p_{yp}Z$. This gives:

$$t = \left(\frac{2.5\,w}{p_{yp}} (a^2 - 0.3b^2) \right)^{0.5}$$

(2) Weld: column to slab

The code states in Section 4.13.3 that where the slab and column end are in tight contact the load is transmitted in direct bearing. The surfaces in contact would be machined in this case. The weld only holds the base slab in position. Where the surfaces are not suitable to transmit the load in direct bearing the weld must be designed to transmit the load.

8.10.4 Axially loaded slab base: example

A column consisting of a 305×305 UC 198 carries an axial dead load of 1600 kN and an imposed load of 800 kN. Adopting a square slab, determine the size and thickness required. The cube strength of the concrete grout is 25 N/mm^2. Use Grade 43 steel.

The area required for the base is calculated first:

Design load $= (1.4 \times 1600) + (1.6 \times 800) = 3520$ kN
Area $= 3520 \times 10^3/(0.4 \times 25)$ $= 35.2 \times 10^4$ mm^2
Make the base 600 mm square
Pressure $w = 3520 \times 10^3/600^2$ $= 9.78$ N/mm^2

The arrangement of the column on the base plate is shown in Figure 8.41. From this:

a—greater projection of base $= 142.95$ mm
b—lesser projection of base $= 130.05$ mm

Assume that the thickness of plate is less than 40 mm. Design strength p_{yp} $= 265$ N/mm^2 (Table 6).
The thickness of the base plate is given by:

$$t = \left(\frac{2.5 \times 9.78}{265} (142.95^2 - 0.3 \times 130.05^2) \right)^{0.5}$$

$= 37.6$ mm

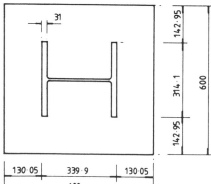

Figure 8.41 Axial loaded base: example

The column flange thickness is 31.4 mm. Make the base plate 40 mm thick. Use 6 mm fillet weld all round to hold the base plate in place. The surfaces are to be machined for direct bearing. The holding-down bolts are nominal but 4 No. 24 mm diameter bolts would be provided.

Base slab: 600 mm × 600 × 40 mm thick.

8.10.5 Eccentrically loaded bases

(1) General considerations

Consider bases subjected to axial load and bending about one axis. Two separate cases occur:

(1) Pressure over the whole base;
(2) Pressure over part of the base and tension in the holding-down bolts

The relative values of moment and axial load determine which case will occur in a given instance.

The code states in Section 4.13.1 that the nominal bearing pressure between the base plate and the support may be determined on the basis of a linear distribution of pressure. Thus the base is designed using elastic analysis.

8.10.6 Compression over the whole of the base

A column base and loading are shown in Figure 8.42 with the pressure distribution under the base. The following terms are defined:

W = total load on the base
M = moment on the base
e = eccentricity of the load = M/W
b = breadth of base
l = length of base
p_{max} = maximum pressures under the base

The middle third rule applies, and if the eccentricity e is less than one sixth, the resultant load lies within the middle third of the length of the base and pressure occurs over the whole of the base:

A = area of base = bl
Z = modulus   = $bl^2/6$

Figure 8.42 Compression over the whole of the base

Maximum pressure on the concrete:

$$p_{max} = (W/A) + (M/Z)$$

The maximum pressure must not exceed the bearing strength of the concrete given in Section 4.13.1 of the code.

The size of base is established by successive trials. If the length is fixed the breadth may be determined so that the design strength of the concrete is not exceeded. The weld size between the base plate and the column is determined using the same requirements that were set out for the axially loaded base in 8.10.3 (2) above.

8.10.7 Slab base—compression over the whole base: example

A column base is subjected to a factored moment of 55 kNm and a factored axial load of 780 kN. The column section is 203 × 203 UC 86. The cube strength of concrete in the foundation f_{cu} is 25 N/mm². Design a slab base and the weld between the column and the plate, assuming that both are machined for tight contact. Use Grade 43 steel.

(1) Size of base

Eccentricity of load

$$e = 55 \times 10^3 / 780 = 70.5 \, mm$$

If the base is made 6e in length there will be pressure over the whole of the base:

$$6e = 423.1 \, mm$$

The breadth required to limit the pressure on the concrete to $0.4 f_{cu} = 10 \, N/mm^2$:

$$b = \frac{2 \times 780 \times 10^3}{423.1 \times 10} = 368.7 \, mm$$

A base 430 mm square could be used. However, a rectangular base 500 mm long by 360 mm wide will be checked. The arrangement of the base is shown in Figure 8.43(a):

Area A $= 500 \times 360$ $= 180 \times 10^3 \, mm^2$
Modulus $Z = 360 \times 500^2/6 = 15 \times 10^6 \, mm^2$

Maximum pressure:

$$p_{max} = \frac{780 \times 10^3}{180 \times 10^3} + \frac{55 \times 10^6}{15 \times 10^6}$$

$$= 4.33 + 3.67 \qquad = 8.0 \, N/mm^2$$

Minimum pressure:
$$p_{min} = 4.33 - 3.67 \qquad = 0.66 \, N/mm^2$$

the pressure distribution is shown in the elevation of the base in Figure 8.42(a).

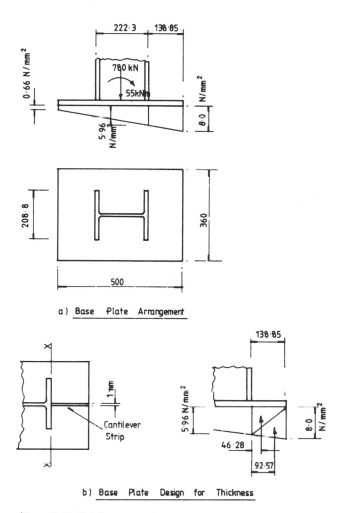

a) Base Plate Arrangement

b) Base Plate Design for Thickness

Figure 8.43 Slab base: example

(2) Thickness of base plate

Consider a 1 mm wide strip as shown in Figure 8.43(b). This acts as a cantilever from the face of the column with the loading caused by pressure on the base. This method gives a conservative design for the thickness of base plate. Plate action due to bending in two directions at right angles is not taken into account.

Base pressure at Section XX:

$$= 0.66 + [(500 - 138.85)/500](8 - 0.66) = 5.96 \, \text{N/mm}^2$$

The trapezoidal pressure diagram loading the cantilever strip is divided into two triangles, as shown in the figure. The moment at Xx is calculated as follows:

$$M_X = (5.96 \times 138.85 \times 46.28)/2 + (8.0 \times 138.85 \times 92.57)/2$$

$$= 70.56 \times 10^3 \, \text{N/mm}$$

The section of plate is 1 mm wide $\times$ t mm thick:

$$Z = t^2/6$$

Assume that the thickness is not greater than 40 mm:

$p_y = 265 \, \text{N/mm}^2$
$t = (6 \times 70.56 \times 10^3/1.2 \times 265)^{0.5} = 36.5 \, \text{mm}$

A slab 40 mm thick is required.

(3) Weld column to base plate

The base slab has been designed on the basis of a linear distribution of pressure. For consistency, the weld will be designed on the same basis: column 203 $\times$ 203 UC 86:

$$A = 110.1 \, \text{cm}^2; \quad Z_x = 851.5 \, \text{cm}^3$$

Axial stress $f_c = 780 \times 10/110.1$ $= 70.84 \, \text{N/mm}^2$
Bending stress $f_{bt} = 55 \times 10^3/851.5$ $= 64.6 \ \text{N/mm}^2$

On the basis of elastic stress distribution there is compressive stress over the whole of the base. The slab and column are machined for tight contact so the weld is required to hold the base slab in position. Use 6 mm fillet weld with continuous weld to give a full seal around the column profile.

8.10.8 Gusseted base

With this type of base part of the load is transmitted from the column through the gussets to the base plate. Gussets and other stiffeners, if provided, support the base plate against bending and thus a thinner plate can be used than with a slab base.

   The gussets are subjected to bending from the upward pressure under the base, as shown in Figure 8.44(a). The top edge of the gusset plates is in compression and must be checked for buckling. To ensure that this does not occur, the limiting proportions should meet those for a semi-compact element in welded construction given in Table 7 of BS 5950: Part 1. Referring to the figure, two requirements must be satisfied:

(1) Gusset between welds to the column flange:

$$D \leqslant 28 \, \varepsilon t$$

(2) Outstand of gusset from column or base plate:

$$S \leqslant 13 \varepsilon t$$
where $\varepsilon = (275/p_{yg})^{0.5}$
    t = thickness of gusset plate
    p_{yg} = design strength of the gusset

The gusset plates are designed to resist shear and bending. The code states in Section 4.13.2.4 that the moment in the gusset should not exceed $p_{yg}Z$, where Z = elastic modulus of the gusset and p_{yg} = design strength of the gusset $\leqslant 270 \, \text{N/mm}^2$.

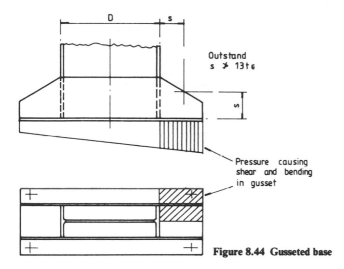

Figure 8.44 Gusseted base

With respect to the welds, the requirements from Section 4.13.3 in the code noted in Section 8.10.3 above apply. These concern whether or not the plates are machined for tight contact in bearing. Weld design is given in the following example.

8.10.9 Gusseted base: example

Redesign the base in the example in Section 8.10.7 using gusseted construction. The parts are not machined for tight contact in bearing, so the welds are to be designed to transmit the whole of the column load and moment to the base plate. The arrangement of the base is shown in Figure 8.45(a).

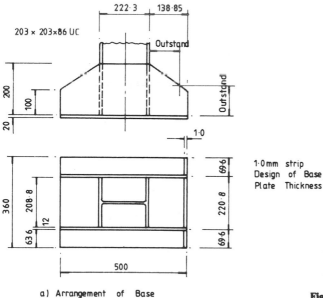

a) Arrangement of Base

Figure 8.45 (contd. overleaf)

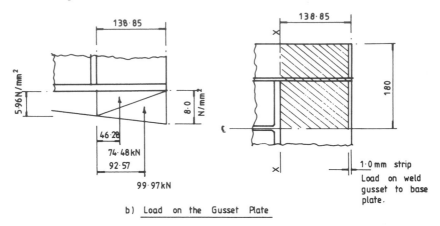

b) Load on the Gusset Plate

Figure 8.45 Gusset base: design example

(1) Gusset plates

The gussets are first checked to ensure that buckling does not occur. Assume 12 mm thick plates:

Design strength $p_{yg} = 270 \, \text{N/mm}^2$ (Table 6)

(1) Gusset between welds to column flange:
Length $= 222.3 \, \text{mm} = 18.52 \, t < 28\varepsilon$

(2) Gusset outstand:

This is not to exceed $13\varepsilon t = 157.4 \, \text{mm}$. In Figure 8.45 the outstand is obviously satisfactory.

The gusset is a semi-compact section. The pressure under the base and the load on a gusset are shown in Figure 8.45(b). The values have been taken from the example in Section 8.10.7.
At section XX the shear on one gusset is:

$$F_v = \frac{5.96 \times 138.85 \times 180}{2 \times 10^3} + \frac{8.0 \times 138.85 \times 180}{2 \times 10^3}$$
$$= 74.48 + 99.97 \qquad\qquad = 174.45 \, \text{kN}$$

The moment is:

$$M_x = [(74.48 \times 46.28) + (99.97 \times 92.57)]10^3 = 12.7 \, \text{kNm}$$

The shear capacity:

$$P_v = 0.9 \times 0.6 \times 275 \times 200 \times 12/10^3 \qquad = 356.4 \, \text{kN}$$
$$F_v = 0.49 \, P_v < 0.6 \, P_v$$

The moment capacity is not reduced by the effect of shear.
The moment capacity:

$$M_C = \frac{270 \times 12 \times 200^2}{6 \times 10^6} \qquad = 21.6 \, \text{kNm}$$

The gusset plates are satisfactory.

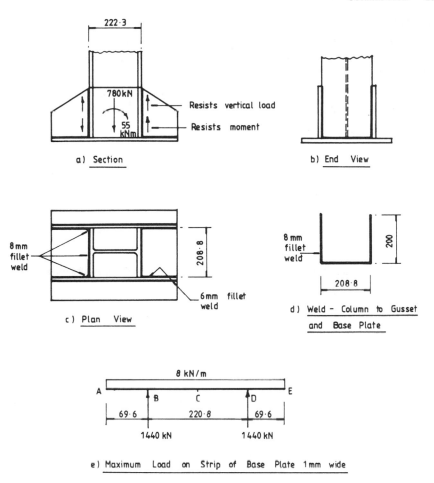

a) Section

b) End View

c) Plan View

d) Weld – Column to Gusset
and Base Plate

e) Maximum Load on Strip of Base Plate 1mm wide

Figure 8.46 Base plate welds and load for calculating thickness

(2) Gusset plate to column weld

The welds between the column, gussets and base plate have to transmit all of
the load to the base plate. The welds are shown in Figure 8.46. The column to
gusset and base plate weld is shown in Figure 8.46(d):

Load per weld $= (780/2) + (55/0.22)$ $= 640 \, \text{kN}$
Length of weld $= 608.8 - 2 \times 8$ $= 592.8$
Load per mm $= 640/592.8$ $= 1.07 \, \text{kN/mm}$
Use 8 mm fillet weld, strength 1.2 kN/mm.

The welds between one gusset plate and the base must support the maximum
pressure under the base. See Figure 8.45(b) and consider a 1.0 mm wide strip at
the edge of the base plate. Load on one weld:

$= 8 \times 180/(2 \times 10^3)$ $= 0.72 \, \text{kN/mm}$

Provide 6 mm fillet weld, strength 0.9 kN/mm. The weld is shown in Figure
8.46(c).

(3) Base-plate thickness

Consider a strip of plate 1.0 mm wide at the edge of the plate, as shown in Figure 8.45(a). This is considered as a beam with over-hanging ends loaded as shown in Figure 8.46(e). The moments are:

$$M_B = 0.5 \times 8 \times 69.6^2 \qquad\qquad = 1.94 \times 10^4 \text{ Nmm}$$
$$M_c = (1440 \times 110.4) - (0.5 \times 8 \times 180^2) \quad = 2.94 \times 10^4 \text{ Nmm}$$

Assume that thickness of plate is greater than 16 mm:

Design strength $p_y = 265 \text{ N/mm}^2$

The moment of resistance of the base plate (see Section 4.13.7.3 of the code)
$$= 1.2 p_{yp} Z \qquad\qquad = 1.2 p_{yp} t^2 / 6$$

Thickness $t = (6 \times 2.94 \times 10^4 / 1.2 \times 265)^{0.5} = 23.6 \text{ mm}$

Provide 25 mm thick base plate.

8.10.10 Compression over part of the base and tension in the holding-down bolts

(1) General considerations and terms

The design method adopted is that used for a reinforced concrete section subjected to axial force and bending moment, assuming a linear distribution of pressure. Elastic theory with a triangular compressive stress block is adopted using factored loads and design strengths for the concrete in compression and bolts in tension. The design could also be based on a rectangular stress block.

A column base is shown in Figure 8.47 with the pressure distribution on the concrete and the tension in the holding-down bolts. The following terms are defined:

p_c = bearing strength of the concrete
p_t = tension strength of the bolts
m = modular ratio
$$= \frac{\text{Young's modulus for steel}}{\text{Young's modulus for concrete}} = \frac{E_s}{E_c} = 15$$
W = axial load on the base
M = moment on the base
e = eccentricity of load $= M/W$
l = length of base
b = breadth of base
x = depth to the neutral axis
d = distance from the centreline of the bolts in tension to the edge of the base plate in compression
z = lever arm, the distance between the compressive force in the concrete and the tensile force in the bolts
A_t = tensile stress area of the holding-down bolts
$2a$ = distance between centres of bolts in the base plate
C = compressive force in the concrete
T = tensile force in the bolts
f = maximum bearing stress in the concrete

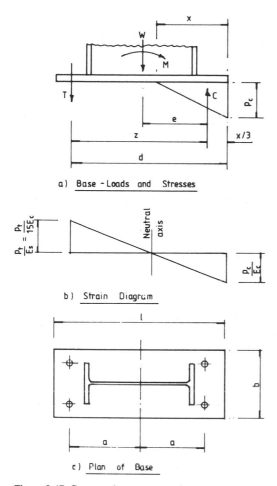

a) Base - Loads and Stresses

b) Strain Diagram

c) Plan of Base

Figure 8.47 Compression over part of the base and tension in holding-down bolts

A direct design method is given first and this is followed by an exact analysis for a given base and loads using elastic theory. The analytical method could be used for checking a base.

(2) Design method

Assume that the maximum design strengths occur simultaneously in the concrete and bolts. The strain diagram with plane sections remaining plane is as shown in Figure 8.47(b). From this diagram:

$$\frac{p_c}{E_c} \bigg/ \frac{P_t}{15E_c} = \frac{x}{d-x}$$

Solve this equation to give the depth to the neutral axis:

$$x = \left(\frac{15p_c}{15p_c + p_t}\right)d$$

The lever arm $z = d - x/3$.

Take moments about the centre line of the bolts in tension:

$$M_1 = M + Wa$$

The compressive force in the concrete:

$$C = M_1/Z$$

The maximum compressive stress in the concrete:

$$f = 2C/bx$$

This must not exceed the bearing strength p_c.

Alternatively, if the length of base is assumed, the breadth can be determined so that the concrete bearing strength is not exceeded. The tensile force in the holding-down bolts is:

$$T = C - W$$

The tensile stress area can then be determined:

$$A_t = T/p_t$$

Suitable bolts can be selected from Table 4.2.

Thus an assumed size of base can be checked and the method gives accurate results if the final stresses in the concrete and steel are close to the design strengths taken to determine the depth to the neutral axis.

(3) Elastic analysis

Consider the base loaded as shown in Figure 8.48. The tensile stress in the bolts:

$$t = 15f(d - x)/x$$

Then the internal forces in bolts and concrete are:

T = tensile force in the holding-down bolts
  $= tA_t$
C = compressive force in the concrete
  $= 1/2\,fbx$

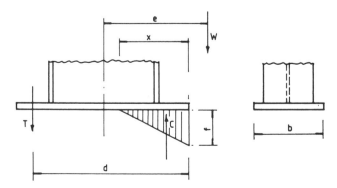

Figure 8.48 Column base: elastic analysis

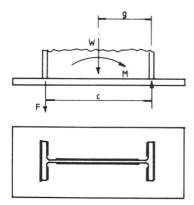

Figure 8.49 Weld: column to base plate

The load on the column:

$$W = C - T$$

The moment on the base:

$$M = We$$
$$= C[(l/2) - (x/3)] + Ta$$
$$= (C - T)e$$

This gives a cubic equation which can be solved for x. Then f and t can be found by back substitution. The method can be used to check a base where the maximum stresses in the steel and concrete are not at the design strengths.

(4) Weld between column and base slab

The weld between the column and base slab is shown in Figure 8.49. The flange welds are assumed to resist the moment and the web welds the shear. The weld force required to resist moment:

$$F = (M - Wg)/c$$

where c = distance centre to center of flanges
       g = one half the depth of the column

Assuming a size of weld, the length required at the tension flange can be found. If heavy moments have to be resisted, a full-strength weld between the column and base slab may be needed.

8.10.11 Tension in holding-down bolts: example

Design the slab base for the side column of a single-storey building for the following conditions at the column base:

Vertical loads
Dead load   = 38.7 kN
Imposed load   = 37.5 kN
Wind load   = − 1.47 kN
Horizontal load = 13.75 kN
Moment (wind) = 32.74 kNm

The column is 406 × 140 UB 39. Use Grade 43 steel, Grade 4.6 bolts and Grade 20 concrete.

(1) Design of base

Design the base for dead + imposed + wind load:

Design load   = 1.2 (38.7 + 37.5 − 1.47)   = 89.7 kN
Design moment = 1.2 × 32.64   = 39.2 kNm
Design strength of the concrete   = 8   N/mm^2
Design strength of the bolts (Table 32)   = 195 N/mm^3

The arrangement of the base is shown in Figure 8.50(a). A length of 560 mm has been selected and the breadth will be calculated to ensure that the bearing stress on the concrete is satisfactory.

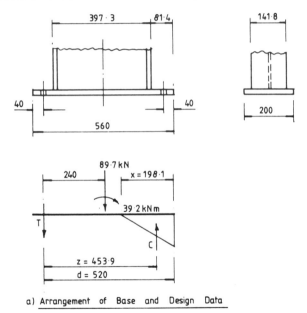

a) Arrangement of Base and Design Data

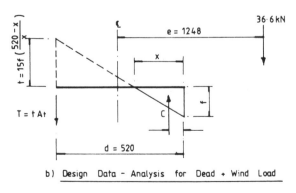

b) Design Data − Analysis for Dead + Wind Load

Figure 8.50 Column base: tension in holding-down bolts

The depth to the neutral axis:

$$x = \left(\frac{15 \times 8}{15 \times 8 + 195}\right) 520 \qquad\qquad = 198.1 \, mm$$

Lever arm $z = 520 - (198.1/3)$ $\qquad\qquad = 453.9 \, mm$

Take moments about the centre line of the holding-down bolts:

$M_1 = 39.2 + (89.7 \times 0.24)$ $\qquad\qquad = 60.73 \, kNm$

The compressive force in the concrete:

$C = 60.73 \times 10^3/453.9$ $\qquad\qquad = 133.8 \, kN$

The base width required to limit the concrete stress to $8 \, N/mm^2$ is:

$$b = \frac{2 \times 133.8 \times 10^3}{8 \times 198.1} \qquad\qquad = 168.9 \, mm$$

Adopt a base width of 200 mm. The maximum concrete stress is:

$f = 168.9 \times 8/200$ $\qquad\qquad = 6.77 \, N/mm^2$

The tension in the holding-down bolts:

$T = 133.8 - 89.7$ $\qquad\qquad = 44.1 \, kN$

The tensile stress area per bolt for two bolts:

$$A_t = \frac{44.1 \times 10^3}{2 \times 195} \qquad\qquad = 113.1 \, mm^2$$

Provide 4 No. 20-mm diameter bolts tensile stress area $= 245 \, mm^2$ per bolt. The bolt size has been selected for practical reasons. The shear stress in the bolts is very small.

(2) Analyse base for dead + wind load

This case will give the maximum tension in the bolts:

Design load	$= 38.7 - (1.4 \times 1.47)$	$= 36.6 \, kN$
Design moment	$= 1.4 \times 32.64$	$= 45.7 \, kNm$
Eccentricity e	$= 45.7 \times 10^3/36.6$	$= 1248 \, mm$

The loads and stress distribution are shown in Figure 8.50(b):

$$W = (0.5fx \times 200 - 15f \times 490)[(520 - x)/x]$$
$$= 100fx - 3.82 \times 10^6 f/x + 7350f$$
$$M = 1248 W$$
$$= 12.48 \times 10^4 fx + 91.73 \times 10^5 f - 47.67 \times 10^8 f/x \qquad (1)$$
$$= 100fx(280 - x/3) + 15f \times 490[(520 - x)/x]240$$
$$= 2.8 \times 10^4 fx - 33.3fx^2 + 9.17 \times 10^8 f/x - 17.64 \times 10^5 f \quad (2)$$

Equate (1) and (2) and reduce, to give

$$x^3 + 2.91 \times 10^3 x^2 + 3.28 \times 10^5 x - 1.71 \times 10^8 = 0$$

Solving by successive trials gives:

$x = 188 \, mm$

The concrete stress f is found by back substitution in the expression for W:

$$36.6 \times 10^3 = 100\, fx \times 188 + 7350f - 3.82 \times 10^6 f/188$$

Concrete: $f = 6.27\, \text{N/mm}^2$
Bolts   : $t = 15 \times 6.27(520 - 188)/188 = 166.1\, \text{N/mm}^2$

(3) Thickness of base plate

The pressure under the edge of the base plate is shown in Figure 8.51(a). Consider a cantilever strip 1 mm × 81.4 mm:

$$M = \frac{3.98 \times 81.4^2}{2 \times 3} + \frac{6.76 \times 81.4^2 \times 2}{2 \times 3} = 1.933 \times 10^4\, \text{Nmm}$$

Assume that the base plate thickness is less than 40 mm:

Design strength $p_y = 265\, \text{N/mm}^2$ (Table 6)

$$t = \left(\frac{6 \times 1.933 \times 10^4}{1.2 \times 265} \right)^{0.5} = 19.1\, \text{m}$$

Use 20 mm thick plate.
    The moment in the base plate due to bolt tension is checked (see Figure 8.51(b)):

Bolt tension $= 166.1 \times 245/10^3 = 40.7\, \text{kN/bolt}$

The section at the column face resisting the bolt load is considered to subtend an angle of 120 degrees with the bolt centre. Spreading the load at 120 degrees would give a width of 143.4 mm at the column face. In this case a width of 100 mm resists the bolt tension. Moment per mm width:

$$M = 41.4 \times 41.38 \times 10^3/100 = 1.71 \times 10^4\, \text{Nmm}$$

This is less than the moment from pressure under the base. The thickness is adequate.

(4) Weld—column to base—Plate

The column and base plate are to be machined for bearing. Design the weld for the maximum overturning case of dead load plus wind load (see Figure 8.51(c)). The tension in the weld:

$$T = \frac{45.7 - 36.6 \times 0.19}{0.38} = 101.96\, \text{kN}$$

Load in weld per mm:
        $= 101.96/141.8$ $= 0.72\, \text{kN/mm}$

A 6 mm fillet weld, strength 0.9 kN/mm, will be satisfactory.
    The shear load to be resisted is 13.75 kN. This requires only 30 mm of weld on the web. The weld is shown in the figure.

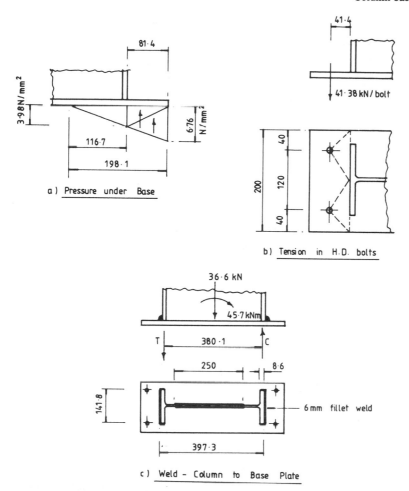

a) Pressure under Base

b) Tension in H.D. bolts

c) Weld - Column to Base Plate

Figure 8.51 Base plate and weld design

8.10.12 Pocket bases: design theory

In this type of base the column is grouted into a pocket in the concrete foundation, as shown in Figure 8.52(a). The axial load is resisted by direct bearing and bond between the steel and concrete. The moment is resisted by compression forces in the concrete acting on the flanges of the column. The forces act on both faces of the flanges of a universal beam. The action in resisting moment is shown in Figure 8.52(b).

To be consistent with the design of the bases above, elastic theory is used for the design of the pocket base. The design strength of the concrete in compression from Section 3.4.4.1 of BS 8110: Part 1 is given as:

$$0.67 f_{cu}/\gamma_m = 0.67 f_{cu}/1.5 = 0.447 f_{cu}$$

where f_{cu} = characteristic cube strength at 28 days. Other terms are defined:

p_y = design strength of the steel
Z_x = elastic modulus of the steel section

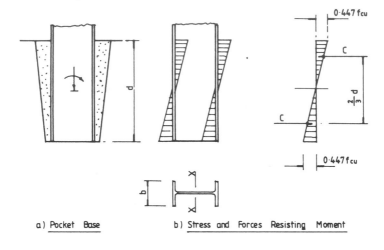

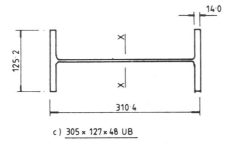

a) Pocket Base   b) Stress and Forces Resisting Moment

c) 305 × 127 × 48 UB

Figure 8.52 Pocket base design

d = depth of embedment of the steel column
b = breadth of flange of the column.

The depth to embed the column to develop its strength can be determined as follows. The compressive force in the concrete:

$$C = (0.447 f_{cu}/2) \times (d/2)xb = 0.112 f_{cu}bd$$

The moment of resistance of the concrete:

$$M = 2 \times C \times 2d/3 \qquad = 0.149 f_{cu}bd^2$$
$$= \text{moment of resistance of the steel section}$$
$$= p_y Z_x$$

The depth of embedment:

$$d = \left(\frac{p_y Z_x}{0.149 f_{cu}b}\right)^{0.5}$$

8.10.13 Pocket base example

Determine the depth that a 305 × 127 UB 48 must be embedded to develop its strength. The section dimensions are shown in Figure 8.52(c). The steel is Grade 43 and the concrete Grade 25.

The elastic modulus $Z_X = 612.4\,\text{cm}^3$
The design strength $p_y = 275\,\text{N/mm}^2$ (Table 6)

$$d = \left(\frac{275 \times 612.4 \times 10^3}{0.149 \times 25 \times 125.2}\right)^{0.5} = 600.9\,\text{mm}$$

If a rectangular stress block in accordance with BS 8110 is used, d is 578 mm.

Problems

8.1 A Grade 43 steel column having 6.0 m effective length for both axes is to carry pure axial loads from the floor above. If a 254×254 UB 89 is available, check the ultimate load that can be imposed on the column. The self weight of the column may be neglected.

8.2 A column has an effective length of 5.0 m and is required to carry an ultimate axial load of 250 kN, including allowance for self weight. Design the column using the following sections:

(1) Universal column section;
(2) Circular hollow section;
(3) Rectangular hollow section.

8.3 A column carrying a floor load is shown in Figure 8.53. The column can be considered as pinned at the top and the base. Near the mid-height it is propped by a strut about the minor axis. The column section provided is an 457×152 UB 60 of Grade 43 steel. Neglecting its self weight, what is the maximum ultimate load the column can carry from the floor above?

8.4 A universal beam, 305×165 UB 54, is cased in accordance with the provisions of Clause 4.14.1 of BS 5950. The effective length of the column is 6.0 m. Check that the section can carry an axial load of 720 kN.

8.5 A Grade 43 steel 457×152 UB 60 used as a column is subjected to uniaxial bending about its major axis. The design data are as follows:

Ultimate axial compression $= 1000\,\text{kN}$
Ultimate moment at top of column $= +200\,\text{kNm}$
Ultimate moment at bottom of column $= -100\,\text{kNm}$
Effective length of column $= 7.0\,\text{m}$

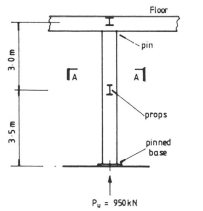
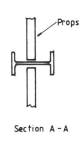

Figure 8.53

Determine the adequacy of the steel section.

8.6 A column between floors of a multi-story building frame is subjected to biaxial bending at the top and bottom. The column member consists of a Grade 43 steel 305 × 305 UC 158 section. Investigate its adequacy if the design load data are as follows:

Ultimate axial compression   = 2300 kN

Ultimate moments,
    Top—about major axis   = 300 kNm
        —about minor axis   =   50 kNm
    Bottom—about major axis   = 150 kNm
        —about minor axis   = − 80 kNm

Effective length of column   = 6.0 m

8.7 A steel tower supports a water tank of size 3 m × 3 m × 3 m. The self weight of the tank is 50 kN when empty. The arrangement of the structure is shown in Figure 8.54. Other design data are given below:

Unit weight of water = 9.81 kN/m³
Design wind pressure = 1.0 kN·m²

Use Grade 43 steel angles for all members. Design the steel tower structure and prepare the steel drawings.

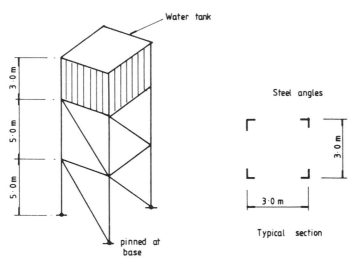

Figure 8.54

9
Trusses and bracing

9.1 Trusses—types, uses and truss members

9.1.1 Types and uses of trusses

Trusses and lattice girders are framed elements resisting in-plane loading by axial forces in either tension or compression in the individual members. They are beam elements but their use gives a large weight saving when compared with a universal beam designed for the same conditions.

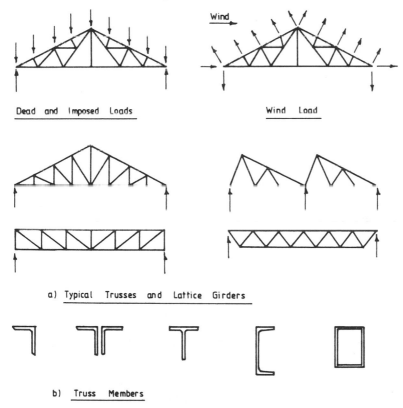

Dead and Imposed Loads   Wind Load

a) Typical Trusses and Lattice Girders

b) Truss Members

Figure 9.1 Roof trusses and lattice girders

The main uses for trusses and lattice girders in buildings are to support roofs and floors and carry wind loads. Pitched trusses are used for roofs while parallel chord lattice girders carry flat roofs or floors.

Some typical trusses and lattice girders are shown in Figure 9.1(a). Trusses in buildings are used for spans of about 10–50 m. The spacing is usually about 5–8 m. The panel length may be made to suit the sheeting or decking used and the purlin spacing adopted. Purlins need not be located at the nodes but this introduces bending into the chord. The panel spacing is usually 1.5–4 m.

9.1.2 Truss members

Truss and lattice girder members are shown in Figure 9.1(b). The most common members used are single and double angles, tees, channels and the structural hollow sections. I and H and compound and built-up members are used in heavy trusses.

9.2 Loads on trusses

The main types of loads on trusses are dead, imposed and wind loads. These are shown in Figure 9.1(a).

9.2.1 Dead loads

The dead load is due to sheeting or decking, insulation, felt, ceiling if provided, weight of purlins and self weight. This load may range from 0.3 to 1.0 kN/m^2. Typical values are used in the worked examples here.

9.2.2 Imposed loads

The imposed load on roofs is taken from BS 6399: Part 1, Section 6. This loading may be summarized as follows:

(1) Where there is only access to the roof for maintenance and repair— $0.75 \, kN/m^2$;
(2) Where there is access in addition to that in (1)—$1.5 \, kN/m^2$.

9.2.3 Wind loads

Wind loads are estimated from CP3: Chapter V: Part 2. The wind load depends on the building dimensions and roof slope, among other factors. The wind blowing over the roof causes a suction or pressure on the windward slope and a suction on the leeward one (see Figure 9.1(a)). The loads act normal to the roof surface.

Wind loads are important in the design of light roofs where the suction can cause reversal of load in truss members. For example, a light angle member is satisfactory when used as a tie but buckles readily when required to act as a strut. In the case of flat roofs with heavy decking the wind uplift will not be greater than the dead load, and it need not be considered in the design.

9.2.4 Application of loads

The loading is applied to the truss through the purlins. The value depends on the roof area supported by the purlin. The purlin load may be at a node point, as shown in Figure 9.1(a), or between nodes, as discussed in Section 9.3.3 below. The weight of the truss is included in the purlin point loads.

9.3 Analysis of trusses

9.3.1 Statically determinate trusses

Trusses may be simply supported or continuous, statically determinate or redundant, pin jointed or rigid jointed. However, the most commonly used truss or lattice girder is single span, simply supported and statically determinate. The joints are assumed to be pinned, though, as will be seen in actual construction, continuous members are used for the chords. This assumption for truss analysis is stated in Section 4.10 of BS 5950: Part 1.
A pin-jointed truss is statically determinate when:

$$m = 2j-3$$
where m = number of members
j = number of joints.

9.3.2 Load applied at the nodes of the truss

When the purlins are located at the node points the following manual methods of analysis are used:[20]

(1) Force diagram. This is the quickest method for pitched roof trusses.
(2) Joint resolution. This the best method for a parallel chord lattice girder.
(3) Method of sections. This method is useful where it is necessary to find the force in only a few members.

The force diagram method is used for the analysis of the truss in Section 9.6. An example of use of the method of sections would be for a light lattice girder where only the force in the maximum loaded member would be found. The member is designed for this force and made uniform throughout.
A matrix analysis program can be used for truss analysis. In Chapter 10 the roof truss for an industrial building is analysed using a computer program.

9.3.3 Loads not applied at the nodes of the truss

The case where the purlins are not located at the nodes of the truss is shown in Figure 9.2(a). In this case the analysis is made in two parts:

(1) The total load is distributed to the nodes as shown in Figure 9.2(b). The truss is analysed to give the axial loads in the members.
(2) The top chord is now analysed as a continuous beam loaded with the normal component of the purlin loads as shown in Figure 9.2(c). The continuous beam is taken as fixed at the ridge and simply supported at the eaves. The beam supports are the internal truss members. The beam is analysed by moment distribution. The top chord is then designed for axial load and moment.

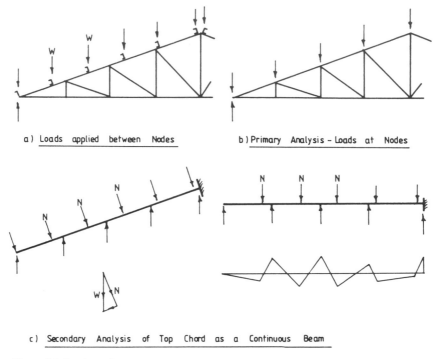

a) Loads applied between Nodes

b) Primary Analysis – Loads at Nodes

c) Secondary Analysis of Top Chord as a Continuous Beam

Figure 9.2 Loads applies between nodes of truss

9.3.4 Eccentricity at connections

BS 5950: Part 1 states in Section 6.1.2 that members meeting at a joint should be arranged so that their centroidal axes meet at a point. For bolted connections the bolt gauge lines can be used in place of the centroidal axes. If the joint is constructed with eccentricity then the members and fasteners must be designed to resist the moment that arises. The moment at the joint is divided between the members in proportion to their stiffness.

9.3.5 Rigid-jointed trusses

Moments arising from rigid joints are important in trusses with short, thick members. BS 5950: Part 1 states in Section 4.10 that secondary stresses from these moments will not be significant if the slenderness of chord members in the plane of the truss is greater than 50 and that of most web members is greater than 100. Rigid jointed trusses may be analysed using a matrix stiffness analysis program.

9.3.6 Deflection of trusses

The deflection of a pin-jointed truss can be calculated using the strain energy method. The deflection at a given node is:[16]

$$\delta = \Sigma PuL/AE$$
$P =$ force in a truss member due to the applied loads

u = force in a truss member due to unit load applied at the node and in the direction of the required deflection.

L = length of a truss member

A = area of a truss member

E = Young's modulus

A computer analysis gives the deflection as part of the output.

9.3.7 Redundant and cross-braced trusses

A cross-braced wind girder is shown in Figure 9.3. To analyse this as a redundant truss it would be necessary to use the computer analysis program.

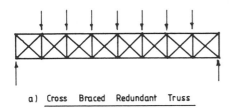

a) Cross   Braced   Redundant   Truss

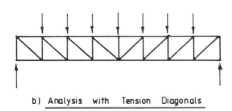

b) Analysis   with   Tension   Diagonals

Figure 9.3 Cross-braced lattice girder

However, it is usual to neglect the compression diagonals and assume that the panel shear is taken by the tension diagonals, as shown in Figure 9.3(b). This idealization is used in design of cross bracing (see Section 9.7).

9.4 Design of truss members

9.4.1 Design conditions

The member loads from the separate load cases—dead, imposed and wind—must be combined and factored to give the critical conditions for design. Critical conditions often arise through reversal of load due to wind, as discussed below. Moments must be taken into account if loads are applied between the truss nodes.

9.4.2 Struts

(1) Maximum slenderness ratios

Maximum slenderness ratios are given in Section 4.7.3.3 of BS 5950: Part 1.

For lightly loaded members these limits often control the size of members. The maximum ratios are:

(1) Members resisting other than wind load—180
(2) Members resisting self weight and wind load—250
(3) Members normally acting as ties but subject to reversal of stress due to wind—350

The code also states that the deflection due to self weight should be checked for members whose slenderness exceeds 180. If the deflection exceeds length/1000, the effect of bending should be taken into account in design.

(2) Limiting proportions of angle struts

To prevent local buckling, limiting width/thickness ratios for single angles and double angles with components separated are given in Table 7 of BS 5950: Part 1. These are shown in Figure 9.4.

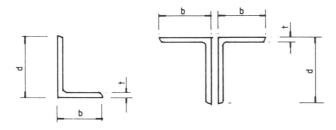

Section	Limiting Proportions	
	b/t and $d/t \leqslant$	$(b+d)/t \leqslant$
Plastic	$8.5\,\epsilon$	–
Compact	$9.5\,\epsilon$	–
Semi – compact	$15.0\,\epsilon$	$23\,\epsilon$

$$\epsilon = (275/p_y)^{0.5}$$

Figure 9.4 Limiting proportions for single- and double-angle struts

(3) Effective lengths for compression chords

The compression chord of a truss or lattice girder is usually a continuous member over a number of panels or, in many cases, its entire length. The chord is supported in its plane by the internal truss members and by purlins at right angles to the plane as shown in Figure 9.5.

The code defines the length of chord members in Section 4.10 as:

(1) In the plane of the truss—panel length L_1 .
(2) Out of the plane of the truss—purlin spacing L_2

The rules from Section 4.7.2 of the code can then be used to determine the effective lengths. The slenderness ratios for single and double angle chords are

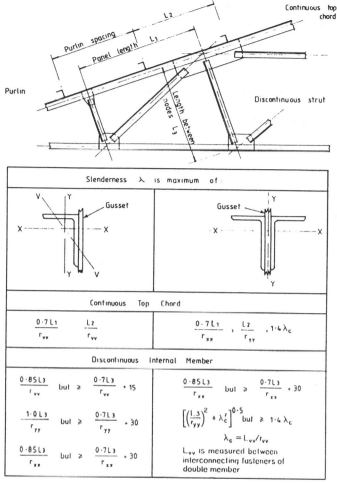

Figure 9.5 Slenderness ratios for truss members

shown in Figure 9.5. Note that truss joints reduce the in-plane value of effective length.

(4) Effective lengths of discontinuous internal truss members

Discontinuous internal truss members, a single angle or double angle connected to a gusset at each end, are shown in Figure 9.5. The effective lengths for the cases where the connections contain at least two fasteners or the equivalent in welding are given in Section 4.7.10 and Table 28 of the code. The slenderness ratios are shown in the figure. The length L_3 is the distance between truss nodes. The effective lengths for other sections are given in the code.

(5) Design procedure

The code states in Section 4.7.10.1 that the end eccentricity for discontinuous struts may be ignored and the design made for an axially loaded member.

For single angles or double angles with members separated, the compression resistance for plastic, compact or semi-compact sections is given by:

$$P_c = A_g p_c$$
$$A_g = \text{gross area}$$

From Table 25 in the code, the strut curve (Table 27(c)) is selected to obtain the compressive strength p_c.

If the section is slender, the design strength is reduced by the lesser of:

$$\frac{11}{(d/t\varepsilon) - 4} \quad \text{or} \quad \frac{19}{[(b+d)/t\varepsilon] - 4}$$

See Table 8 and Figure 3 in BS 5950: Part 1 and Figure 9.4 here.

9.4.2 Ties

The effective area is used in the design of discontinuous angle ties. This was set out in Section 7.4 above. Tension chords are continuous throughout all or the greater part of their length. Checks will be required at end connections and splices.

9.4.3 Members subject to reversal of load

In light roofs the uplift from wind can be greater than the dead load. This causes a reversal of load in all members. The bottom chord is the most seriously affected member and must be supported laterally by a lower chord bracing system, as shown in Figure 1.2. It must be checked for tension due to dead and imposed loads and compression due to wind load.

9.4.4 Chords subjected to axial load and moment

Angle top chords of trusses may be subjected to axial load and moment, as discussed in Section 9.3.3 above. The buckling capacity for axial load is calculated in accordance with Section 9.4.2 (3) above.

The buckling resistance moment for a single angle is given in Section 4.3.8 of BS 5950: Part 1. This depends on the slenderness about the weakest axis, for example:

$$M_b = 0.8 \, p_y Z \text{ when } L/r_v < 100$$

where Z = elastic modulus about the appropriate axis
   r_v = radius of gyration about the weakest axis
   L = unrestrained length

The same expression may be used for a double angle chord as shown in Figure 9.5 using the maximum of L_1/r_x and L_2/r_y. The design check is elastic, so the value of Z to be used depends on where the check is made. For example, for the hogging moment at a node, the minimum value of Z_x is used. It is conservative not to take moment gradient into account, i.e. $m = n = 1$.

The interaction expression for the overall buckling check is:

$$F/(A_g p_c) + M/M_b < 1$$

where F = axial load

A_g = gross area of chord

p_c = compressive strength

M = applied moment

M_b = buckling resistance moment

9.5 Truss connections

9.5.1 Types

The following types of connections are used in trusses:

(1) Column cap and end connections;
(2) Internal joints in welded construction;
(3) Bolted site joints—internal and external.

The internal joints may be made using a gusset plate or the members may be welded directly together. Some typical connections using gussets are shown in Figure 9.6. In these joints all welding is carried out in the fabrication shop. The site joints are bolted.

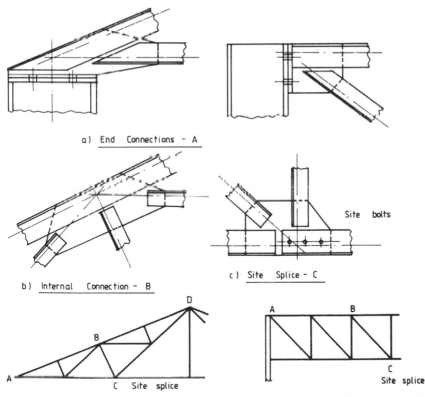

a) End   Connections - A

b) Internal   Connection - B

c) Site   Splice - C

Site   bolts

C   Site   splice

C
Site splice

Figure 9.6 (contd. overleaf)

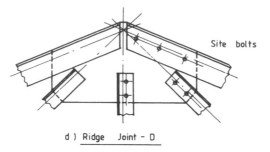

d) Ridge   Joint - D

Figure 9.6 Truss and lattice girder connections

9.5.2 Joint design

Joint design consists of designing the bolts, welds and gusset plate.

(1) Bolted joints

The load in the member is assumed to be divided equally between the bolts. The bolts are designed for direct shear and the eccentricity between the bolt gauge line and the centroidal axis is neglected (see Figure 9.7(a)). The bolts and gusset plate are checked for bearing.

(2) Welded joints

In Figure 9.7(a) the weld groups can be balanced as shown. That is, the centroid of the weld group is arranged to coincide with the centroidal axis of the angle in the plane of the gusset. The weld is designed for direct shear.

If the angle is welded all round, the weld is loaded eccentrically, as shown in Figure 9.7(b). However, the eccentricity is generally not considered in practical design because much more weld is provided than is needed to carry the load.

(3) Gusset plate

The gusset plate transfers loads between members. The thickness is usually

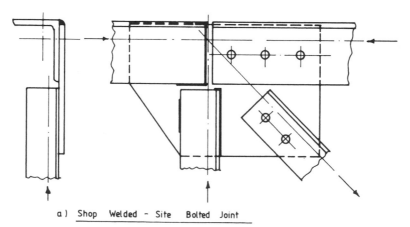

a)   Shop   Welded - Site   Bolted   Joint

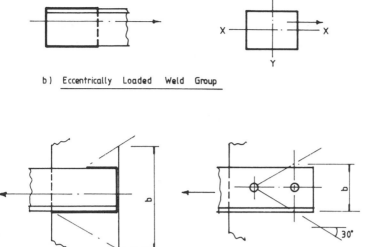

b) Eccentrically Loaded Weld Group

c) Effective Width of Gusset Plate

Figure 9.7 Truss connections and gusset plate design

selected from experience but it should be at least equal to that of the members to be connected.

The actual stress conditions in the gusset are complex. The direct stress in the plate can be checked at the end of the member assuming that the load is dispersed at 30 degrees as shown in Figure 9.7(c). The direct load on the width of dispersal b should not exceed the design strength of the gusset plate. In joints where members are close together it may not be possible to disperse the load. In this case a width of gusset equal to the member width is taken for the check.

9.6 Design of a roof truss for an industrial building

9.6.1 Specification

A section through an industrial building is shown in Figure 9.8(a). The frames are at 5-m centres and the length of the building is 45 m. The purlin spacing on the roof is shown in Figure 9.8(b). The loading on the roof is as follows:

(1) Dead load—measured on the slope length
    Sheeting and insulation board $= 0.25\,\text{kN/m}^2$
    Purlins $= 0.1\ \ \text{kN/m}^2$
    Truss $= 0.1\ \ \text{kN/m}^2$

    Total dead load $= 0.45\,\text{kN/m}^2$

(2) Imposed load—measured on plan $= 0.75\,\text{kN/m}^2$
    measured on slope $= 0.75 \times 10/10.77$ $= 0.7\ \ \text{kN/m}^2$
(3) Wind load

This is calculated using CP3: Chapter V: Part 2. The derivation of the wind loading is set out below.

Design the roof truss using angle members and gusseted joints. The truss is to be fabricated in two parts for transport to site. Bolted site joints are to be provided at A, B and C as shown in the figure.

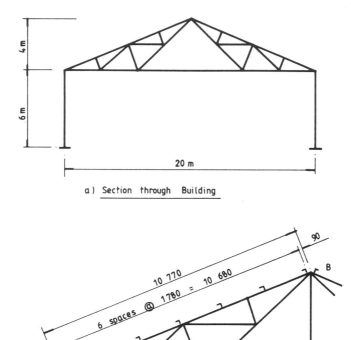

a) Section through Building

b) Arrangement of Purlins

Figure 9.8 A pitched-roof truss

9.6.2 Truss loads

(1) Dead and imposed loads

Because of symmetry only one half of the truss is considered.

Dead loads:
End panel point $= 1/8 \times 0.45 \times 10.77 \times 5$ $= 3.03\,kN$
Internal panel points $= 2 \times 3.03$ $= 6.06\,kN$

Imposed loads:
End panel point $= 3.03 \times 0.7/0.45$ $= 4.71\,kN$
Internal panel points $= 2 \times 4.71$ $= 9.42\,kN$

The dead loads are shown in Figure 9.10.

(2) Wind loads

Location—north-east England

Basic wind speed		$= 45 \, \text{m/s}$
Topography factor	$S_1 = 1.0$	

The building is sited on the outskirts of a city with obstructions up to 10 m high

Ground roughness—category		$= 3$
Building size class		$= \text{B}$
Height to the top of the truss	$H = 10 \, \text{m}$	
Factor from Table 3	$S_2 = 0.74$	
Statistical factor	$S_3 = 1.0$	
Design wind speed $V_s = 0.74 \times 45$		$= 33.3 \, \text{m/s}$
Dynamic pressure $q = 0.613 \times 33.3^2/10^3$		$= 0.68 \, \text{kN/m}^2$

The external pressure coefficients C_{pe} from Table 8 of the wind code are shown in Figure 9.9(a). The internal pressure coefficients C_{pi} are taken from Appendix E in the code. The values used are where there is only a negligible probability of a dominant opening occurring during a severe storm. C_{pi} is taken as the more onerous of the values $+0.2$ or -0.3.

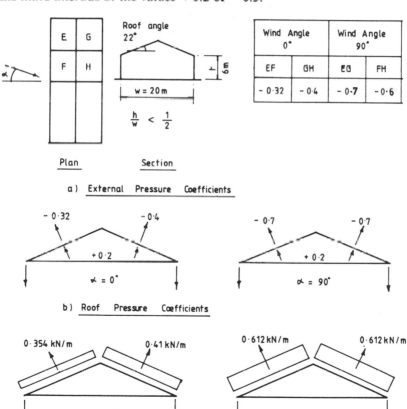

a) External Pressure Coefficients

b) Roof Pressure Coefficients

c) Roof Wind Loads

Figure 9.9 Wind loads on the roof truss

For the design of the roof truss the condition of maximum uplift is the only one that need be investigated. A truss is selected from Section FH of the roof shown in Figure 9.9(a), where C_{pe} is a maximum and C_{pi} is taken as $+0.2$, the case of internal pressure. The wind load normal to the roof is:

$$0.68\,(C_{pe} - C_{pi})$$

The wind loads on the roof are shown in Figure 9.9(c) for the two cases of wind transverse and longitudinal to the building.

The wind loads at the panel points normal to the top chord for the case of wind longitudinal to the building are:

End panel points $= 1/8 \times 0.612 \times 10.77 \times 5 \quad = 4.12\,\text{kN}$
Internal panel points $\qquad\qquad\qquad\qquad\qquad = 8.24\,\text{kN}$

The wind loads are shown in Figure 9.11

9.6.3 Truss analysis

(1) Primary forces in truss members

Because of symmetry of loading in each case only one half of the truss is considered. The truss is analysed by the force diagram method and the analyses are shown in Figure 9.10 and 9.11. Note that members 4–5 and 5–6 must

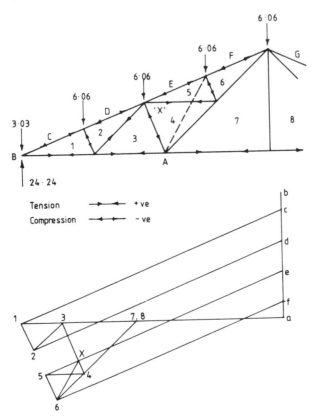

Figure 9.10 Dead-load analysis

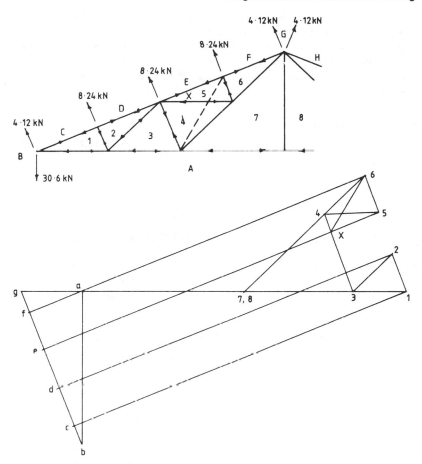

Figure 9.11 Wind-load analysis

Table 9.1 Forces in members of roof truss (kN)

Member		Dead load	Imposed load	Wind load
C–1		− 57.1	− 88.8	72.1
D–2	Top chord	− 54.9	− 85.4	72.1
E–5		− 52.6	− 81.8	74.9
F–6		− 50.3	− 78.2	74.9
A–1		53.0	82.4	− 66.9
A–3	Bottom chord	45.5	− 70.8	− 55.9
A–7		30.3	47.1	− 32.1
1–2		− 5.6	− 8.7	·8.2
3–4	Struts	− 11.3	− 17.6	17.6
5–6		− 5.6	− 8.7	8.2
2–3		7.6	11.8	− 11.1
4–5		7.6	11.8	− 14.1
4–7	Ties	15.2	23.6	− 23.7
6–7		22.7	35.3	− 34.8
7–8		0.0	0.0	0.0

be replaced by the fictitious member 6–X to locate point 6 on the force diagram. Then point 6 is used to find points 4 and 5.

The dead load case is analysed and the forces due to the imposed loads are found by proportion. The case for maximum wind uplift is analysed. The forces in the members of the truss are tabulated for dead, imposed and wind load in Table 9.1.

(2) Moments in the top chord

The top chord is analysed as a continuous beam for moments caused by the normal component of the purlin load from dead load:

Purlin load	$= 1.78 \times 5 \times 0.35$	$= 3.12\,\text{kN}$
Normal component	$= 3.12 \times 10/10.77$	$= 2.89\,\text{kN}$
End purlin L	$= 2.89 \times 0.98/1.78$	$= 1.59\,\text{kN}$

The top chord loading is shown in Figure 9.12(a). The fixed end moments are:

Span AB
$$M_A \quad = \qquad\qquad\qquad\qquad\qquad\qquad\qquad = 0$$
$$M_{BA} = 2.89 \times 1.78(2.69^2 - 1.78^2)/2 \times 2.69^2 \quad = 1.44\,\text{kNm}$$

Span BC
$$M_{BC} = 2.89[(0.87 \times 1.82^3) + (2.65 \times 0.04^2)]/2.69^2 = 1.15\,\text{kNm}$$
$$M_{CB} = 2.89\cdot[(1.82 \times 0.87^2) + (0.04 \times 2.65^2)]/2.69^2 = 0.66\,\text{kNm}$$

Span CD
$$M_{CD} = 2.89 \times 1.74 \times 0.95^2/2.69^2 \qquad\qquad = 0.63\,\text{kNm}$$
$$M_{DC} = 2.89 \times 0.95 \times 1.74^2/2.69^2 \qquad\qquad = 1.15\,\text{kNm}$$

Span DE
$$M_{DE} = 2.89 \times 0.82 \times 1.87^2/2.69^2$$
$$+ 1.59 \times 2.6 \times 0.09^2/2.69^2 \qquad = 1.15\,\text{kNm}$$
$$M_{ED} = 2.89 \times 1.87 \times 0.82^2/2.69^2$$
$$+ 1.59 \times 2.6 \times 0.09^2/2.69^2 \qquad = 0.51\,\text{kNm}$$

The distribution factors are:

| Joint B | BA: BC = 0.75:1 | $= 0.43: 0.57$ |
| Joints C and D | | $= 0.5: 0.5$ |

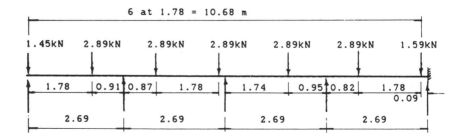

a) Top Chord Dead Loads

0.43	0.57		0.5	0.5		0.5	0.5	
0.0	1.44 -0.12	-1.15 -0.17	0.66 -0.02	-0.63 -0.02	1.15 0.0	-1.15 0.0	0.63
	0.0 0.0	-0.01 0.0	-0.09 0.05	0.0 0.05	0.0 0.0	0.0 0.0	0.0
0.0	1.32	-1.32	0.6	-0.6	1.15	-1.15	0.63

b) Moment Distribution, Top Chord Analysis

Figure 9.12 Top-chord analysis

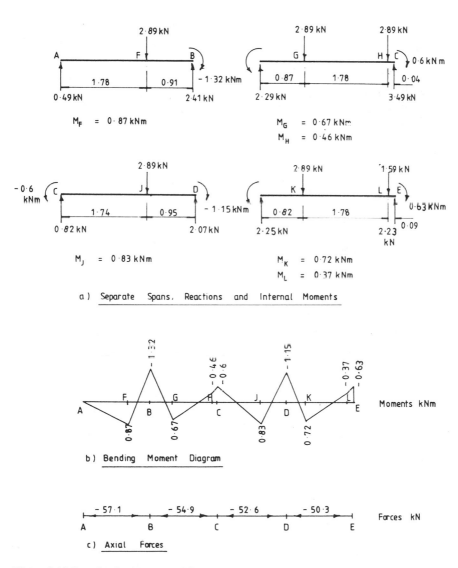

a) Separate Spans, Reactions and Internal Moments

b) Bending Moment Diagram

c) Axial Forces

Figure 9.13 Top-chord moments and forces

The moment distribution is shown in Figure 9.12(b) and the reactions and internal moments for the separate spans are shown in Figure 9.13(a). The bending-moment diagram for the complete top chord is shown in Figure 9.13(b) and the axial loads from the force diagram in Figure 9.13(c).

9.6.4 Design of truss members

Part of the member design depends on the joint arrangement. The joint detail and design are included with the member design. Full calculations for joint design are not given in all cases.

(1) Top chord

The top chord is to be a continuous member with a site joint at the ridge.

Member C-1 at eaves

The maximum design conditions are at B (Figure 9.13)

Dead + imposed load
Compression $F = -(1.4 \times 57.1) - (1.6 \times 88.8)$ $= -222.0\,\text{kN}$
Moment M_B $= -(1.4 \times 1.32) - (1.6 \times 1.32 \times 0.7/0.35) = -6.07\,\text{kNm}$

Dead + wind load
Tension F $= -57.1 + (1.4 \times 72.1)$ $= 43.8\,\text{kN}$

Try 2 No. $100 \times 65 \times 10$ angles with 10 mm thick gusset plate. The gross section is shown in Figure 9.14(a). The properties are:

$A = 31.2\,\text{cm}^4$; $r_x = 3.14\,\text{cm}$, $r_y = 2.79\,\text{cm}$
$Z_X = 46.4\,\text{cm}^3$ (minimum)
Design strength $p_y = 275\,\text{N/mm}^3$ (Table 6)

Check the section classification using Table 7.

d/t $= 100/10 = 10$ < 15
$(b + d)/t \times 165/10 = 16.5 < 23$
The section is semi-compact.

The slenderness is the maximum of:
$L_E/r_x = 0.7 \times 2690/31.4$ $= 60.0$
$L_E/r_y = 1780/27.9$ $= 63.8$
Compressive strength p_c $= 193.4\,\text{N/mm}^2$ (Table 27(c))

The buckling resistance moment for a single angle is given in Section 4.3.8 of the code. The two angles act together and the slenderness is less than 100. Thus:

$M_b = 0.8 \times 275 \times 46.4/10^3 = 10.21\,\text{kNm}$

Interaction expression for overall buckling:

$$\frac{222.02 \times 10}{31.2 \times 193.4} + \frac{6.07}{10.21} = 0.96 \ < 1.0$$

Provide 2 No. $100 \times 65 \times 10$ angles.

The member will be satisfactory for the case of dead plus wind load.

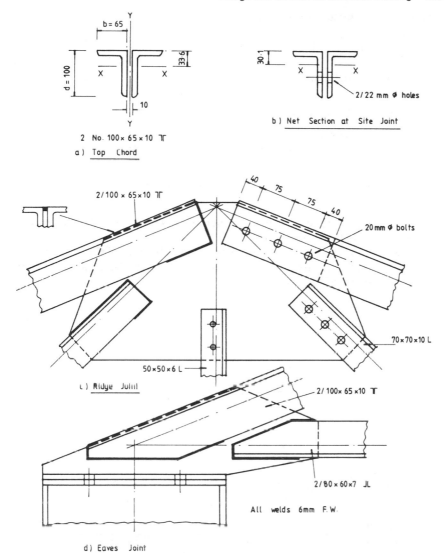

a) Top Chord

2 No. 100 × 65 × 10 ⊤

b) Net Section at Site Joint

2/22 mm ∅ holes

Figure 9.14 Top-chord design details

Member F–6 at ridge

The ridge joint is shown in Figure 9.14(c). A bolted site joint is provided in the chord on one side. The design conditions are:

Compression $F = (1.4 \times 50.3) + (1.6 \times 78.2)$ $= 195.54\,\text{kN}$

Moment $M_E = -(1.4 \times 0.57) - (1.6 \times 0.57 \times 0.07/0.35) = 2.62\,\text{kNm}$

The net section is shown in Figure 9.14(b):

Net area $= 26.8\,\text{cm}^2$; $Z_x = 40.69\,\text{cm}^3$ (minimum)

$M_b = 0.8 \times 275 \times 40.69/10^3 = 8.95\,\text{cm}^3$

Interaction expression for local capacity is:

$$\frac{195.54 \times 10}{275 \times 26.8} + \frac{2.62}{8.95} = 0.56 < 1.0$$

The section is satisfactory.

Eaves joint (see Figure 9.14(d)

The member is connected to both sides of the gusset so the gross section is effective in resisting load (see Section 4.6.3.3 of the code).

Compression $F = 222.02$ kN
Length of 6 mm fillet, strength 0.9 kN/mm, required:
$= 222.02/0.9 = 246.6$ mm

This may be balanced around the member as shown in the figure. More weld has been provided than needed.

The bearing capacity of the gusset is checked at the end of the member on a width of 100 mm. No dispersal of the load is considered because of the compact arrangement of the joint:

Bearing capacity $= 275 \times 100 \times 10/10^3$
$= 275$ kN

The gusset is satisfactory.

Ridge joint (see Figure 9.14(c))

Try three No. 20-mm diameter Grade 4.6 bolts at the centres shown on the figure. From Table 4.2 in the code, the double shear value is 76.2 kN. The bolts resist:

Direct shear	$= 195.54$ kN	
Moment	$= 2.62$ kN	
Direct shear per bolt	$= 195.54/3$	$= 65.18$ kN
Shear due to moment	$= 2.62/0.15$	$= 17.47$ kN
Resultant shear	$= (65.18^2 + 17.47^2)^{0.5}$	$= 67.4$ kN

The bolts are satisfactory.

The shop-welded joint must be designed for moment and shear. The weld is shown in the figure. The gusset will be satisfactory.

(2) Bottom chord

The bottom chord is to have two site joints at P and R, as shown in Figure 9.15.

Member A–1

The design conditions are:

Dead + imposed load		
Tension	$F = (1.4 \times 53) + (1.6 \times 82.4)$	$= 206.04$ kN
Dead + wind load		
Compression	$F = 53 - (1.4 \times 66.9)$	$= -40.66$ kN

Try two No. $80 \times 60 \times 7$ angles. The section is shown in Figure 9.15(b). The properties are:

$A = 18.76\,\text{cm}^2$; $r_x = 2.51\,\text{cm}$; $r_y = 2.34\,\text{cm}$.

The section is semi-compact.

When the bottom chord is in compression due to uplift from wind, lateral supports will be provided at P, Q and R by the lower chord bracing shown in Figure 9.15(a). The effective length for buckling about the YY axis is 5800 mm:

$$L_E/r_Y = 5800/23.4 \qquad = 248$$
$$p_c = 28.4\,\text{N/mm}^2 \text{ (Table 27(c))}$$
$$P_c = 18.76 \times 28.4/10 = 53.2\,\text{kN}$$

At end A the angles are connected to both sides of the gusset:

$$P_t = 275 \times 18.76/10 \qquad = 515.9\,\text{kN}$$

Provide 2 No. $80 \times 60 \times 7$ angles. The wind load controls the design.

The connection to the gusset is shown in Figure 9.14(d). The length of 6 mm weld required:

$$= 206.04/0.9 = 228.9\,\text{mm}$$

The weld is placed as shown in the figure. The gusset is satisfactory.

Member A–7

Dead + imposed load
Tension $F \qquad = (1.4 \times 30.3) + (1.6 \times 47.1) = 117.8\,\text{kN}$
Dead + wind load
Compression $= 30.3 - (1.4 \times 32.1) \qquad = -14.64\,\text{kN}$

Member A–7 is connected to the gusset by 2 No. 20-mm diameter bolts in double shear as shown in Figure 9.15(d):

$$\text{Net area} = 18.76 - 2 \times 22 \times 7/10^2 = 15.68\,\text{cm}^2$$
$$P_t = 15.68 \times 275/10 \qquad = 431.2\,\text{kN}$$

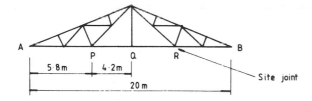

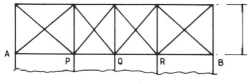

a) Lower Chord Bracing

Figure 9.15 (contd. overleaf)

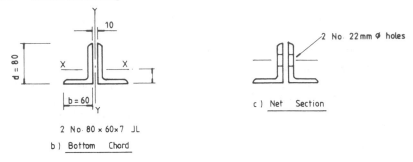

c) Net   Section

2 No. 80 × 60×7   JL

b)   Bottom   Chord

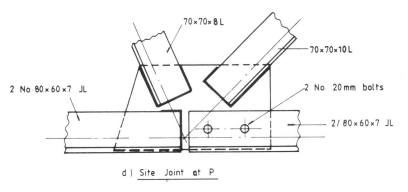

d) Site   Joint   at   P

Figure 9.15 Bottom-chord design details

The member will also be satisfactory when acting in compression due to wind. The shear capacity of the joint:

$$= 2 \times 76.2 = 152.4\,\text{kN} > 117.8\,\text{kN}$$

(3) Internal members

Members 4–7, 6–7

Design for the maximum load in members 6–7:

Dead + imposed load
Tension F $= (1.4 \times 22.7) + (1.6 \times 35.3) =$ 88.26 kN
Dead + wind load
Compression load $= 22.7 - (1.4 \times 34.8)$ $= -26.02\,\text{kN}$

Try $70 \times 70 \times 10$ angle. The member lengths and section are shown in Figure 9.16. The properties are:

$A = 13.1\ \text{cm}^2;\ r_x = 2.09\ \text{cm};\ r_y = 1.36\ \text{cm}.$

The slenderness values are calculated below (see Figure 9.5).

Members 6–7 buckling about the VV axis:
$\lambda = 0.85 \times 2900/13.6 = 181.2$ but $\geqslant 0.7 \times 2900/13.6 + 15 = 164.3$
$\lambda = 2900/20.9 = 138.8$ but $\geqslant 0.7 \times 2900/20.9 + 30 = 127.1$

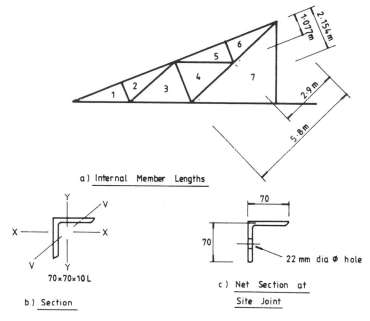

a) Internal Member Lengths

70×70×10 L

b) Section

c) Net Section at Site Joint

22 mm dia φ hole

Figure 9.16 Design for members 4–7, 6–7 and 3–4

Members 6–7 and 4–7 buckling laterally:
$\lambda = 5800/20.9 = 277.5$ but $\geqslant 0.7 \times 5800/20.9 + 30 = 224.2$
Compressive strength for $\lambda = 277.5$:
$p_c = 23.2$ N/mm^2 (Table 27(c))
Compression resistance:
$P_c = 23.2 \times 13.1/10 = 30.4$ kN

This is satisfactory.
    The end of member 6–7 is connected at the ridge by bolts, as shown in Figure 9.14(c). The net section is shown in Figure 9.16(c):

Net area of connected leg = $10(65 - 22)$ = 430 mm^2
Area of unconnected leg = 10×65 = 650 mm^2

Net area = $430 + \left(\dfrac{3 \times 430}{(3 \times 430) + 650} \right) 650$ = 862.2 mm^2

Tension capacity P_t = $275 \times 862.2/10^3$ = 237.1 kN

This is satisfactory.

Joints for Members 4–7, 6–7

Use 20-mm bolts in clearance holes.

Single shear value   = 38.1 kN (Table 4.2)
Number of bolts required = 88.26/38.1 = 2.31. Use three bolts.

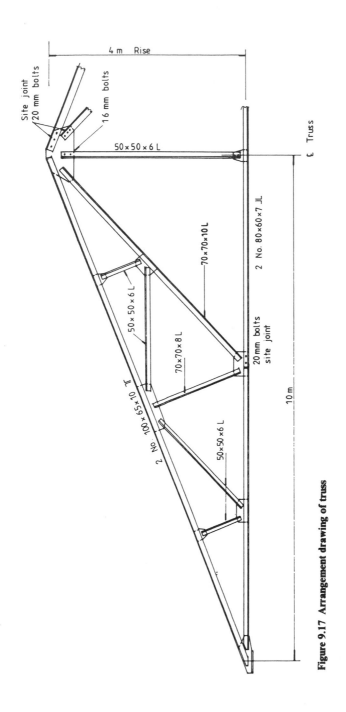

Figure 9.17 Arrangement drawing of truss

The bolts are shown in Figure 9.14(c) and the welded connection is also shown on the figure. The connection to the site joint at P is shown in Figure 9.15(d).

Members 3–4

The design loads are:

Dead + imposed load
Compression $F = -(1.4 \times 11.3) - (1.6 \times 17.6) = -43.98\,\text{kN}$

Dead + wind load
Tension $F \quad = -11.3 + (1.4 \times 17.6) \quad = 13.34\,\text{kN}$

Try 70 ×70 × 8 angle. The properties are:
  $A = 10.6\,\text{cm}^2$; $r_x = 2.11\,\text{cm}$; $r_v = 1.36\,\text{cm}$
See Figure 9.16.
  $\lambda = 0.85 \times 2154/13.6 = 134.6$ but $\geqslant 0.7 \times 2154/13.6 + 15 = 125.9$
    $= 2154/21.1 = 102.1$ but $\geqslant 0.7 \times 2154/21.1 + 30 = 101.5$
  $p_c = 81.4\,\text{N/mm}^2$
  $P_c = 81.4 \times 10.6/10 \quad = 86.2\,\text{kN}$

A smaller angle could be used but this section will be adopted for uniformity.

Other internal members

All other members are to be $50 \times 50 \times 6$ angles. The design for these members is not given.

(4) Truss arrangement

A drawing of the truss is shown in Figure 9.17 and details for the main joints are shown in Figures 9.14 and 9.15.

9.7 Bracing

9.7.1 General considerations

Bracing is required to resist horizontal loading in buildings designed to the simple design method. The bracing also generally stabilizes the building and ensures that the framing is square. It consists of the diagonal members between columns and trusses and is usually placed in the end bays. The bracing carries the load by forming lattice girders with the building members.

9.7.2 Bracing for single-storey industrial buildings

The bracing for a single-storey building is shown in Figure 9.18(a). The internal frames resist the transverse wind load by bending in the cantilever columns. However, the gable frame can be braced to resist this load, as shown. The wind blowing longitudinally causes pressure and suction forces on the windward and leeward gables and wind drag on the roof and walls. These forces are resisted by the roof and wall bracing shown.

   If the building contains a crane an additional load due to the longitudinal crane surge has to be taken on the wall bracing. A bracing system for this case is shown in Figure 9.18(b).

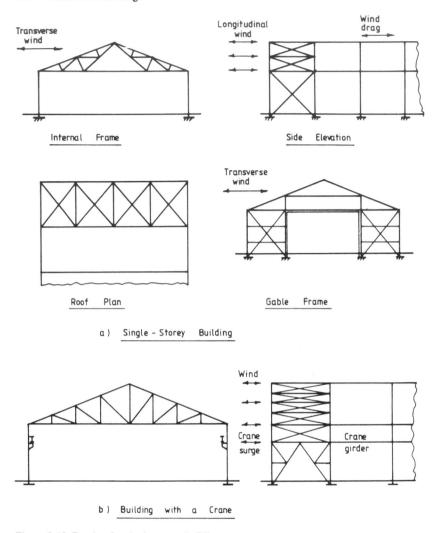

Figure 9.18 Bracing for single-storey building

9.7.3 Bracing for a multi-storey building

The bracing for a multi-storey building is shown in Figure 9.19. Vertical bracing is required on all elevations to stabilize the building. The wind loads are applied at floor level. The floor slabs transmit loads on the internal columns to the vertical lattice girders in the end bays. If the building frame and cladding are erected before the floors are constructed, floor bracing must be provided, as shown in Figure 9.19(c). Floor bracing is also required if precast slabs not effectively tied together are used.

9.7.4 Design of bracing members

The bracing can be single diagonal members or cross members. The loading is generally due to wind or crane surge and is reversible. Single bracing members

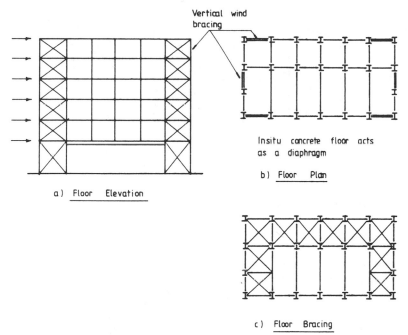

Figure 9.19 Bracing for multi-storey building

must be designed to carry loads in tension and compression. With cross-bracing, only the members in tension are assumed to be effective and those in compression are ignored.

The bracing members are the diagonal web members of the lattice girder formed with the main building members, the column, building truss chords, purlins, eaves and ridge members, floor beams and roof joists. The forces in the bracing members are found by analysing the lattice girder. The members are designed as ties or struts, as set out in Section 9.4 above. Bracing members are often very lightly loaded and minimum-size sections are chosen for practical reasons.

9.7.5 Example: bracing for a single-storey building

The gable frame and bracing in the end bay of a single-storey industrial building are shown in Figure 9.20. The end bay at the other end of the building is also braced. The length of the building is 50 m and the truss and column frames are at 5 m centres. Other building dimensions are shown in the figure. Design the roof and wall bracing to resist the longitudinal wind loading using Grade 43 steel.

(1) Wind load (refer to CP3: Chapter V: Part 2)

Basic wind speed   = 45 m/s
Topography factor $S_1 = 1.0$
Ground roughness—category 3
Building size—Class B

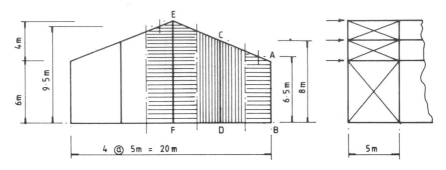

Sheeting rails and door are not shown.

Figure 9.20 Gable frame and bracing in end bay

Height to ridge $H = 10\,\text{m}$
Factor $S_2 = 0.74$ (Table 3)
Statistical factor $S_3 = 1.0$
Design wind speed $V_s = 45 \times 0.74 = 33.3\,\text{m/s}$
Dynamic pressure $q = 0.613 \times 33.3^2/10^3 = 0.68\,\text{kN/m}^2$

The pressure coefficients on the end walls from Table 7 in the code are set out in Figure 9.21(a). The overall pressure coefficient is 0.8. This is not affected by the internal pressure:

Wind pressure $= 0.8 \times 0.68 = 0.544\,\text{kN/m}^2$

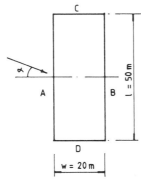

Wind Angle α	C_{pe} for Surface	
	C	D
0°	−0·6	−0·6
90°	+0·7	−0·1

a) External Pressure Coefficients – End Walls

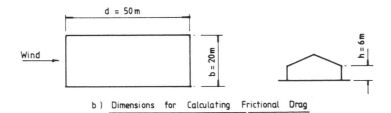

b) Dimensions for Calculating Frictional Drag

Figure 9.21 Data for calculating wind loads

The method for calculating the frictional drag is given in Clause 7.4 of the code:

Cladding—corrugated plastic coated steel sheet.
Factor $C_f' = 0.02$ for surfaces with corrugations across the wind direction.
Refer to Figure 9.21(b).
$d/h = 50/6 \quad = 8.33 > 4$
$d/b = 50/20 \quad = 2.5 \ < 4$

The ratio d/h is greater than 4, so the drag must be evaluated.
For $h < b$, the frictional drag is given by:

F' roof $= C_f'qbd(d - 4h)$
$\qquad = 0.02 \times 0.68 \times 20(50 - 24) \ = 7.1\,\text{kN}$
F' walls $= C_f'q\,2h(d - 4h)$
$\qquad = 0.02 \times 0.68 \times 12(50 - 24) \ = 4.24\,\text{kN}$

The total load is the sum of the wind load on the gable ends and the frictional drag. The load is divided equally between the bracing at each end of the building.

(2) Loads on bracing (see Figure 9.20)

Point E

Load from the wind on the end gable column EF	
$= 9.5 \times 5 \times 0.5 \times 0.544$	$= 12.92\,\text{kN}$
Reaction at E, top of the column	$= 6.46\,\text{kN}$
Load at E from wind drag on the roof	
$= 0.5 \times 7.1 \times 0.25$	$= 0.9\ \text{kN}$
Total load at E	$= 7.36\,\text{kN}$

Point C

Load from the wind on the end gable column CD	
$= 8 \times 5 \times 0.5 \times 0.544$	$= 10.88\,\text{kN}$
Reaction at C, top of the column	$= 5.44\,\text{kN}$
Load C from wind drag on the roof	$= 0.9\,\text{kN}$
Total load at C	$= 6.34\,\text{kN}$

Point A

Load from the wind on building column AB	
$= 2.5 \times 6.5 \times 0.5 \times 0.544$	$= 4.42\,\text{kN}$
Reaction at A, top of the column	$= 2.21\,\text{kN}$
Load at A from wind drag on roof and wall	
$= (0.125 \times 7.1 \times 0.5) + (0.5 \times 4.24 \times 0.25)$	$= 0.97\,\text{kN}$
Total load at A	$= 3.18\,\text{kN}$

(3) Roof bracing

The loading on the lattice girder formed by the bracing and roof members and the forces in the bracing members are shown in Figure 9.22(a). Note that the

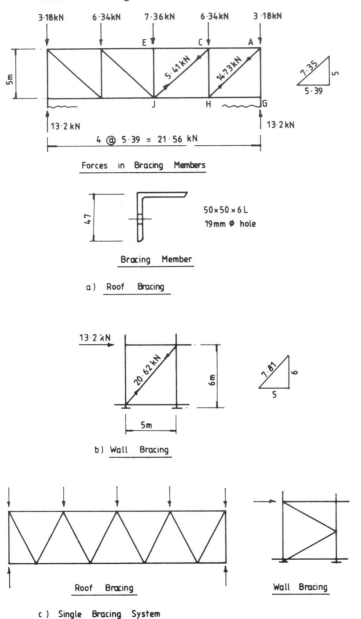

Figure 9.22 Bracing trusses and member forces

members of the cross bracing in compression have not been shown. Forces are transmitted through the purlins in this case. The maximum loaded member is AH:

Design load = 1.4 × 14.17   = 20.62 kN

Try 50 × 50 × 6 angle with 2 No. 16-mm diameter Grade 4.6 bolts in the end connections.

Bracing   295

Bolt capacity $= 2 \times 25.1 = 50.2\,\text{kN}$ (Table 4.2).
Referring to Figure 9.22(a):

$$\text{Net area} = (47 - 18)6 + \left(\frac{3 \times 174}{3 \times 174 \times 282}\right) 282 = 357.1\,\text{mm}^2$$

Tension capacity $P_t = 275 \times 357.1/10^3 = 98.2\,\text{kN}$
Make all the members the same section.

(4) Wall bracing

The load on the wall bracing and the force in the bracing member are shown in Figure 9.22(b).

Design load $= 1.4 \times 20.62 = 28.9\,\text{kN}$
Provide $50 \times 50 \times 6$ angle

(5) Further considerations

Design for load on one gable

In a long building the bracing should be designed for the maximum load at one end. This is the external pressure and internal suction on the gable plus one half of the frictional drag on the roof and walls. In this case if the design is made on this basis:

Roof bracing: Design load in $A\,H - 25.1\,\text{kN}$
Wall bracing: Design load $\quad = 39.2\,\text{kN}$

The $50 \times 50 \times 6$ angles will be satisfactory.

Single bracing system

In the above analysis cross bracing is provided and the purlins form part of the bracing lattice girder. This arrangement is satisfactory when angle purlins are used. However, if cold-rolled purlins are used the bracing system should be independent of the purlins. A suitable system is shown in Figure 9.22(c), where the roof-bracing members support the gable columns. Circular hollow sections are often used for the bracing members.

9.7.6 Example: bracing for a multi-storey building

The framing plans for an office building are shown in Figure 9.23. The floors and roof are cast *in situ* reinforced concrete slabs which transmit the wind load from the internal columns to the end bracing. Design the wind bracing using Grade 43 steel.

(1) Wind loads

The data for the wind loading are:

Basic wind speed $\qquad\qquad V = 45\,\text{m/s}$
Topography factor $\qquad\qquad S_1 = 1.0$
Ground roughness—Category $\quad = 4$
Building size $\qquad$ Class $\qquad = B$

Factor S_2
Roof $H = 13$ m $S_2 = 0.66$
Second floor $H = 11$ m $S_2 = 0.63$
First floor $H = 7$ m $S_2 = 0.58$

The height of the building has been divided into parts for measurement of H as follows:

Roof   —to the top of the building
Second floor—mid-way between second floor and roof
First floor   —mid-way between the first and second floor (see Figure 9.23).
Statistical factor $S_3 = 1.0$

The design wind speeds and dynamic pressures are:

Roof $V_s = 45 \times 0.66 = 29.7$ m/s
      $q = 0.613 \times 29.7^2/10^3 = 0.54$ kN/m^2
Second floor $V = 28.4$ m/s
      $q = 0.49$ kN/m^2
First floor $V = 26.1$ m/s
      $q = 0.42$ kN/m^2

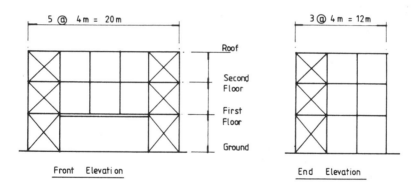

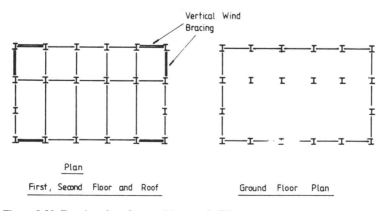

Figure 9.23 Framing plans for a multi-storey building

The force coefficients C_f for wind on the building as a whole are taken from Table 10 of the wind code. These are shown in Figure 9.24 for transverse and longitudinal wind:

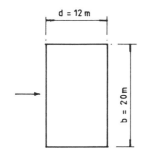

l = length of building   = 20 m
w = width   of building   = 12 m
h = height of building   = 13 m

Wind	Transverse	
l/w	b/d	Cf for h/b = 0·65
1·66	1·66	0·981

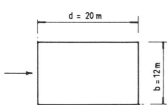

Wind	Longitudinal	
l/w	b/d	Cf for h/b = 1·08
1·66	0·6	0·816

Figure 9.24 Wind loads: force coefficients

$$\text{Force} = C_f q A_e$$

where A_e = effective frontal area under consideration

Wind drag on the roof and walls need not be taken into account because neither the ratio d/h nor d/b is greater than 4.

(2) Transverse bracing

The loads at the floor levels are:

$P = 0.981 \times 0.54 \times 2 \times 10 \qquad = 10.6\,\text{kN}$
$Q = 0.981 \times 0.49 \times 4 \times 10 \qquad = 19.2\,\text{kN}$
$R = 0.981 \times 0.42 \times 4.5 \times 10 \quad = 18.5\,\text{kN}$

The loads are shown in Figure 9.25(a) and the forces in the bracing members in tension are also shown in the figure. The member sizes are selected (see Figure 9.25(c)).

Member QT, RU

Design load $= 1.4 \times 42.1 \qquad = 58.9\,\text{kN}$
Provide 50 × 50 × 6 angle. The tension capacity allowing for one No. 18-mm diameter holes was calculated in Section 9.7.5 (2) above. This is 98.2 kN.

Using 16 mm diameter Grade 4.6 bolts:
Capacity 25.1 kN in single shear:

Member QT—provide two bolts each end
Member RU—provide three bolts each end

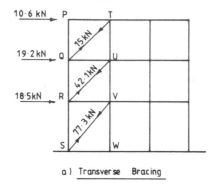

a) Transverse   Bracing

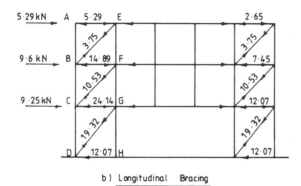

b) Longitudinal   Bracing

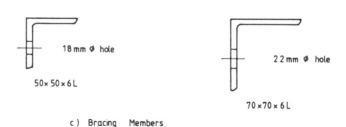

c) Bracing   Members

Figure 9.25 Bracing member design

Member SV

Design load = 1.4 × 77.3   = 108.22 kN
Try 70 × 70 × 6 angle with 20-mm diameter bolts with capacity in single shear of 39.2 kN per bolt.
No. of bolts required at each end = 3
For the angle
    Net area   = 538.7 mm^2
    Tension capacity = 148.1 kN

Note that a 60 × 60 × 6 angle with 16 mm bolts could be used but five bolts would be required in the end connection.

(3) Longitudinal bracing

The loads at the floor levels are:

$A = 0.816 \times 0.54 \times 2 \times 6 \qquad = 5.29\,\text{kN}$
$B = 0.816 \times 0.49 \times 4 \times 6 \qquad = 9.6\ \ \text{kN}$
$C = 0.816 \times 0.42 \times 4.5 \times 6 \quad = 9.25\,\text{kN}$

The loads are shown in Figure 9.25(b) and are divided between the bracing at each end of the building. Those in the bracing members in tension are shown in the figure.

The maximum design load for member DG is

$= 1.4 \times 19.32 \qquad\qquad = 27.1\,\text{kN}$

Provide $50 \times 50 \times 6$ angles for all members. Tension capacity 98.2 kN with one 18 mm diameter hole. 2 No. 16-mm diameter bolts are required at the ends of all bracing member.

Problems

9.1 A flat roof building of 18 m span has 1.5 m deep trusses at 4-m centres. The trusses carry purlins at 1.5-m centres. The total dead load is $0.7\,\text{kN/m}^2$ and the imposed fload is $0.75\,\text{kN/m}^2$:

(1) Analyse the truss by joint resolution.
(2) Design the truss using angle sections with welded internal joints and bolted field splices.

9.2 A roof truss is shown in Figure 9.26. The trusses are at 6-m centres, the length of the building is 36 m and the height to the eaves is 5 m. The roof loading is:

Dead load $= 0.4\,\text{kN/m}^2$ (on slope)
Imposed load $= 0.75\,\text{kN/m}^2$ (on plan)

The wind load is to be estimated using CP3: Chapter V: Part 2. The building is located on the outskirts of a city and the basic wind speed is 45 m/s.

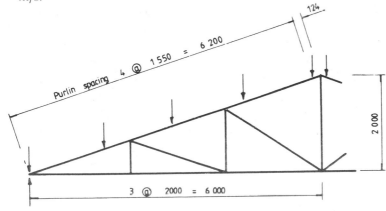

Figure 9.26 Roof truss

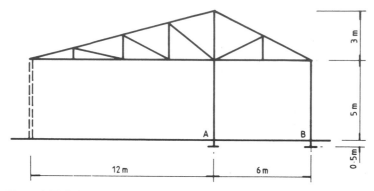

Figure 9.27 Building frame with cantilever truss

(1) Analyse the truss for the roof loads.
(2) Analyse the top chord for the loading due to the purlin spacing shown. The dead load from the roof and purlins is 0.32 kN/m²
(3) Design the truss.

9.3 A section through a building is shown in Figure 9.27. The roof trusses are supported on columns at A and B and cantilever out to the front of the building. The front has roller doors running on tracks on the floor. The frames are at 6-m centres and the length of the building is 48 m. The roof load is:

Dead load   = 0.45 kN/m² (on slope)
Imposed load = 0.75 kN/m² (on plan)

The wind loads are to be in accordance with CP3: Chapter V: Part 2. The basic wind speed is 45 m/s and the location is in the suburbs of a city. The structure should be analysed for wind load for the two conditions of doors opened and closed. Analyse and design the truss.

9.4 The end framing and bracing for a single-storey building are shown in Figure 9.28. The location of the building is on an industrial estate on the outskirts of a city in the north-east of England. The length of the building is 32 m. The wind loads are to be in accordance with CP3: Chapter V: Part 2. Design the bracing.

9.5 The framing for a square tower building is shown in Figure 9.29. The bracing is similar on all four faces. The building is located in a city centre in an area where the basic wind speed is 50 m/s. Design the bracing.

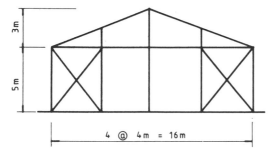
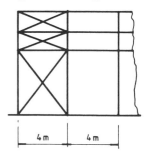

Figure 9.28 Bracing for a factory building

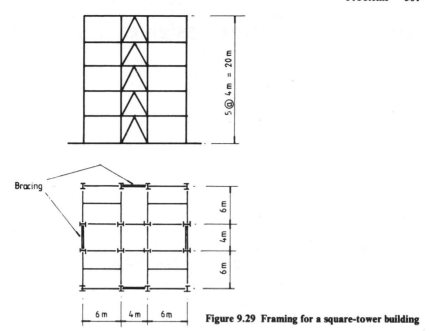

Figure 9.29 Framing for a square-tower building

10

Workshop steelwork design example

10.1 Introduction

An example giving the design of the steel frame for a workshop is presented here and illustrates the following steps in the design process:

(1) Preliminary considerations and estimation of loads for the various load cases;
(2) Computer analysis for the structural frame;
(3) Design of the truss and crane column;
(4) Sketches of the steelwork details.

The framing plans for the workshop with overhead crane are shown in Figure 10.1. The frames are spaced at 6.0 m centres and the overall length of building is 48.0 m. The crane span is 19.1 m and the capacity is 50 kN.

Design the structure using Grade 43 steel. Structural steel angle sections are used for the roof truss and universal beams for the columns.

Computer analysis is used because plane frame programs are now generally available for use on microcomputers. (The reader should consult references 16 and 22 for particulars of the matrix stiffness method analysis. The plane frame manual for the particular software package used should also be consulted.)

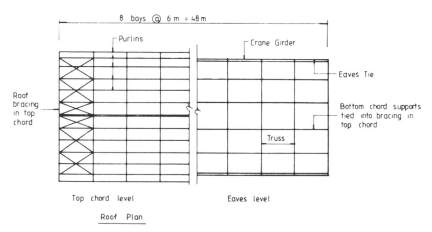

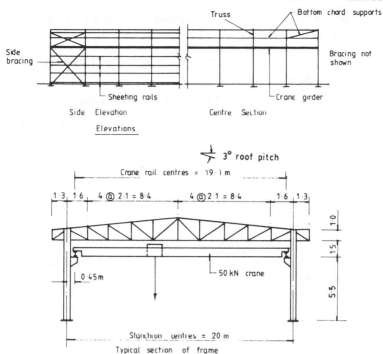

Figure 10.1 Arrangement of workshop steel frame

A manual method of analysis could also be used and the procedure is as follows:

(1) The roof truss is taken to be simply supported for analysis;
(2) The columns are analysed for crane and wind loads assuming portal action with no change of slope at the column top. The portal action introduces forces into the chords of the truss which should be added to the forces in (1) for design.

The reader should consult references 15 and 21 for further particulars of the manual method of analysis.

10.2 Basic design loads

Details of sheeting and purlins used are given below:

Sheeting   Cellactite 11/3 corrugated sheeting, type 800 thickness 0.8 mm. Dead load = 0.1 kN/m². The loads and estimated self weight on plan are as follows:

Roof dead load	(kN/m²)
Sheeting	0.11
Insulation and lighting	0.14
Purlin self weight	0.03
Truss and bracing	0.10
Total load on plan	0.38

Imposed load on plan = 0.75 kN/m²

Purlins Purlins are spaced at 2.1 m centres and span 6.0 m between trusses
Purlin loads $= 0.11 + 0.14 + 0.75 = 1.0 \, kN/m^2$
Provide Ward Building Components Purlin A200/180
Safe load $= 1.07 \, kN/m^2$

Walls— Cladding, insulation, sheeting rails and bracing $= 0.3 \, kN/m^2$

Stanchion— Universal beam section, say 457 × 191 UB 67 for self-weight estimation

Crane data—Hoist capacity $= 50 \, kN$
Bridge span $= 19.1 \, m$
Weight of bridge $= 35 \, kN$
Weight of hoist $= \; 5 \, kN$
End clearance $= 220 \, mm$
End-carriage wheel centres $= 2.2 \, m$
Minimum hook approach $= 1.0 \, m$

Wind data— Wind loads are calculated using CP3: Chapter V: Part 2.
Basic wind speed, $V = 45 \, m/s$
Location—North-east England: the structure is sited on the outskirts of a city with obstructions up to 10 m height. Ground roughness is Category 3 and building size is Class B.

10.3 Computer analysis data

10.3.1 Structural geometry and properties

The computer model of the steel frame is shown in Figure 10.2 with numbering for the joints and members. The joint coordinates are shown in Table 10.1. The column bases are taken as fixed at the floor level and the truss joints and connections of the truss to the columns are taken as pinned. The structure is analysed as a plane frame. The steel frame resists horizontal load from wind and crane surge by cantilever action from the fixed based columns and portal action from the truss and columns. The other data and member properties are shown in Table 10.2.

Table 10.1 Joints coordinates of structure

Joint	X-dist.	Y-dist.	Joint	X-dist.	Y-dist.
1	0.00	0.000	14	12.1	8.415
2	0.00	5.500	15	12.1	7.000
3	0.00	7.000	16	14.2	8.305
4	0.00	8.000	17	14.2	7.000
5	1.60	7.000	18	16.3	8.195
6	1.60	8.085	19	16.3	7.000
7	3.70	7.000	20	18.4	8.085
8	3.70	8.195	21	18.4	7.000
9	5.80	7.000	22	20.0	8.000
10	5.80	8.305	23	20.0	7.000
11	7.90	7.000	24	20.0	5.500
12	7.90	8.415	25	20.0	0.000
13	10.0	8.525	11a	10.0	7.000

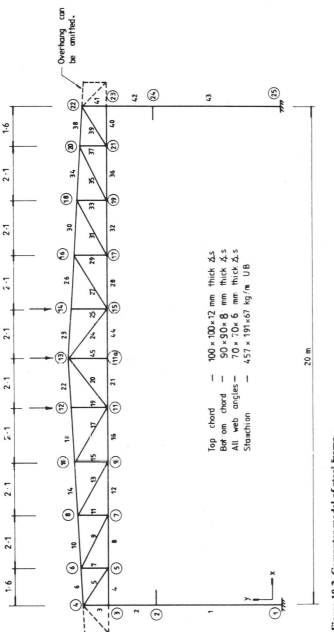

Overhang can be omitted.

20 m

Top chord	—	100 × 100 × 12 mm thick ∠s
Bottom chord	—	90 × 90 × 8 mm thick ∠s
All web angles	—	70 × 70 × 6 mm thick ∠s
Stanchion	—	457 × 191 × 67 kg/m UB

Figure 10.2 Computer model of steel frame

Table 10.2 Structural control data

Number of joints	= 26
Number of members	= 45
Number of joint loaded	= 13
Number of member loaded	= 4

Elastic modulus E = 205 000 N/mm²
Columns: try 457 × 191 UB67

  $A = 85.4\,\text{cm}^2$ $E \times A = 1751\,\text{MN}$

  $I_x = 29401\,\text{cm}^4$ $E \times I_x = 60.3\,\text{MN/m}^2$

Top chord angles: try 100 × 100 × 12 mm angles

  $A = 22.7\,\text{cm}^2$ $E \times A = 465\,\text{MN}$

Bottom chord angles: try 90 × 90 × 8 mm angles

  $A = 13.9\,\text{cm}^2$ $E \times A = 285\,\text{MN}$

All web members: try 80 × 80 × 6 mm angles

  $A = 9.35\,\text{cm}^2$ $E \times A = 192\,\text{MN}$

Note: All $E \times I$ values of the steel angles used in the truss are set to nearly zero.

10.3.2 Roof truss: dead and imposed loads

For the steel truss the applied loads are considered as concentrated at the purlin node points. With the purlin spacing of 2.1 m the applied joint loads at the top chord are:

Dead loads per panel point $= 0.38 \times 6 \times 2.1 = 4.8\,\text{kN}$
Imposed loads per panel point $= 0.75 \times 6 \times 2.1 = 9.45\,\text{kN}$

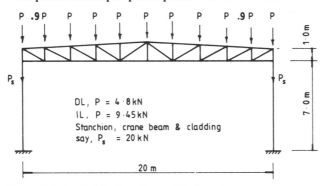

Figure 10.3 Applied dead and imposed loads on truss

Figure 10.3 shows the dead and imposed loads applied on the steel truss at the top chord node points. The dead load of crane girder, side rail, cladding and vertical bracing are assumed to load directly on to the column without affecting the roof truss. The estimated load is 18 kN per leg.

10.3.3 Crane loads

The maximum static wheel load from the manufacturer's table is

Maximum static load per wheel = 35 kN
Add 25% for impact $35 \times 1.25 = 43.8\,\text{kN}$

The location of wheels to obtain the maximum reaction on the column leg is shown in Figure 10.4, with one of the wheels directly over the support point:

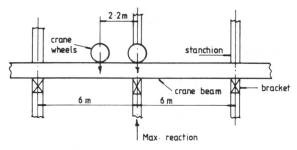

Figure 10.4 Wheels location for maximum reaction

Maximum reaction on the column through the crane bracket
$$= 43.8 + 43.8 \times (6 - 2.2)/6$$
$$= 43.8 \times 1.63 \qquad\qquad = 71.5$$

Corresponding reaction on the opposite column from the crane
$$= 8.7 \times 1.25 \times 1.63 \qquad = 17.7 \, \text{kN}$$

Transverse surge per wheel is 10 per cent of hoist weight plus hook load
$$= 0.1 \times (50 + 5)/4 \qquad = 1.4 \, \text{kN}$$

The reaction on the column is
$$= 1.4 \times 71.4/43.8 \qquad = 2.28 \, \text{kN}$$

Crane load eccentricity from centre line of column assuming 457 × 191 UB 67:
$$e = 220 + 457/2 = 448.5 \text{ say } 450 \, \text{mm}$$

Figure 10.5 shows the crane loads acting on the frame. The two applied moments are due to the vertical loads multiplied by the eccentricities.

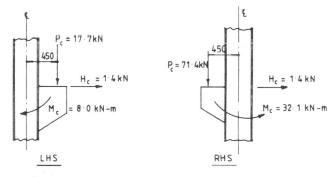

Figure 10.5 Crane loading on steel frame

10.3.3 Wind loads, wind angle 0 degrees to the structure

Design wind speed = basic wind speed $\times S_1 \times S_2 \times S_3$. Take the topographical and statistical factors S_1 and S_2 as unity. For a building of size class B and ground roughness category 3, for a given height, the wind factor S_2 is determined:

Design wind speed $\quad V_s = 45 \times 1 \times 1 \times S_2 = 45 \times S_2$ m/s
Dynamic wind pressure $q = 0.613 \times (45 \times S_2)^2$ N/m^2

Table 10.3 Wind pressures on roof and walls

	Height (m)	S_2	V_s (m/s)	q (kN/m^2)
Roof	9	0.74	33.3	0.68
Walls	7	0.66	29.7	0.54

Table 10.3 gives the dynamic wind pressures for the roof and the walls.
For roof surface with pitch of 3 degrees, wind angle 0 degrees and

$$h/w = 6/20 = 0.3 \ < 1/2$$

External pressure coefficient on roof surface:

$$\begin{aligned} \text{EF} \quad & C_{pe} = -0.9 \\ \text{GH} \quad & = C_{pe} = -0.4 \end{aligned}$$

For walls $\quad h/w < 0.5; \ 3/2 \ < 1/w = 48/20 = 2.4 \ < 4$

$$\begin{aligned} \text{Wall} \quad \text{A} \quad & C_{pe} = 0.7 \\ \text{B} \quad & C_{pe} = -0.25 \end{aligned}$$

From Appendix E, CP 3: Chapter V: Part 2 for the case where there is a negligible probability of a dominant opening occurring during a severe storm C_{pi} is taken as the more onerous of:

Internal suction $= -0.3$
Internal pressure $= +0.2$

The net pressure coefficients acting on the surface of the structure are shown in Figures 10.6(a) and (b).
Internal suction, wind loads:

$$\begin{aligned} \text{on wall } \text{AB} &= 0.54 \times (0.7 + 0.3) \times 6 & &= 3.25 \text{ kN/m} \\ \text{DE} &= 0.54 \times (0.25 - 0.3) \times 6 & &= -0.16 \text{ kN/m} \\ \text{on roof } \text{BC} &= 0.68 \times (-0.9 + 0.3) \times 6 \times 2.1 &&= -5.14 \text{ kN per node} \\ \text{CD} &= 0.68 \times (-0.4 + 0.3) \times 6 \times 2.1 &&= -0.86 \text{ kN per node} \end{aligned}$$

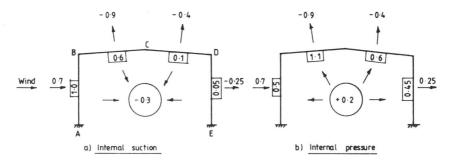

a) Internal suction   b) Internal pressure

Figure 10.6 Wind-pressure coefficients: $\alpha = 0$ degrees

Internal pressure, wind loads:

on wall $AB = 0.54 \times (0.7 - 0.2) \times 6$ $= 1.62\,kN/m$
         $DE = 0.54 \times (-0.25 - 0.2) \times 6$ $= -1.46\,kN/m$
on roof $BC = 0.68 \times (-0.9 - 0.2) \times 2.1$ $= -9.43\,kN$ per node
         $CD = 0.68 \times (-0.4 - 0.2) \times 6 \times 2.1 = -5.15\,kN$ per node

The calculated wind forces for the computer analysis are given in Figures 10.7(a) and (b) for the internal pressure and suction cases, respectively.

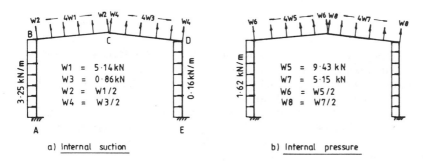

Figure 10.7 Wind loads on structure: $\alpha = 0$ degrees

10.3.5 Wind loads, wind angle 90 degrees to the structure

For roof surface with 3 degree pitch, wind angle 90 degrees and $h/w = 6/20 = 0.3 < 1/2$:

External pressure coefficients on both roof slopes are -0.8
External pressure coefficients on both walls are -0.6
Internal pressure coefficient is $+0.2$

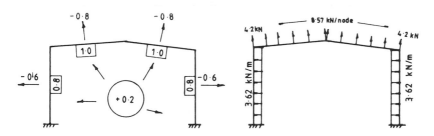

Figure 10.8 Wind loads on structure: $\alpha = 90$ degrees

The wind pressure coefficients and calculated wind loads are shown in Figure 10.8.

10.4 Results of computer analysis

A total of seven computer runs were carried out with one run each for the following load cases:

Case (1) Dead loads   —(DL)
Case (2) Imposed loads—(IL)
Case (3) Wind loads at 0 degrees angle, internal suction—(WL, IS)
Case (4) Wind loads at 0 degrees angle, internal pressure—(WL, IP)
Case (5) Wind loads at 90 degrees angle, internal pressure—(WL, IM)
Case (6) Crane loads, when maximum wheel loads occur—(CRWL)
Case (7) Crane surge loads—(CRSL)

Table 10.4 gives the summary of the truss member axial forces extracted from the computer output. There are no bending moments in the truss members. It was found that the crane loads do not produce any axial force in truss members listed in Table 10.4, except for those truss members connected directly to the column legs, which have some forces due to crane loads. They are members 3–4, 3–5, 4–5, 4–6, 20–22, 21–22 and 22–23.

Table 10.5 shows the member axial forces and moments for the column legs and the truss members connected directly on to it. There are no bending moments for members 3–4, 3–5, 4–5, 4–6, 20–22, 21–22 and 22–23.

The following five critical loads combinations are computed:

(1) 1.4 DL + 1.6 LL
(2) 1.4 DL + 1.6 LL + 1.6 CRWL
(3) 1.0 DL + 1.4 WL (wind at 0°, I.S.)
(4) 1.0 DL + 1.4 WL (wind at 0°, I.P.)
(5) 1.0 DL + 1.4 WL (wind at 90°, I.P.)

The maximum values from the above load combinations are tabulated in Tables 10.4 and 10.6. These will be used later in the design of members.

Note that design conditions arising from notional horizontal loads specified in Clause 2.4.2.3 of BS 5950: Part 1 are not as severe as those in cases 2–5 in Table 10.4. The displacements at every joint are computed in the analyses, but only the critical values are of interest. They are summarized in Table 10.7 and compared with the maximum allowable values.

Table 10.4 Summary of member forces (kN) for truss

Member	Dead load (DL)	Imposed load (IL)	Wind at 0° (IP)	Wind at 90° (IP)	1.4 DL + 1.6 IL	DL + 1.4 × wind at 0°	DL + 1.4 × wind at 90°
4– 6	– 1.3	– 2.56	17.5	3.0	– 5.92	23.2	– 1.3
6– 8	– 30.7	– 60.45	63	55.6	– 139.70	57.5	47.14
8–10	– 47.5	– 93.53	85.6	85.4	– 216.14	72.34	72.06
10–12	– 54.5	– 107.31	90.7	97.9	– 248.00	72.48	82.56
12–13	– 54.5	– 107.31	90.7	– 97.9	– 248.00	72.48	82.56
13–14	– 54.5	– 107.31	90.7	97.9	– 248.00	72.48	82.56
14–16	– 54.5	– 107.31	90.7	97.9	– 248.00	72.48	82.56
16–18	– 47.5	– 93.53	85.6	85.4	– 216.14	72.34	72.06
18–20	– 30.7	– 60.45	63	55.6	– 139.70	57.5	47.14
20–22	– 1.3	– 2.56	17.5	3.0	– 5.92	23.2	– 1.3
3– 5	40.5	79.74	– 50.3	– 63	184.29	– 29.92	– 47.7
5– 7	6.1	12.01	– 6.0	– 1.6	27.76	6.1	– 38.6
7– 9	23.3	45.88	– 51.4	– 50.9	106.02	– 48.66	– 47.96
9–11	40	78.76	– 74	– 80.6	182.02	– 63.6	– 72.84
11–11a	46.4	91.36	– 70.4	– 91.9	211.14	– 52.16	– 82.26
11a–15	46.4	91.36	– 70.4	– 91.9	211.14	– 52.16	– 82.26
15–17	40	78.76	– 47.3	– 80.6	182.02	– 26.22	– 72.84
17–19	23.3	45.88	– 18.9	– 50.9	106.02	– 3.16	– 47.96
19–21	6.1	12.01	– 25.1	– 1.6	27.76	– 29.04	3.86
21–23	40.5	79.74	– 73.4	– 63	184.29	– 62.26	– 47.7
4– 5	40.6	79.94	– 66.4	– 72.3	184.75	52.36	60.62
5– 6	– 21.5	– 42.33	35.2	38.4	– 97.83	27.78	32.26
4– 5	40.6	79.94	– 66.4	– 72.3	184.75	– 52.36	– 60.62
5– 6	– 21.5	– 42.33	35.2	38.4	– 97.83	27.78	32.26
6– 7	33.1	65.17	– 51	– 59	150.62	– 38.3	– 49.5
7– 8	– 15.2	– 29.93	23.5	27.1	– 69.17	17.7	22.74
8– 9	19.3	38.00	– 25.9	– 34.3	87.82	– 16.96	– 28.72
9–10	– 9.6	– 18.90	12.8	16.9	– 43.68	8.32	14.06
10–11	8.3	16.34	6.0	– 14.7	– 37.77	8.3	– 12.28
11–12	– 4.8	9.45	9.4	8.6	– 21.84	8.36	7.24
11–13	.8	9.45	– 10.7	– 1.4	3.64	– 14.18	– 1.16
13–11a	0	0	0	0	0	0	0
13–15	.8	1.58	8.5	– 1.4	3.64	– 2.7	– 1.16
14–15	– 4.8	– 9.45	5.1	8.6	– 21.84	2.34	7.24
15–16	8.3	16.34	– 19.2	– 14.7	37.77	– 18.58	– 12.28
16–17	– 9.6	– 18.90	16.1	16.9	– 43.68	12.94	14.06
17–18	19.3	38.00	– 32.6	– 34.3	87.82	– 26.34	– 28.72
18–19	– 15.2	– 29.93	22.8	27.1	– 69.17	16.72	22.74
19–20	33.1	65.17	– 49.6	– 59	150.62	– 36.34	– 49.5
20–21	– 21.5	– 42.33	30.2	38.4	– 97.83	20.78	32.26
21–22	40.6	79.94	– 57	– 72.3	184.75	– 39.2	– 60.2

Notations: DL = dead load
   IL = imposed load
   IP = internal pressure
   (–) = compression

Table 10.5 Summary of member forces (kN) and moments (kNm) for columns and truss members connected to them

Member		Dead load (DL)	Imposed load (IL)	Crane load (CRWL)	Crane load (CRSL)	Wind at 0° (IS)	Wind at 0° (IP)	Wind at 90° (IP)
1–2		− 26.4	− 51.9	− 16.9	0.4	9.4	40.9	42.7
(M)	Top	− 22.1	− 43.5	− 7.1	− 4	6.5	− 24.1	− 47.7
	Bottom	18.6	36.6	11.3	8.6	− 35.1	8.3	54.1
2–3		− 26.4	− 51.9	− 0.7	0.4	8.8	38.5	40.7
(M)	Top	− 33.1	− 64.9	4.2	4	− 3.5	− 38.9	− 58.4
	Bottom	− 22.1	− 43.5	0.9	− 4	6.5	− 24.1	− 47.7
3–4		− 26.4	− 51.9	− 0.7	− 0.4	8.8	38.5	40.7
(M)	Top	0	0	0	0	0	0	4
	Bottom	− 33.1	− 64.9	4.2	4	− 3.5	− 38.9	− 58.4
3–5		− 40.5	− 79.7	7.5	4	13.2	50.3	63
4–5		40.6	79.9	− 1.7	− 1	− 14.5	− 66.4	− 72.3
4–6		− 1.3	− 2.6	− 2.7	− 3.1	− 8.8	17.5	3
20–22		− 1.3	− 2.6	9.1	3.1	− 7.2	17.5	3
21–22		40.6	79.9	2.4	1	− 10.1	− 57	− 72.3
21–23		− 40.5	− 79.7	− 7.8	− 4	25.5	− 73.4	− 63
22–23		− 26.4	− 51.9	− 0.7	− 0.4	5	30.8	40.7
(M)	Top	0	0	0	0	0	0	0
	Bottom	− 33.1	− 64.9	11	4	− 17.9	− 61.9	− 58.4
23–24		− 26.4	− 51.9	− 0.7	− 0.4	5	30.8	40.7
(M)	Top	− 33.1	− 64.9	11	4	− 17.9	− 61.9	− 58.4
	Bottom	− 22.1	− 43.5	16	0.4	6.9	− 42.7	− 47.2
24–25		− 26.4	− 51.9	− 72.2	− 0.4	5.2	32.1	42.7
(M)	Top	− 22.1	− 43.5	− 15.9	− 4	6.9	− 42.7	− 47.7
	Bottom	18.6	36.6	2.4	8.6	− 17.4	63.6	54.2

Table 10.6 Loads combination for Table 10.5

	1.4 DL + 1.6 DL	1.4 DL + 1.6 LL + 1.6 CRWL	1.4 DL + 1.6 LL + 1.4 TCL	DL + 1.4 × Wind at 0° (IS)	DL + 1.4 × Wind at 0° (IP)	DL + 1.4 × Wind at 90° (IP)	DL + 1.4 × 1.2(DL + TC + 1.2 Wind at 0° (IS)
	− 120.00	− 147.04	− 146.4	− 13.24	30.86	33.38	− 40.20
Top	− 100.54	− 111.90	− 118.3	− 13.00	− 55.84	− 88.88	− 32.04
Bottom	84.60	102.68	116.68	− 30.54	30.22	94.34	4.08
	− 120.00	− 121.00	− 120.48	− 14.08	27.50	30.58	− 21.48
Top	− 150.18	− 143.46	− 137.1	− 38.00	− 87.56	− 114.86	− 34.08
Bottom	− 100.54	− 99.00	− 105.5	− 13.00	− 55.84	− 88.88	− 22.44
	− 120.00	− 121.00	− 121.76	− 14.08	27.50	30.58	− 22.44
Bottom	− 150.18	− 143.46	− 137.06	− 38.00	− 87.56	− 114.86	− 34.08
	− 184.22	− 172.22	− 165.82	− 22.02	29.92	47.70	− 18.96
	184.68	182.00	− 180.36	20.30	− 52.36	− 60.62	28.08
	− 5.98	− 10.30	− 15.26	− 13.62	23.20	2.90	− 19.08
	− 5.98	8.58	13.54	− 11.38	23.20	2.90	4.44
	184.68	188.52	190.12	26.46	− 39.20	− 60.62	40.68
	− 184.22	− 196.70	− 197.34	− 4.80	− 143.26	− 128.70	− 27.84
	− 120.00	− 121.12	− 121.76	− 19.40	16.72	30.58	− 27.00
Top	0	0	0	0	.00	.00	.00
Bottom	− 150.18	− 132.58	− 126.18	− 58.16	− 119.76	− 114.86	− 43.20
	− 120.00	− 121.10	− 121.76	− 19.40	30.58		− 27.00
Top	− 150.18	− 132.58	− 126.18	− 56.16	119.76	− 114.86	− 43.20
Bottom	− 100.54	− 74.94	− 68.54	− 12.44	− 81.88	88.18	5.76
	− 120.00	− 235.52	− 236.16	− 19.12	18.54	33.38	− 112.56
Top	− 100.54	− 125.98	− 132.38	− 12.44	− 81.88	− 88.88	− 42.12
Bottom	84.60	88.44	102.2	− 5.76	107.64	94.48	14.64

Read member number from Table 10.5

Table 10.7 Critical joint displacements

Load case: 1.0 DL + 1.0 LL
Max. vertical deflection at joint 13	= 32.1 mm
Max. horizontal deflection at joints 2 and 24	= 3.7 mm
Horizontal deflection at joint 22	= 1.8 mm

Load case: 1.0 CRWL + 1.0 CRSL
Max. horizontal deflection at joint 22	= 2.62 mm

Load case: 1.0 CRWL = 1.0 Wind at 0° I.S.
Max. horizontal deflection at joint 22	= 7.86 mm

Allowable vertical deflection	= L/200 = 20 × 1000/200
	= 100 mm > 32.1 (Satisfactory)
Allowable horizontal deflection	= L/300 = 8 × 1000/300
	= 27 mm
Max. horizontal deflection	= 1.8 + 7.86
	= 9.66 mm < 16 (Satisfactory)

10.5 Structural design of members
10.5.1 Design of the truss members

Using Grade 43 steel with a design strength of 275 N/mm², the truss members are designed using structural steel angle sections.

Top chord members 10–12, 12–13, 13–14 and 14–16, etc.

Maximum compression from loads combination (2) = -248 kN
Maximum tension from loads combination (5)   = 82.6 kN
Try $100 \times 100 \times 12.0$ mm angle
    $r_v = 1.94$ cm, $A = 22.7$ cm^2
The section is plastic. Lateral restraint is provided by the purlins and web members at the node points.

Slenderness $L/r_v = 2110/19.4 = 108$
$p_c = 113$ N/mm^2 (Table 27(c))

Compression resistance:

$P_c = 113 \times 22.7/10 = 256$ kN > 248 kN (Satisfactory)
The section will also be satisfactory in tension.

Bottom chord members 9–11, 11–11a, 11a–15 and 15–17 etc.
    Maximum tension from loads combination (1)   = 211.1 kN
    Maximum compression from loads combination (5) = -82.3 kN
    Try $90 \times 90 \times 8.0$ mm angle
        $r_v = 1.76$ cm, $A = 13.9$ cm^2
    Tension capacity = $275 \times 13.9/10 = 382$ kN > 211.1 kN
    Lateral support for the bottom chord are shown in Figure 10.1. The slenderness values are:
    $L_E/r_v = 2110/17.6 = 119$
    $L_E/r_x = 4200/27.4 = 153$
    $p_c = 66$ N/mm^2 (Table 27(c))
    Compression resistance = $66 \times 13.9/10 = 92$ kN
                                    > 82 kN (Satisfactory)

All web members
    Maximum tension = 184.8 kN (members 4–5)
    Maximum compression = -97.8 kN (members 5–6)

    Try $70 \times 70 \times 6.0$ mm angles
        $A = 8.13$ cm^2
        $r_v = 1.37$ cm, $r_x = 2.13$ cm

The slenderness is the maximum of:
    $L_E/r_v = 0.85 \times 1085/13.7 = 67.3$
    $L_E/r_x = (0.7 + 1085/21.3) + 30 = 65.6$
    $p_c = 186.4$ N/mm^2 (Table 27(c))
    $P_c = 186.4 \times 8.13/10 = 151.5$ kN > 97.8 kN (Satisfactory)

The angle is connected through one leg to a gusset by welding.
Net area = 7.77 cm^2
Tension capacity = $275 \times 7.77/10 = 214$ kN
                                > 184.8 kN (Satisfactory)

10.5.2 Column design
The worst loading condition for the column is that due to ultimate loads from dead load, imposed load and maximum crane load on members 24–25. The design loads are extracted from Table 10.6 as follows:

Maximum column compressive load $= -236\,kN$
Maximum moment at top of column $= -126\,kNm$
Corresponding moment at bottom $=\quad 88\,kNm$

Try 457×191 UB67, the properties of which are:

$A = 85.4\,cm^2$, $r_x = 18.55\,cm$,
$u = 0.873$ $r_y = 4.12\,cm$,
$x = 37.9$ $S_x = 1470\,cm^3$
                         $Z_x = 1296\,cm^3$

Design strength $p_y = 275\,N/mm^2$ (Table 6 of code)
Effective lengths $L_x = 1.5 \times 7000 = 10\,500\,mm$ (Appendix D of code)
                   $L_y = 0.85 \times 5500 = 4675\,mm$ (Appendix D of code)

Maximum slenderness ratio:

$\lambda = L_y/r_y = 4675/41.2 = 113.4$
from Table 27(c) the value of $p_c = 105.6\,N/mm^2$
Value of $M_{cx} = S_x \times p_y = 1470 \times 275/1000 = 404\,kNm$
    Check for $1.2 \times 275 \times 1296/1000 = 427\,kNm > 404$
$\beta = M_2/M_1 = -88/126 = -0.698$
From Table 18 equivalent moment factor $m = 0.43$
Equivalent moment $\bar{M} = m \times M_1 = 0.43 \times 126 = 54.2\,kNm$

Determine the value of buckling resistance moment M_b:

$\lambda_{LT} = nuv\,\lambda$
take $n = 1.0$ and $N = 0.5$
with $\lambda/x = 113.4/37.9 = 2.99$; from Table 14, $v = 0.91$
$\lambda_{LT} = 1.0 \times 0.873 \times 0.91 \times 113.4 = 90.1$
From Table 11, the value of $p_b = 143.8\,N/mm^2$
$M_b = S_x \times p_b = 1470 \times 143.8/1000 = 211.4\,kNm$

Check for local capacity at top of column:

$$\frac{F}{A_u \times p_y} + \frac{M_x}{M_{cx}} < 1$$

i.e. $\dfrac{236 \times 10}{85.4 \times 275} + \dfrac{126}{404} = 0.1 + 0.31 = 0.41 \; < 1$ (Satisfactory)

Similarly, check for local capacity at bottom of column. The interaction criteria is equal to 0.317, which is satisfactory. Check the stanchion for overall buckling:

$$\frac{F}{A_g \times p_c} + \frac{\bar{M}}{M_b} < 1$$

i.e. $\dfrac{236 \times 10}{85.4 \times 105.6} + \dfrac{54.2}{211.4} = 0.26 + 0.26 = 0.52 < 1$ (Satisfactory)

The 457×191 UB67 provided is satisfactory. The reader may try with a 457×152 UB60 to increase the stress ratio and achieve greater economy. For the design of the crane girder the reader should refer to Chapter 5.

10.6 Steelwork detailing

The details for the main frame and connections are presented in Figure 10.9.

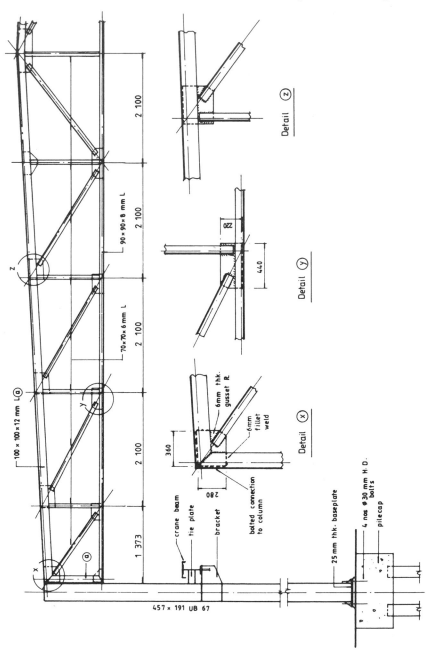

Figure 10.9 Steel frame details

Microcomputer applications

11.1 Application of computers in design

The use of the computer in structural design greatly increases speed of calculation and numerical accuracy. It also makes possible the selection of a number of alternative sections, all of which have been subjected to comprehensive checks to ensure that they meet the relevant requirements. The final choice of section is made on the basis of economic or functional criteria.

The design process of a structure consists of four main stages, as discussed in Chapter 1. These are:

(1) Preliminary considerations and design;
(2) Structural analysis;
(3) Detailed design;
(4) Preparation of drawings.

The application of the computer differs in character in the various stages due to the nature and requirements of the particular process under consideration. These aspects are discussed below.

11.1.1 Preliminary design

Preliminary design is carried out to evaluate alternatives and come to a decision regarding the materials and structural system to be used. The effect of the other considerations, including environmental factors, construction methods, etc., will also need to be assessed.

At this stage many of the decisions are based on the engineer's experience, but a computer, with suitable software, can be useful in the appraisal of different design proposals. In order to reduce the work and computation time only a simplified structural model using typical loads should be used. Calculation of preliminary quantities and costs of alternative systems to enable comparisons to be made can also be computerized.

11.1.2 Structural analysis

Following the preliminary design, the material and structural system to be used can be selected and the design loads can be closely estimated. The structural model can be built up using dimensions and member sizes from the preliminary design. Several runs may be required before arriving at the final analysis.

In rigid frame analysis the computer sets up the stiffness matrix for the structure, inverts this matrix and multiplies by the load matrix to give the deformation matrix and subsequently all the member forces.[22] In this process very large matrices must be stored, accessed and manipulated mathematically.

Moderate-sized structures can be analysed using a plane-frame program on a microcomputer. Large structures will require the use of a mini- or mainframe computer.

Part of the analysis is to combine the separate load cases and introduce the appropriate load factors to give the critical values of axial load, shear and moment for design of each element and connection.

In most of the examples in this book the design actions can be determined manually. However, in the example in Chapter 10 a plane frame matrix stiffness program has been used in the analysis.

The output of structural analysis is conveniently stored in large sequential files on magnetic disk. In this form it is very convenient for subsequent retrieval and input as data into the design program.

11.1.3 Detailed design

Detailed design consists of selecting suitable member sizes and designing connections using design actions from the analysis program. All design programs follow through the procedure, checking strength (including stability and serviceability) to ensure that the section meets all the requirements set out in the code of practice.

Microcomputers of not less than 64K RAM are capable of handling most design programs and programs for beams, columns and column bases are given later in this chapter.

11.1.4 Preparation of drawings

Drawings are the final product of a design office. Computer-aided drafting (CAD) can make a considerable saving in the time required for preparation of drawing. As graphics require large storage memories and fast computing speed, CAD was previously possible only on mainframe and minicomputers. With recent improvements some CAD software can be run on a microcomputer.

The basic equipment required for CAD is as follows:

(1) The basic computer system—this includes the processor, keyboard, text display screen and disk drives.
(2) The workstation—a high-resolution graphics monitor to display the drawing on the screen.
(3) The digitizer—also termed the 'tablet', used as a pointing device for inputting graphics into the computer.
(4) The plotter—used to plot the final drawing on paper.

It is now possible to link computer analysis, design and drafting together into one package system, known as Computer Aided Design and Drafting (CADD). In such a fully computerized system no manual work is required in transferring data from one stage to another. Repeated designs can be performed effortlessly until the designer is satisfied with the final results.

11.2 General aspects of microcomputers

The 8-bit microcomputers are now very cheap. Some cost less than the monthly salary of an engineer for a complete system, including a printer. The microprocessors employed in these 8-bit computers include the Zilog – Z80 chips and the MOS – 6502 chips, which have clock speeds of 2.2 MHz. With 8-bit chips the maximum random access memory that can be directly accessed at a time is roughly 64K bytes, equal to the size of a 25-page document.

The more powerful 16-bit microcomputers commonly employ Intel 8088 and 8086 chips, which have clock speeds of 4.77 MHz and 8.0 MHz, respectively. These computers with faster processing speeds are capable of running programs requiring up to 512K RAM memories. The IBM-PC and many other similar 16-bit microcomputers are currently in general use in industry. When 'number crunching' is important, an 8087 numeric data coprocessor can be installed. This increases the operating speed by four to fifteen times, and improves the numerical accuracy of the computations.

For heavy users, installing a hard disk with the PC will speed up data input/output and loading of programs. Hard disks are now low in cost and are available in 20, 30 M byte capacities. They are thin and fit neatly into the drive slot. Directories and Paths for programs and data filing make file searching extremely easy. At the time of writing, a 32-bit processing board using the NS-32081 chip has become available for attachment to the PC. A clock speed of 10 MHz is achieved with this chip.

Higher-end microcomputers using 32-bit chips are now available. These include microcomputers based on Intel 80386 and Motorola 6800 series chips. Many small design firms that may not be able to afford a minicomputer will find these machines a good substitute. The high computing speed and large RAM and storage capacity make them suitable for many engineering and business applications. Two common disk-operating systems, DOS, that are currently employed in the 16-bit microcomputers include the MS-DOS and the CP/M DOS.

It is very important to keep abreast of computer developments. In the near future automated design and drafting will be available in most design offices.

11.3 Considerations in writing design programs

In writing a design program it is much easier if the code of practice follows a logical sequence of requirements and checks and all the design stresses and factors are given in closed-form mathematical expressions. This increases the speed of calculation and avoids unecessary wastage of memory space or the need for additional storage data files. In the above aspects, the new limit state steel code is much superior to BS 449.

Programming languages such as FORTRAN, BASIC and PASCAL can be used on the IBM-PC. Microsoft BASIC has been found to be very satisfactory. It allows interactive input and ease of input/output control and the

BASIC statements and functions are well documented in the IBM manual. When speed is of great importance, the BASIC compiler version may be used. This increases computation speed several times, and also gives the capability for processing batch files.

When the requirements of a design program are known the overall approach to the solution is decided first. This usually requires successive trials using a file of section properties. Then the complete programming process involves the following three main stages:

(1) Flow charting. This is the systematic presentation of the precise logical procedures and steps required in the program. The flow charts are independent of the particular computer and programming language
(2) Program coding. The procedures given in the flow charts are translated into a form which is acceptable to the computer.
(3) Computer operation. The program coding is typed into the computer memory and the entire program is saved as files in a permanent storage device. The program is then throughly tested and 'de-bugged' and, when all errors have been eliminated, it is ready for production runs.

It is essential to structure the program before constructing flow charts and starting coding. This is carried out by systematically breaking up the overall program into separate subprograms and subroutines. This simplifies the program writing and 'de-bugging' and reduces the length.

It should be noted that the design output format is important and should present all the necessary information clearly. The printed results should include the input data, design strengths, the key parameters and the selected section. Graphic output could enhance the clarity of the results. The graphic capability depends on the type of computer used and its available graphic commands. In some cases it is possible to pipe data through drafting software, resulting in considerable time saving in programming for graphic output.

11.4 Steel beam design program

11.4.1 Program discussion

Simply support steel beam design is chosen first here to illustrate computer programming to the new limit state code. The beam arrangement is shown in Figure 11.1. The program designs beams of compact and plastic section and only simple loadings are used to reduce the program length. The flow chart of the program is shown in Figure 11.2 and the subroutine 'PB' for computing the bending strength p_b is shown in Figure 11.3. The notations used are the same as that given in BS 5950: Part 1 or defined within the flow charts. The salient points are discussed below:

(1) There are two sets of input data in the design program. The first is the structural data consisting of span, laterally unsupported length and the design loads. The second set is the input of the trial section and its properties. If a section properties data files is created then the trial section can be selected by a search process in the program.
(2) The program computes the values of the maximum moment, shear force and the equivalent slenderness ratio. The bending strength p_b, moment

capacity M_b and shear capacity P_v are then calculated. Both bending and shear are checked for adequacy.

(3) The beam must also comply with the serviceability limit state for deflection. The unfactored imposed loads are used to calculate the deflection which should not exceed the limit given by the code. These limits are $L/180$ for cantilever, $L/360$ for beams carrying brittle finishes and $L/200$ for other beams.

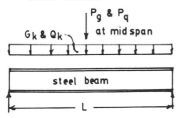

Figure 11.1 Steel beam design

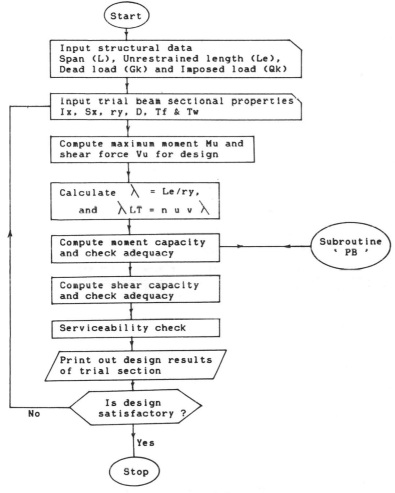

Figure 11.2 Flow chart for simple steel beam design

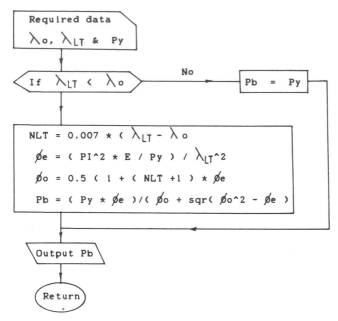

Figure 11.3 Subroutine 'PB'

11.4.2 Program "S-Beam" source listing

```
100 CLS   :REM  ROLLED STEEL BEAM - SIMPLY SUPPORTED SPAN"
110 PRINT "**************************************************"
120 PRINT "*    LIMIT STATE DESIGN OF ROLLED STEEL BEAMS    *"
130 PRINT "*         ( TO B.S. 5950, 1985 )                 *"
140 PRINT "**************************************************"
150 PRINT
160 INPUT "SPAN UNIFORM DISTRIBUTED DEAD LOAD    Gk  (KN/M) =";GK
170 INPUT "SPAN UNIFORM DISTRIBUTED LIVE LOAD    Qk  (KN/M) =";QK
180 INPUT "MIDSPAN CONCENTRATED DEAD LOAD        Pg  ( KN ) =";PG
190 INPUT "MIDSPAN CONCENTRATED LIVE LOAD        Pq  ( KN ) =";PQ
200 INPUT "SPAN LENGTH OF BEAM (BETWEEN SUPPORT)     ( M ) =";LS
210 INPUT "EFFECTIVE LATERALLY UNSUPPORTED LENGTH    ( M ) =";LE
220 PRINT
230 INPUT "STEEL BEAM SIZE (TRIAL SECTION) IS  ( DxBxkg/m )=";B$
240 INPUT "BEAM MOMENT OF INERTIA                Ix  (cm^4) =";IX
250 INPUT "BEAM PLASTIC MODULUS                  Sx  (cm^3) =";SP
260 INPUT "BEAM RADIUS OF GYRATION (MINOR AXIS) Ry  ( cm ) =";RY
270 INPUT "ACTUAL DEPTH OF BEAM SECTION          D   ( mm ) =";DA
280 INPUT "BEAM FLANGE THICKNESS                 Tf  ( mm ) =";TF
290 INPUT "BEAM WEB THICKNESS                    tw  ( mm ) =";TW
300 PRINT
310 NS=.5 : U=.9 :REM SET MONOSYMMETRY & BUCKLING PARAMETERS
320 PRINT
330 INPUT "ACTUAL VALUE OF BUCKLING PARAMETER (DEFAULT=0.9) =";U
340 PRINT
350 INPUT "YIELD STRESS OF STEEL MATERIAL USED  Py (N/mm2) =";PY
360 XO=DA/TF
370 WU=1.4*GK+1.6*QK
380 PU=1.4*PG+1.6*PQ
390 MU=WU*LS^2/8+PU*LS/4
400 VU=(WU*LS+PU)/2
410 LAMD=LE*100/RY
420 PI=3.1416  :E=206000!
430 LAMDO=.4*SQR(PI^2*E/PY)
440 IF NS<.5 THEN 460
450 PSI=.8*(2*NS-1)   :GOTO 470
460 PSI=1*(2*NS-1)
470 V=((4*NS*(1-NS)+.05*(LAMD/XO)^2+PSI^2)^.5+PSI)^(-.5)
480 U=.9
```

```
 490 N=(.94*WU+.86*PU)/(WU+PU)
 500 LAMDLT=N*U*V*LAMD
 510 GOSUB 2000   :REM TO COMPUTE BENDING STRENGTH, PB
 520 MB=SP*PB/1000
 530 KO=MU/MB
 540 IF ABS(KO)<1! THEN 570
 550 C$="MOMENT CAPACITY OF TRIAL SECTION IS INADEQUATE"
 560 GOTO 580
 570 C$="MOMENT CAPACITY OF TRIAL SECTION IS ADEQUATE"
 580 VB=.6*PY*DA*TW/1000
 590 IF VB>=VU THEN 620
 600 E$="SHEAR  CAPACITY OF TRIAL SECTION IS INADEQUATE"
 610 GOTO 630
 620 E$="SHEAR  CAPACITY OF TRIAL SECTION IS ADEQUATE"
 630 REM SERVICEABILITY CHECK
 640 DI=((5*QK*LS^4)/384+PQ*LS^3/48)/(E*IX)*(1E+08)
 650 DM=LS*1000/360
 660 IF DM>=DI THEN 680
 670 D$="BEAM IMPOSED LOAD DEFLECTION IS EXCESSIVE"  :GOTO 690
 680 D$="BEAM IMPOSED LOAD DEFLECTION IS WITHIN LIMIT"
 690 IF ABS(KO)<=1! AND VB>=VU AND DM>=DI THEN 710
 700 F$="ANALYSIS SHOWS THAT TRIAL SECTION IS NOT ADEQUATE" :GOTO 1000
 710 F$="ANALYSIS SHOWS THAT TRIAL SECTION IS ADEQUATE"
1000 PRINT "-----------------------------------------------------------"
1010 PRINT
1020 PRINT "BEAM DESIGN DATA AS FOLLOW :"
1030 PRINT "==========================================================="
1C40 PRINT USING"SPAN ULTIMATE DISTRIBUTED LOAD      Wu  = +####.## (KN/M)";WU
1050 PRINT USING"SPAN ULTIMATE CONCENTRATED LOAD     Pu  = +#####.# ( KN )";PU
1060 PRINT USING"MAXIMUM DESIGN MOMENT ON BEAM       Mmax= +#####.# (KN-M)";MU
1070 PRINT USING"MAXIMUM DESIGN SHEAR  ON BEAM       Vu  = +#####.# ( KN )";VU
1080 PRINT USING"SPAN LENGTH OF BEAM BETWEEN SUPPORTS Ls = ###.### ( M  )";LS
1090 PRINT USING"LATERALLY UNSUPPORTED LENGTH OF BEAM Le = ###.### ( M  )";LE
1100 PRINT
1110 PRINT "-----------------------------------------------------------"
1120 PRINT "TRIAL SECTION SIZE :"; :PRINT B$
1130 PRINT "-----------------------------------------------------------"
1140 PRINT USING"BEAM SECTION MOMENT OF INERTIA      Ix  = ####### (cm^4)";IX
1150 PRINT USING"BEAM SECTION PLASTIC MODULUS        Sx  = ####### (cm^3)";SP
1160 PRINT USING"BEAM SECTION RADIUS OF GYRATION (MINOR AXIS) Ry = ####.## ( cm )";RY
1170 PRINT USING"OVERALL DEPTH OF BEAM SECTION       D   = ####.## ( mm )";DA
1180 PRINT USING"BEAM FLANGE THICKNESS               Tf  = ####.## ( mm )";TF
1190 PRINT USING"BEAM WEB THICKNESS                  tw  = ####.## ( mm )";TW
1200 PRINT USING"BEAM SECTION RATIO OF  ( D/T )      X   = ####.## ";XO
1210 PRINT
1220 PRINT "-----------------------------------------------------------"
1230 PRINT
1240 PRINT "RESULTS OF TRIAL SECTION DESIGN :"
1250 PRINT "==========================================================="
1260 PRINT
1270 PRINT USING"EQUIVALENT SLENDERNESS RATIO     LAMDA(LT)= ####.## ";LAMDLT
1280 PRINT USING"MAXIMUM DESIGN BEAM BENDING MOMENT  Mmax= +#####.# (KN-M)";MU
1290 PRINT USING"COMPUTED BENDING STRENGTH           Pb  = +#####.# N/mm2";PB
1300 PRINT USING"ULTIMATE SECTION MOMENT CAPACITY    Mb  = +#####.# (KN-M)";MB
1310 PRINT USING"BENDING MOMENT STRESS RATIO      Mmax/Mb = ####.## ";KO
1320 PRINT USING"MAXIMUM SHEAR FORCE ON BEAM         Vu  = ####.## ( KN )";VU
1330 PRINT USING"ULTIMATE SHEAR CAPACITY OF BEAM     Pv  = ####.## ( KN )";VB
1340 PRINT USING"DEFLECTION DUE TO IMPOSED ON BEAM   Di  = ####.## ( mm )";DI
1350 PRINT USING"MAXIMUM ALLOWABLE DEFLECTION (L/360) Dm = ####.## ( mm )";DM
1360 PRINT
1370 PRINT "-----------------------------------------------------------"
1380 PRINT "STATUS REPORT"
1390 PRINT "============="
1400 PRINT "FOR BENDING : "; :PRINT C$
1410 PRINT "FOR SHEAR   : "; :PRINT E$
1420 PRINT "DEFLECTION  : "; :PRINT D$
1430 PRINT
1440 PRINT "- REMARK -  : "; :PRINT F$
1450 PRINT "==========================================================="
1460 PRINT
1470 INPUT "IS REDESIGN OF SECTION REQUIRED, Y = YES, N = NO ";ANS$
1480 IF ANS$=Y THEN 220
1490 END
2000 REM SUBROUTINE TO COMPUTE PB
2010 IF (LAMDLT>LAMDO) THEN 2030
2020 PB=PY :GOTO 2070
2030 NLT=.007*(LAMDLT-LAMDO)
3040 RHOE=(PI^2*E/PY)/LAMDLT^2
2050 PHIO=.5*(1+(NLT+1)*RHOE)
2060 PB=(PY*RHOE)/(PHIO+SQR(PHIO^2-RHOE))
2070 RETURN
```

11.4.3 Program "S-Beam" statements and functions

The source listing of the computer program for the design of hot-rolled steel beams is given in Section 11.4.2. The notation used corresponds to that in BS 5950: Part 1. For clarity in illustrating the design program, the input of section properties from a data file is replaced by interactive input. The functions of the various program blocks are as follows:

Statements	*Functions*
100– 150	Clear screen and display title.
160– 220	Input beam load data G_k, Q_k, P_g, P_q, span and effective length.
230– 360	Input trial beam size, properties of the section and the material design strength.
370– 400	Compute ultimate design moment M_u and shear V_u.
410– 500	Compute λ, n, u, v, and λ_{LT}.
510– 570	Call subroutine 'PB', calculate and check the buckling resistance moment.
580– 620	Calculate and check the shear capacity of beam.
630– 680	Check serviceability of trial beam.
690– 710	Select status string variable from the results of all the three checks performed.
1000–1230	Print out beam design input data.
1240–1370	Print out design results.
1380–1450	Print status report of trial section.
1460–1490	If design is satisfactory then end: otherwise go to 230 for redesign.
2000–2070	Subroutine 'PB'.

11.4.4 Sample run of the steel beam program

A sample run is made to show the use of the steel beam design program. The beam span is 6.0 m and the load arrangement is shown in Figure 11.4. A trial beam section 610 × 305 UB 149 is checked for adequacy.
    The sample interactive input data for the design problem is as follows:

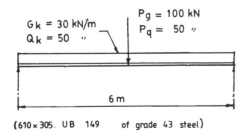

G_k = 30 kN/m
Q_k = 50 ,,

P_g = 100 kN
P_q = 50 ,,

6 m

(610 × 305. UB 149   of grade 43 steel)

Figure 11.4 Example of steel beam design

```
RUN
****************************************************
*    LIMIT STATE DESIGN OF ROLLED STEEL BEAMS     *
*           ( TO B.S. 5950, 1985 )                *
****************************************************

SPAN UNIFORM DISTRIBUTED DEAD LOAD     Gk  (KN/M) =? 30
SPAN UNIFORM DISTRIBUTED LIVE LOAD     Qk  (KN/M) =? 50
MIDSPAN CONCENTRATED DEAD LOAD         Pg  ( KN ) =? 100
MIDSPAN CONCENTRATED LIVE LOAD         Pq  ( KN ) =? 50
SPAN LENGTH OF BEAM (BETWEEN SUPPORT)      ( M  ) =? 6
EFFECTIVE LATERALLY UNSUPPORTED LENGTH     ( M  ) =? 6


STEEL BEAM SIZE (TRIAL SECTION) IS   ( DxBxkg/m )=? 610x305 UB 149
BEAM MOMENT OF INERTIA             Ix  (cm^4) =? 125000
BEAM PLASTIC MODULUS               Sx  (cm^3) =? 4570
BEAM RADIUS OF GYRATION (MINOR AXIS) Ry ( cm ) =? 6.99
ACTUAL DEPTH OF BEAM SECTION       D   ( mm ) =? 609.6
BEAM FLANGE THICKNESS              Tf  ( mm ) =? 19.7
BEAM WEB THICKNESS                 tw  ( mm ) =? 11.9


ACTUAL VALUE OF BUCKLING PARAMETER (DEFAULT=0.9) =? .886

YIELD STRESS OF STEEL MATERIAL USED  Py (N/mm2) =? 265
------------------------------------------------------------
```

The complete results from the computer run are given below:

```
RUN
****************************************************
*    LIMIT STATE DESIGN OF ROLLED STEEL BEAMS     *
*           ( TO B.S. 5950, 1985 )                *
****************************************************


------------------------------------------------------------


BEAM DESIGN DATA AS FOLLOW :
============================================================
SPAN ULTIMATE DISTRIBUTED LOAD        Wu  =  +122.00 (KN/M)
SPAN ULTIMATE CONCENTRATED LOAD       Pu  =  +220.0 ( KN )
MAXIMUM DESIGN MOMENT ON BEAM         Mmax=  +879.00 (KN-M)
MAXIMUM DESIGN SHEAR  ON BEAM         Vu  =  +476.0 ( KN )
SPAN LENGTH OF BEAM BETWEEN SUPPORTS Ls  =   6.000 ( M  )
LATERALLY UNSUPPORTED LENGTH OF BEAM Le  =   6.000 ( M  )


------------------------------------------------------------

TRIAL SECTION SIZE :610x305 UB 149
------------------------------------------------------------
BEAM SECTION MOMENT OF INERTIA       Ix  =   125000 (cm^4)
BEAM SECTION PLASTIC MODULUS         Sx  =    4570 (cm^3)
BEAM RADIUS OF GYRATION (MINOR AXIS) Ry  =    6.99 ( cm )
OVERALL DEPTH OF BEAM SECTION        D   =  609.60 ( mm )
BEAM FLANGE THICKNESS                Tf  =   19.70 ( mm )
BEAM WEB THICKNESS                   tw  =   11.90 ( mm )
BEAM SECTION RATIO OF  ( D/T )       X   =   30.94


------------------------------------------------------------


RESULTS OF TRIAL SECTION DESIGN :
============================================================

EQUIVALENT SLENDERNESS RATIO    LAMDA(LT)=    63.28
MAXIMUN DESIGN BEAM BENDING MOMENT    Mmax=  +879.0 (KN-M)
COMPUTED BENDING STRENGTH             Pb  =  +199.9 (N/mm2)
ULTIMATE SECTION MOMENT CAPACITY      Mb  =  +913.3 (KN-M)
BENDING MOMENT STRESS RATIO      Mmax/Mb  =    0.96
MAXIMUM SHEAR FORCE ON BEAM           Vu  =  476.00 ( KN )
ULTIMATE SHEAR CAPACITY OF BEAM       Pv  = 1153.42 ( KN )
DEFLECTION DUE TO IMPOSED ON BEAM     Di  =    4.15 ( mm )
MAXIMUM ALLOWABLE DEFLECTION (L/360) Dm  =   16.67 ( mm )


------------------------------------------------------------
STATUS REPORT
=============
FOR BENDING : MOMENT CAPACITY OF TRIAL SECTION IS ADEQUATE
FOR SHEAR   : SHEAR  CAPACITY OF TRIAL SECTION IS ADEQUATE
DEFLECTION  : BEAM IMPOSED LOAD DEFLECTION IS WITHIN LIMIT

- REMARK -  : ANALYSIS SHOWS THAT TRIAL SECTION IS ADEQUATE
============================================================
```

11.5 Column design program

11.5.1 Program discussion

Column design using hot-rolled sections is considered in this program. The three possible cases for bending actions that can occur at the column ends are included. These are:

Type (1)—Uniaxial bending about major axis.
Type (2)—Uniaxial bending about minor axis.
Type (3)—Biaxial bending.

In all cases axial compression, taken as positive, is present. The sign convention for the moments at the top and bottom ends of the column is positive for clockwise applied moments and negative for anti-clockwise moments.

Hot-rolled I or H plastic, compact or semi-compact sections can be used. The appropriate column curve is selected by inputting the Robertson constant 'a' which depends on the section and axis of buckling. This is used in the subroutine, 'PC' for computing the compressive strength p_c. Two local capacity checks at the top and bottom of the column and an overall buckling check are made. In the last check, computed equivalent moments are used.

11.5.2 Program "S-COLUMN' source listing

```
100 CLS    :REM LIMIT STATE DESIGN OF ROLLED STEEL COLUMNS
110 PRINT "****************************************************"
120 PRINT "*    LIMIT STATE DESIGN OF ROLLED STEEL COLUMNS   *"
130 PRINT "*             ( TO B.S. 5950, 1985 )              *"
140 PRINT "****************************************************"
150 PRINT
160 PRINT "TYPE OF COLUMN DESIGN CONSIDERED HERE "
170 PRINT "----------------------------------------------------------"
180 PRINT
190 PRINT "TYPE (1) - AXIAL LOAD & BENDING ABOUT MAJOR AXIS"
200 PRINT
210 PRINT "TYPE (2) - AXIAL LOAD & BENDING ABOUT MINOR AXIS"
220 PRINT
230 PRINT "TYPE (3) - AXIAL LOAD & BIAXIAL  BENDING OF COLUMN"
240 PRINT
250 PRINT "----------------------------------------------------------"
260 PRINT
270 INPUT "YOUR CHOICE OF STEEL COLUMN DESIGN =";CHOICE
280 CLS   :PRINT
290 INPUT "ULTIMATE AXIAL LOAD ON COLUMN           Fu  ( KN ) =";FU
300 MX1=MX2=MY1=MY2=M0=.0001
310 IF CHOICE=2 THEN 350
320 PRINT :PRINT "MAJOR AXIS ULTIMATE MOMENTS  ( CLOCKWISE + )"
330 INPUT "TOP & BOTTOM MOMENTS     ( Mux1 & Mux2 )  (KN-M) =";MX1,MX2
340 IF CHOICE=1 THEN 370
350 PRINT :PRINT "MINOR AXIS ULTIMATE MOMENTS  ( CLOCKWISE + )"
360 INPUT "TOP & BOTTOM MOMENTS     ( Muy1 & Muy2 )  (KN-M) =";MY1,MY2
370 PRINT
380 PRINT "-------------------------------------------"
390 PRINT "TRIAL STEEL COLUMN DESIGN   -   INPUT DATA"
400 PRINT "-------------------------------------------"
410 INPUT "MINOR AXIS, COLUMN EFFECTIVE HEIGHT  LX  ( M ) =";LX
420 INPUT "MAJOR AXIS, COLUMN EFFECTIVE HEIGHT  LY  ( M ) =";LY
425 PRINT
430 PRINT "TRIAL COLUMN SECTION PROPERTIES :"
435 PRINT
440 INPUT "SERIAL SIZE OF TRIAL STEEL COLUMN SECTION      =";S$
445 INPUT "SERIAL SIZE OF TRIAL STEEL COLUMN SECTION      =";S$
450 INPUT "SECTION OVERALL DEPTH (mm) & X-AREA (cm^2) D,A =";DO,AG
460 INPUT "THICKNESS OF FLANGE (mm) &  WEB (mm)       T,t =";TF,TW
470 INPUT "RADIUS OF GYRATION (cm)            Rxx & Ryy =";RX,RY
480 INPUT "PLASTIC MODULUS   (cm^3)           Spx & Spy =";SPX,SPY
490 INPUT "ELASTIC MODULUS ABOUT MINOR AXIS   (cm^3)   Zy =";ZY
500 INPUT "YIELD STRENGTH OF COLUMN MATERIAL (N/mm2)   Py =";PY
510 M1=MX1 :M2=MX2
```

```
520 GOSUB 2000
530 MX3=MEQV
540 M1=MY1 :M2=MY2
550 GOSUB 2000
560 MY3=MEQV
570 XO=DO/TF
580 LAMDX=LX*100/RX
590 LAMDY=LY*100/RY
600 U=.9
610 IF LAMDX>LAMDY THEN 630
620 LAMD1=LAMDY :GOTO 640
630 LAMD1=LAMDX
640 GOSUB 5000
650 FC=FU*10/AG
660 MCX=SPX*PY/1000
670 MCY=SPY*PY/1000
680 GOSUB 3000
690 MB=SPX*PB/1000
700 C1=FC/PY
710 C3=FC/PC
720 X1=ABS(MX1/MCX)
730 Y1=ABS(MY1/MCY)
740 X2=ABS(MX2/MCX)
750 Y2=ABS(MY2/MCY)
760 X3=ABS(MX3/MB)
770 Y3=ABS(MY3/PY/ZY*1000)
780 R1=C1+X1+Y1
790 R2=C1+X2+Y2
800 R3=C3+X3+Y3
810 ON CHOICE GOTO 820,830,840
820 T$="AXIAL LOAD & BENDING ABOUT MAJOR AXIS" :GOTO 850
830 T$="AXIAL LOAD & BENDING ABOUT MINOR AXIS" :GOTO 850
840 T$="AXIAL LOAD & BIAXIAL  BENDING ABOUT BOTH AXES"
850 PRINT "***************************************************"
860 PRINT "*    LIMIT STATE DESIGN OF ROLLED STEEL COLUMN   *"
870 PRINT "*          ( TO B.S. 5950, 1985 )                *"
880 PRINT "***************************************************"
890 PRINT
1000 PRINT "TITLE : "; :PRINT T$
1010 PRINT "-----------------------------------------------------------"
1020 PRINT
1030 PRINT "DESIGN DATA FOR STEEL COLUMN AS FOLLOWS :"
1040 PRINT "==========================================================="
1050 PRINT USING"ULT. COLUMN AXIAL COMPRESSIVE LOAD    Fu  = ####.## ( KN )";FU
1060 PRINT USING"MAJOR AXIS ULTIMATE MOMENT (TOP)    Mux1 =+####.## (KN-M)";MX1
1070 PRINT USING"     - do -          MOMENT (BOTTOM) Mux2 =+####.## (KN-M)";MX2
1080 PRINT USING"MINOR AXIS ULTIMATE MOMENT (TOP)    Muy1 =+####.## (KN-M)";MY1
1090 PRINT USING"     - do -          MOMENT (BOTTOM) Muy2 =+####.## (KN-M)";MY2
1100 PRINT
1110 PRINT USING"MAJOR AXIS COLUMN EFFECTIVE HEIGHT    LX  = ####.## ( M )";LX
1120 PRINT USING"MINOR AXIS COLUMN EFFECTIVE HEIGHT    LY  = ####.## ( M )";LY
1130 PRINT
1140 PRINT "TRIAL COLUMN SECTION PROPERTIES INPUT :"
1150 PRINT "==========================================================="
1155 PRINT "SERIAL SIZE OF SECTION  : ";S$
1160 PRINT USING"SECTION OVERALL DEPTH              D   = #####.# ( mm )";DO
1170 PRINT USING"SECTION GROSS X-AREA               Ag  = #####.# (cm^2)";AG
1180 PRINT USING"THICKNESS OF FLANGE                T   = #####.# ( mm )";TF
1190 PRINT USING"THICKNESS OF WEB                   t   = #####.# ( mm )";TW
1200 PRINT USING"RADIUS OF GYRATION                 Rxx = #####.# ( cm )";RX
1210 PRINT USING"RADIUS OF GYRATION                 Ryy = #####.# ( cm )";RY
1220 PRINT USING"PLASTIC MODULUS ABOUT MAJOR AXIS   Spx = #####.# (cm^3)";SPX
1230 PRINT USING"PLASTIC MODULUS ABOUT MINOR AXIS   Spy = #####.# (cm^3)";SPY
1240 PRINT USING"ELASTIC MODULUS ABOUT MINOR AXIS   Zy  = #####.# (cm^3)";ZY
1250 PRINT USING"YIELD STRENGTH OF COLUMN MATERIAL  Py  = #####.# (N/mm2)";PY
1260 PRINT "-----------------------------------------------------------"
1270 PRINT
1280 PRINT "STEEL COLUMN MEMBER DESIGN        - OUTPUT RESULTS -"
1290 PRINT "==========================================================="
1300 PRINT USING"VALUE OF (FC)  = +####.## ";FC;
1310 PRINT USING"      AND (PC)  = +####.## ";PC
1320 PRINT USING"VALUE OF (MCX) = +####.## ";MCX;
1330 PRINT USING"      AND (MCY) = +####.## ";MCY
1340 PRINT USING"VALUE OF (MeqX)= +####.## ";MX3;
1350 PRINT USING"      AND (MeqY)= +####.## ";MY3
1360 PRINT USING"VALUE OF (MB)  = +####.## ";MB
1370 PRINT
1380 PRINT "STRESS RATIOS COMPUTED AS FOLLOWS :"
1390 PRINT "--------------------------------------";
1400 PRINT
```

```
1410 PRINT USING"RATIO OF        FC/PY           [C1]   = +###.### ";C1
1420 PRINT USING"RATIO OF        FC/PC           [C3]   = +###.### ";C3
1430 PRINT USING"RATIO OF        MX1/MCX         [X1]   = +###.### ";X1
1440 PRINT USING"RATIO OF        MY1/MCY         [Y1]   = +###.### ";Y1
1450 PRINT USING"RATIO OF        MX2/MCX         [X2]   = +###.### ";X2
1460 PRINT USING"RATIO OF        MY2/MCY         [Y2]   = +###.### ";Y2
1470 PRINT USING"RATIO OF        MeqvX/MB        [X3]   = +###.### ";X3
1480 PRINT USING"RATIO OF        MeqvY/(Py*Zy)   [Y3]   = +###.### ";Y3
1490 PRINT
1500 PRINT "LOCAL CAPACITY CHECK"
1510 PRINT "--------------------"
1520 PRINT USING"    AT TOP    OF COLUMN,    STRESS RATIO = +###.### ";R1
1530 PRINT USING"    AT BOTTOM OF COLUMN,    STRESS RATIO = +###.### ";R2
1540 PRINT
1550 PRINT USING"OVERALL BUCKLING CHECK,     STRESS RATIO = +###.### ";R3
1560 PRINT "======================================================="
1570 END

2000 REM SUBROUTINE TO COMPUTE MEQV (REQ. M1,M2)
2010 BETA=-(M2+.0000001)/(M1+.0000001)
2020 IF BETA>.5 THEN 2040
2030 MBETA=.43 :GOTO 2060
2040 MBETA=.57+.33*BETA+.1*BETA^2
2050 PRINT
2060 IF M1<0! THEN 2080
2070 MEQV=MBETA*M1+M0 :GOTO 2120
2080 MEQV=MBETA*M1
2090 M3=    M1*(1+BETA)/2
2100 IF (ABS(MEQV)>ABS(M3)) THEN 2120
2110 MEQV=M3
2120 RETURN

3000 REM SUBROUTINE TO COMPUTE PB
3010 LAMDO=.4*SQR(PI^2*E/PY)
3020 V=(1+.05*(LAMD1/X0)^2)^(-.25)
3030 LAMDLT=U*V*LAMD1
3040 IF (LAMDLT>LAMDO) THEN 3060
3050 PB=PY :GOTO 3100
3060 NLT=.007*(LAMDLT-LAMDO)
3070 RHOE=(PI^2*E/PY)/LAMDLT^2
3080 PHIO=.5*(1+(NLT+1)*RHOE)
3090 PB=(PY*RHOE)/(PHIO+SQR(PHIO^2-RHOE))
3100 RETURN

5000 REM SUBROUTINE TO COMPUTE PC
5010 PI=3.1416 :E=206000!
5020 LAMDOC=.2*SQR(PI^2*E/PY)
5030 A=5.5
5040 IF LAMD1>LAMDOC THEN 5060
5050 PC=PY :GOTO 5100
5060 NC=.001*A*(LAMD1-LAMDOC)
5070 PE=PI^2*E/LAMD1^2
5080 PHIC=(PY+(NC+1)*PE)/2
5090 PC=PE*PY/(PHIC+SQR(PHIC^2-PE*PY))
5100 RETURN
```

11.5.3 Program "S-COLUMN" statements and functions

The source listing of the column design programme is given in Section 11.5.2. The various program statements and their corresponding functions are explained below:

Statements	Functions
100– 50	Clear screen and display title.
160– 280	Display options menu and input "Choice" of column type.
290– 370	Input axial load and all respective values of end moments.
380– 500	Input properties of the trial section for column design.
510– 570	Call Subroutine 'MEQV' to compute equivalent moments for major and minor axes.
580– 640	Determine critical slenderness ratio and call Subroutine 'PC' to calculate the compressive resistance p_c.

650– 690	Compute F_c, M_{CX}, M_{CY}. Call Subroutine 'PB' for bending strength p_b and compute buckling resistance moment capacity M_b.
700– 800	Compute strength ratios.
810–1020	Select and print title.
1030–1270	Print out all input data.
1280–1370	Print out various strength values for the trial stanchion section.
1380–1560	Print out in detail all the strength ratios and interaction criteria for the local capacity and overall buckling checks.
1570	End.
2000–2130	Subroutine 'MEQV'.
3000–3100	Subroutine 'PB'.
5000–5100	Subroutine 'PC'.

11.5.4 Sample run of the column design programme

The design of a column subjected to biaxial bending is given as an example to show the use of the design program. The column height and the values of axial load and bending moments at the ends are shown in Figure 11.5. The moments M_{ux1} and M_{ux2} at the top and bottom ends produce double curvature bending about the major axis. The minor axis is in single curvature bending due to moments M_{uy1} and M_{uy2}. The sign convention for moment is positive in the clockwise direction.

All input data are prepared from Figure 11.5, where the loads and moments are factored ultimate values. The complete output results are given below, which include the various values of compressive strength, bending strength, relevant strength ratios and interaction criteria for each case checked. The reader may wish to verify the results by manual calculation.

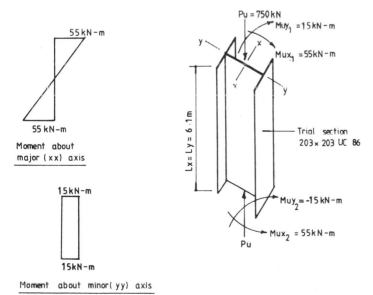

Figure 11.5 Example of column design

```
RUN

*******************************************************
*    LIMIT STATE DESIGN OF ROLLED STEEL COLUMN    *
*         ( TO B.S. 5950, 1985 )                  *
*******************************************************

TITLE : AXIAL LOAD & BIAXIAL BENDING ABOUT BOTH AXES
-------------------------------------------------------

DESIGN DATA FOR STEEL COLUMN AS FOLLOWS :
===========================================================
ULT. COLUMN AXIAL COMPRESSIVE LOAD   Fu  =   750.00 ( KN )
MAJOR AXIS ULTIMATE MOMENT (TOP)     Mux1 =   +55.00 (KN-M)
   - do -       MOMENT (BOTTOM)      Mux2 =   +55.00 (KN-M)
MINOR AXIS ULTIMATE MOMENT (TOP)     Muy1 =   +15.00 (KN-M)
   - do -       MOMENT (BOTTOM)      Muy2 =   -15.00 (KN-M)

MAJOR AXIS COLUMN EFFECTIVE HEIGHT   LX  =     6.00 ( M )
MINOR AXIS COLUMN EFFECTIVE HEIGHT   LY  =     6.00 ( M )

TRIAL COLUMN SECTION PROPERTIES INPUT :
----------------------------------------
SERIAL SIZE OF SECTION  : 203 x 203 UB 86
SECTION OVERALL DEPTH               D   =   209.0 ( mm )
SECTION GROSS X-AREA                Ag  =   110.0 (cm^2)
THICKNESS OF FLANGE                 T   =    20.5 ( mm )
THICKNESS OF WEB                    t   =    10.9 ( mm )
RADIUS OF GYRATION                  Rxx =     9.3 ( cm )
RADIUS OF GYRATION                  Ryy =     5.3 ( cm )
PLASTIC MODULUS ABOUT MAJOR AXIS    Spx =   979.0 (cm^3)
PLASTIC MODULUS ABOUT MINOR AXIS    Spy =   456.0 (cm^3)
ELASTIC MODULUS ABOUT MINOR AXIS    Zy  =   299.0 (cm^3)
YIELD STRENGTH OF COLUMN MATERIAL   Py  =   265.0 (N/mm2)
-------------------------------------------------------

STEEL COLUMN MEMBER DESIGN      -  OUTPUT RESULTS  -
===========================================================
VALUE OF (FC)  =    +68.18    AND  (PC)  =  +104.94
VALUE OF (MCX) =   +259.44    AND  (MCY) =  +120.84
VALUE OF (MeqX)=    +23.65    AND  (MeqY)=   +15.00
VALUE OF (MB)  =   +198.29

STRESS RATIOS COMPUTED AS FOLLOWS :
------------------------------------
RATIO OF          FC/PY         [C1]   =  +0.257
RATIO OF          FC/PC         [C3]   =  +0.650
RATIO OF          MX1/MCX       [X1]   =  +0.212
RATIO OF          MY1/MCY       [Y1]   =  +0.124
RATIO OF          MX2/MCX       [X2]   =  +0.212
RATIO OF          MY2/MCY       [Y2]   =  +0.124
RATIO OF          MeqvX/MB      [X3]   =  +0.119
RATIO OF          MeqvY/(Py*Zy) [Y3]   =  +0.189

LOCAL CAPACITY CHECK
--------------------
     AT TOP    OF COLUMN,   STRESS RATIO =  +0.593
     AT BOTTOM OF COLUMN,   STRESS RATIO =  +0.593

OVERALL BUCKLING CHECK,     STRESS RATIO =  +0.958
===========================================================
Ok
```

11.6 Column-base design program

11.6.1 Program discussion

Three load cases for rectangular bases are considered in this program. They are:

Type (1)—Axially loaded rectangular column bases.
Type (2)—Eccentrically loaded column bases with no tension in the holding-down bolts.
Type (3)—Eccentrically loaded column bases with tension in the holding-down bolts

11.6.2 Program "C-BASE" source listing

```
100 CLS    :REM COLUMN BASE DESIGN
110 PRINT "*****************************************"
120 PRINT "*    COLUMN BASE DESIGN TO B.S. 5950    *"
130 PRINT "*****************************************"
140 PRINT :PRINT
150 PRINT "3 TYPE OF COLUMN BASES CONSIDERED HERE "
160 PRINT "-------------------------------------------------------"
170 PRINT
180 PRINT "TYPE (1) - AXIALLY LOADED RECTANGULAR COLUMN BASE"
190 PRINT "TYPE (2) - ECCENTRICALLY LOADED RECTANGULAR COLUMN BASE"
200 PRINT "         o  ( HOLDING-DOWN BOLTS     UNDER TENSION )"
210 PRINT "         o  ( HOLDING-DOWN BOLTS NOT UNDER TENSION )"
220 PRINT "-------------------------------------------------------"
230 PRINT
240 INPUT "YOUR CHOICE OF COLUMN BASE";CHOICE
250 PRINT
260 INPUT "REFERENCE TITLE ( <40 CHARACTERS )";T$
270 PRINT  :CLS
280 ON CHOICE GOTO 300,1000
290 PRINT
300 PRINT "======================================================"
310 PRINT "#    AXIALLY LOADED  -  RECTANGULAR COLUMN BASE    #"
320 PRINT "======================================================"
330 PRINT
340 INPUT "REQUIREMENT ?  ( 1 = CHECKING, 2 = DESIGN PLATE SIZE )";REQ
350 PRINT
360 IF REQ<>1 THEN 380
370 INPUT "BASE PLATE OVERALL LENGTH & WIDTH  D, B  ( mm ) =";LP,BP
380 INPUT "COLUMN EFFECTIVE SIZE              d, b  ( mm ) =";D2,B2
390 IF REQ=1 THEN 410
400 INPUT "BASE PLATE INTENDED ASPECT RATIO   R = B/L     =";R
410 INPUT "AXIALLY APPLIED DEAD LOAD ON BASE  Pg  ( KN ) =";PG
420 INPUT "AXIALLY APPLIED LIVE LOAD ON BASE  Pk  ( KN ) =";PK
430 INPUT "CONCRETE BASE CUBE STRENGTH AT 28 DAYS  (N/mm2)=";FCU
440 INPUT "MATERIAL STRENGTH OF STEEL PLATE   Pyg (N/mm2)=";PYG
450 REM PRINT
460 FB=.4*FCU
470 PU=1.4*PG+1.6*PK
480 IF REQ=1 THEN 550
490 L=(PU*1000/R/FB)^.5   :B=R*L
500 TEMPL=L/100   :TEMPB=B/100
510 TEMPL=TEMPL+.99
520 TEMPB=TEMPB+.99
530 LP=INT(TEMPL)*100
540 BP=INT(TEMPB)*100
550 WU=(PU*1000)/(LP*BP)
560 AAP=(LP-D2)/2   :BBP=(BP-B2)/2
570 IF AAP<BBP THEN 590
580 AT=AAP   :BT=BBP   :GOTO 600
590 AT=BBP   :BT=AAP
600 TH=(2.5*WU*(AT^2-.3*BT^2)/PYG)^.5
610 RP=BP/LP
620 GOSUB 2000
630 PRINT "-------------------------------------------------------"
640 PRINT "#  RECTANGULAR COLUMN BASE PLATE :  - RESULTS -    #"
650 PRINT "-------------------------------------------------------"
660 PRINT
670 PRINT USING"MINIMUM THICKNESS OF PLATE        T = ##.### ( mm ) ";TH
680 PRINT USING"LIMITING  CONC. BEARING PRESSURE Fb= ###.## (N/mm2)";FB
690 PRINT USING"ACTUAL CONCRETE BEARING PRESSURE Wu= ###.## (N/mm2)";WU
700 PRINT "-------------------------------------------------------"
710 END
1000 CLS   :REM  ECCENTRICALLY LOADED RECTANGULAR COLUMN BASE
1010 PRINT "======================================================"
1020 PRINT "#    ECCENTRICALLY LOADED RECTANGULAR COLUMN BASE   #"
1030 PRINT "======================================================"
1040 PRINT
1050 INPUT "BASE PLATE OVERALL LENGTH AND WIDTH D,B ( mm ) =";D,B
1060 INPUT "COLUMN EFFECTIVE SIZE              d,b  ( mm ) =";D2,B2
1070 INPUT "AXIALLY APPLIED DEAD LOAD ON BASE  Pg  ( KN ) =";PG
1080 INPUT "AXIALLY APPLIED LIVE LOAD ON BASE  Pk  ( KN ) =";PK
1090 INPUT "ECCENTRICITY OF ALL APPLIED LOADS  Ec  ( mm ) =";EC
1100 INPUT "HORIZONTAL IMPOSED SHEAR FORCE     Hs  ( mm ) =";HS
1110 INPUT "CONCRETE BASE CUBE STRENGTH AT 28 DAYS (N/mm2)=";FCU
1120 INPUT "HOLDING-DOWN BOLTS TENSILE AREA    As ( mm^2)=";A1
1130 INPUT "BOLT DIAMETER & EDGE DISTANCE      Do,a ( mm ) =";DD,A2
1140 INPUT "BOLT TYPE  ( 1=GRADE 4.6, 2=GRADE 8.8 )        =";TYPE
1150 INPUT "MATERIAL STRENGTH OF STEEL PLATE   Pyg (N/mm2)=";PYG
```

```
1160 PU=1.4*PG+1.6*PK
1170 MU=PU*EC/1000
1180 HU=1.6*HS
1190 FB=.4*FCU   :N1=15   :RP=B/D
1200 IF TYPE=2 THEN 1230
1210 IF TYPE<>1 THEN 1240
1220 PS=160   :PT=195   :PBB=435   :GOTO 1250
1230 PS=375   :PT=450   :PBB=970   :GOTO 1250
1240 INPUT "STRENGTH VALUES FOR BOLTS USED  Ps, Pt, Pbb =";PS,PT,PBB
1250 D1=D-A2
1260 D6=D1/6
1270 IF (EC-A2/2)>D6 THEN 1390
1280 T=0!  :R1=0!  :R2=0!  :R3=0!
1290 XN=3*(D/2-EC)
1300 FC1=PU*1000/(B*D)
1310 FC2=MU*6000000!/(B*D^2)
1320 FC=FC1+FC2
1330 FCO=FC1-FC2
1340 IF XN<D THEN 1630
1350 XN=D*(1+FC1/FC2)
1360 IF XN<10000 THEN 1630
1370 XN=10000
1380 GOTO 1630
1390 K1=EC-D/2
1400 K2=D/2-A2+EC
1410 K3=6*N1*A1*K2/B
1420 AO=3*K1
1430 BO=K3
1440 CO=-(K3*D1)
1450 REM SOLVE CUBIC EQUATION
1460 PX=BO/3-AO^2/9
1470 QX=CO-AO*BO/3+2*AO^3/27
1480 QO=3*PX
1490 RO=QX
1500 YO=50+AO/3
1510 Y1=(-(QO*YO+RO))^(1/3)
1520 IF ABS(Y1-YO)<1 THEN 1540
1530 YO=Y1  :GOTO 1510
1540 XN=Y1-AO/3
1550 DN=XN
1560 T=PU*(DN+AO)/(3*D1-DN)
1570 FC=2*(PU+T)*1000/B/DN
1580 FQ=HU*1000/(2*A1/.8)
1590 FT=T*1000/A1
1600 R1=FQ/PS
1610 R2=FT/PT
1620 R3=R1+R2
1630 WU=FC  :LP=D   :BP=B
1640 GOSUB 2000
1650 PRINT
1660 PRINT USING"ECCENTRICITY OF APPLIED LOAD     Ec = ###### ( mm ) ";EC
1670 PRINT USING"ULTIMATE MOMENT             Pu*Ec = ###### (KN-M) ";MU
1680 PRINT USING"ULTIMATE HORIZONTAL SHEAR       Hu = ###.## ( KN ) ";HU
1690 PRINT USING"TOTAL HOLDING-DOWN BOLT AREA    As = ###### (mm^2) ";A1
1700 PRINT USING"HOLDING-DOWN BOLT DIAMETER      Dd = ###### ( mm ) ";DD
1710 PRINT USING"BOLT TYPE (1=GRADE 4.6,2=GRADE 8.8)= ###### (TYPE) ";TYPE
1720 PRINT USING"EDGE DISTANCE OF BOLT CENTRE     a = ###### ( mm ) ";A2
1730 PRINT
1740 PRINT "---------------------------------------------------"
1750 PRINT "#   RECTANGULAR COLUMN BASE PLATE :   - RESULTS -   #"
1760 PRINT "==================================================="
1770 PRINT
1780 PRINT USING "NEUTRAL AXIS DISTANCE           Xn = ####.# ( mm ) ";XN
1790 PRINT USING "MAX. HOLDING-DOWN BOLT TENSION Tb =+####.# ( KN ) ";T
1800 PRINT USING"LIMITING CONC. BEARING PRESSURE Fb = ###.## (N/mm2)";FB
1810 PRINT USING "MAXIMUM CONCRETE PRESSURE       Fc = ####.# (N/mm2)";FC
1820 PRINT USING "BOLT; STRESS/STRENGTH RATIO Fs/Ps =+####.##";R1
1830 PRINT USING "BOLT; STRESS/STRENGTH RATIO Ft/Pt =+####.##";R2
1840 PRINT USING "COMBINED  STRESS/STRENGTH  RATIO  =+####.##";R3
1850 PRINT
1860 PRINT "BASE PLATE MIN. THICKNESS"
1870 PRINT "------------------------------------------------"
1880 C2=.5*(D-D2)
1890 FC4=FC*(XN-C2)/XN
1900 FC5=FC-FC4
1910 M2=.5*FC4*C2^2+FC5*C2^2/3
1920 T2=SQR(6*M2/PYG)
1930 QCR=.6*PYG
1940 T3=(FC4*C2+.5*FC5*C2)/QCR
1950 PRINT USING "PLATE MAX. MOMENT               Mm = ####.# (KN-M/M)";M2/1000
1960 PRINT USING "MIN. PLATE THICKNESS (BENDING)     = ###.## ( mm ) ";T2
```

```
1970 PRINT USING "MIN. PLATE THICKNESS (SHEAR  )   = ###.## ( mm ) ";T3
1980 PRINT "=================================================="
1990 END
2000 CLS    :REM SUBROUTINE TO PRINT COMMON DATA
2010 PRINT "REFERENCE TITLE : "; :PRINT T$
2020 PRINT
2030 PRINT "--------------------------------------------------"
2040 PRINT "#    RECTANGULAR COLUMN BASE PLATE :    - INPUT -   #"
2050 PRINT "=================================================="
2060 PRINT
2070 PRINT USING"BASE PLATE OVERALL LENGTH        L = ###### ( mm ) ";LP
2080 PRINT USING"BASE PLATE OVERALL WIDTH         B = ###### ( mm ) ";BP
2090 PRINT USING"BASE PLATE ASPECT RATIO          R = ###.##";RP
2100 PRINT USING"COLUMN EFFECTIVE DEPTH           d = ###### ( mm ) ";D2
2110 PRINT USING"COLUMN EFFECTIVE WIDTH           b = ###### ( mm ) ";B2
2120 PRINT USING"AXIALLY APPLIED ULTIMATE LOAD    Pu= ###### ( KN ) ";PU
2130 PRINT USING"CONCRETE 28 DAYS CUBE STRENGTH  Fcu= ###.## (N/mm2)";FCU
2140 PRINT USING"MATERIAL STRENGTH, STEEL PLATE  Pyp= ###.## (N/mm2)";PYG
2150 PRINT
2160 RETURN
```

11.6.3 Program "C-BASE" statements and functions

The source listing of the column base design programme is given in Section 11.6.2. The program statements and functions are explained as follows:

Statements	Functions
100– 140	Clear screen and display title.
150– 260	Display menu of column base design, input "Choice" and reference title.
270– 290	Clear screen and route selected choice to the appropriate location.
300– 450	For an axially loaded stanchion base, input the necessary data.
460– 500	Computation of plate length L and breadth B.
510– 540	Round off plate size to the upper 100 mm.
550– 610	Compute bearing stress and the required minimum base plate thickness.
620	Call Subroutine to print out the common input data.
630– 710	Print all output results and end.
1000–1150	For an eccentrically loaded stanchion base, input all the required data.
1160–1240	Calculation of P_u, M_u, H_u and select strength value for the bolt type to be used.
1250–1440	Determine the neutral axis location; if outside the base then compute stresses for the case of no tension in holding-down bolts.
1450–1630	If neutral axis is inside the base, set up cubic equation for the tension case and determine all the values of stresses.
1640	Call Subroutine to print out the common input data.
1650–1730	Print out remaining input data.
1740–1990	Output all computed results and end.
2000–2160	Subroutine to print out common input data.

11.6.4 Sample run of the column base design program

As an example to show program execution, an eccentrically loaded column base with tension in the holding-down bolts (type) 3 is chosen. The detailed arrangement is shown in Figure 11.6 with the material strengths indicated. The applied loads are as follows:

(1) Vertical dead and imposed loads, each   = 152 kN
(2) Eccentricity of applied loads   = 630 mm
(3) Horizontal imposed shear force at the base   = 10 kN

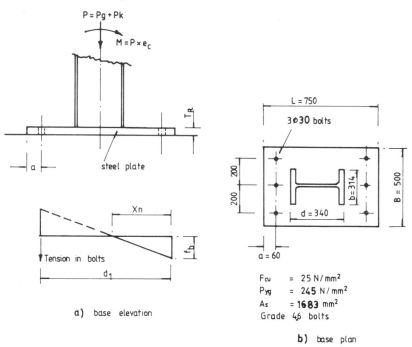

a) base elevation

b) base plan

Figure 11.6 Example of column base design

The input data for a sample column base design are prepared and fed into the program. The output results include the neutral axis distance, X_n, the holding-down bolt tension T_b, the maximum applied bearing pressure F_c and the stress in the bolts. The complete computer output is given below:

```
RUN
*******************************************
*    COLUMN BASE DESIGN TO B.S. 5950      *
*******************************************

3 TYPE OF COLUMN BASES CONSIDERED HERE
-----------------------------------------------------------

TYPE (1) - AXIALLY LOADED RECTANGULAR COLUMN BASE
TYPE (2) - ECCENTRICALLY LOADED RECTANGULAR COLUMN BASE
          o   ( HOLDING-DOWN BOLTS      UNDER TENSION )
          o   ( HOLDING-DOWN BOLTS NOT UNDER TENSION )
-----------------------------------------------------------

YOUR CHOICE OF COLUMN BASE? 2

REFERENCE TITLE ( <40 CHARACTERS )? COLUMN BASE EXAMPLE

===========================================================
#    ECCENTRICALLY LOADED RECTANGULAR COLUMN BASE   #
===========================================================
```

```
BASE PLATE OVERALL LENGTH AND WIDTH D,B ( mm ) =? 750,500
COLUMN EFFFCTIVE SIZE                d,b ( mm ) =? 340,314
AXIALLY APPLIED DEAD LOAD ON BASE    Pg  ( KN ) =? 152
AXIALLY APPLIED LIVE LOAD ON BASE    Pk  ( KN ) =? 152
ECCENTRICITY OF ALL APPLIED LOADS    Ec  ( mm ) =? 630
HORIZONTAL IMPOSED SHEAR FORCE       Hs  ( mm ) =? 10
CONCRETE BASE CUBE STRENGTH AT 28 DAYS (N/mm2)=? 25
HOLDING-DOWN BOLTS TENSILE AREA      As  ( mm^2)=? 1683
BOLT DIAMETER & EDGE DISTANCE      Do,a ( mm ) =? 30,60
BOLT TYPE  ( 1=GRADE 4.6, 2=GRADE 8.8 )    =? 1
MATERIAL STRENGTH OF STEEL PLATE    Pyg (N/mm2)=? 245


----------------------------------------------------------
#   RECTANGULAR COLUMN BASE PLATE :   - INPUT -   #
==========================================================

BASE PLATE OVERALL LENGTH         L =    750 ( mm )
BASE PLATE OVERALL WIDTH          B =    500 ( mm )
BASE PLATE ASPECT RATIO           R =    0.67
COLUMN EFFECTIVE DEPTH            d =    340 ( mm )
COLUMN EFFECTIVE WIDTH            b =    314 ( mm )
AXIALLY APPLIED ULTIMATE LOAD     Pu=    456 ( KN )
CONCRETE 28 DAYS CUBE STRENGTH    Fcu=  25.00 (N/mm2)
MATERIAL STRENGTH, STEEL PLATE    Pyp= 245.00 (N/mm2)


ECCENTRICITY OF APPLIED LOAD      Ec =    630 ( mm )
ULTIMATE MOMENT             Pu*Ec =    287 (KN-M)
ULTIMATE HORIZONTAL SHEAR         Hu =  16.00 ( KN )
TOTAL HOLDING-DOWN BOLT AREA      As =   1683 (mm^2)
HOLDING-DOWN BOLT DIAMETER        Dd =     30 ( mm )
BOLT TYPE (1=GRADE 4.6,2=GRADE 8.8)=     1 (TYPE)
EDGE DISTANCE OF BOLT CENTRE      a =     60 ( mm )


----------------------------------------------------------
#  RECTANGULAR COLUMN BASE PLATE :  - RESULTS -   #
==========================================================

NEUTRAL AXIS DISTANCE            Xn =  315.2 ( mm )
MAX. HOLDING-DOWN BOLT TENSION Tb = +280.7 ( KN )
LIMITINC CONC. BEARINC PRESSURE Fb -  10.00 (N/mm2)
MAXIMUM CONCRETE PRESSURE        Fc =    9.3 (N/mm2)
BOLT; STRESS/STRENGTH RATIO Fs/Ps =   +0.02
BOLT; STRESS/STRENGTH RATIO Ft/Pt =   +0.86
COMBINED  STRESS/STRENGTH  RATIO  =   +0.88

BASE PLATE MIN. THICKNESS
----------------------------------------------------
PLATE MAX. MOMENT                Mm =  153.9 (KN-M/M)
MIN. PLATE THICKNESS (BENDING)      =  61.39 ( mm )
MIN. PLATE THICKNESS (SHEAR )       =   8.80 ( mm )
==========================================================
Ok
```

11.7 Computer programs disclaimer

The inclusion of the above computer programs is for the convenience of readers. Because of the diversity of conditions and hardware under which these programs may be used, neither the publisher nor the author provide warranty for their use to solve design problems. The user is advised to test the programs throughly before relying on them.

Steelwork detailing

12.1 Drawings

Drawings are the means by which the requirements of architects and engineers are communicated to the fabricators and erectors, and must be presented in an acceptable way. Detailing is given for selected structural elements.

Drawings are needed to show the layout and to describe and specify the requirements of a building. They show the location, general arrangement and details for fabrication and erection. They are also used for estimating quantities and cost and for making material lists for ordering materials.

Sufficient information must be given on the designer' sketches for the draughtsman to make up the arrangement and detail drawings. A classification of drawings is set out below:

Site or location plans

These show the location of the building in relation to other buildings, site boundaries, streets, roads, etc.

General arrangement

This consists of plans, elevations and sections to set out the function of the building. These show locations and leading dimensions for offices, rooms, work areas, machinery, cranes, doors, services, etc. Materials and finishes are specified.

Marking plans

These are the framing plans for the steel-frame building showing the location and mark numbers for all steel members in the roof, floors and various elevations.

Foundation plans

These show the setting out for the column bases and holding-down bolts and should be read in conjunction with detail drawings of the foundations.

Sheeting plans

These show the arrangement of sheeting and cladding on building.

Key plan

If the work is set out on various drawings, a key plan may be provided to show the portion of work covered by the particular drawing.

Detail drawings

These show the details of structural members and give all information regarding materials, sizes, welding, drilling, etc. for fabrication. The mark number of the detail refers to the number on the marking plan.

Detail drawings and marking plans will be dealt with here.

12.2 General recommendations

12.2.1 Scales, drawing sizes and title blocks

The following scales are recommended:

Site, location, key plans	1:500
	1:200
General arrangement	1:200
	1:100
	1:50
Marking plans	1:200
	1:100
Detail drawings	1:25
	1:10
	1:5
Enlarged details	1:5
	1:2
	1:1

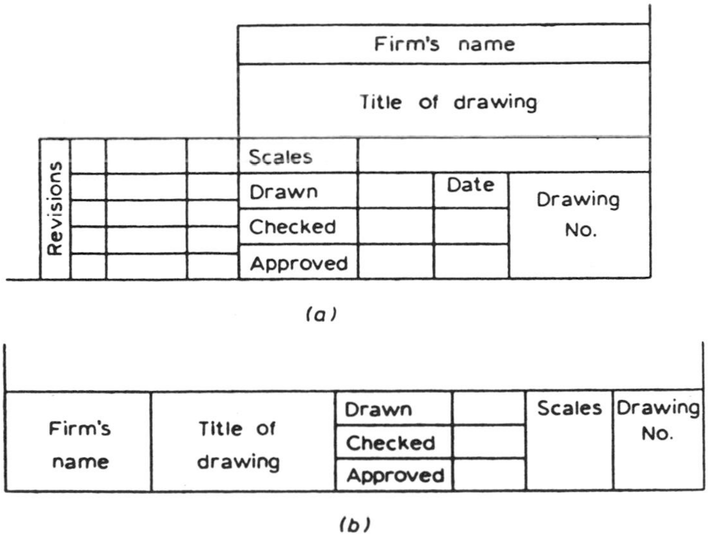

(a)

(b)

Figure 12.1 Typical title blocks

The following drawing sizes are used:

A4 210 × 297—Sketches
A3 297 × 420—Details
A2 420 × 595—General arrangement, details
A1 594 × 841—General arrangement, details
A0 841 × 1189—General arrangement, details

Title blocks on drawings vary to suit the requirements of individual firms and authorities. Typical title blocks are shown in Figure 12.1.

Materials lists can either be shown on the drawing or on separate A4 size sheets. These generally give the following information:

Item or mark number
Description
Material
Number off
Weight
Etc.

12.2.2 Lines, sections, dimensions and lettering

Recommendations regarding lines, sections and dimensions are shown in Figure 12.2.

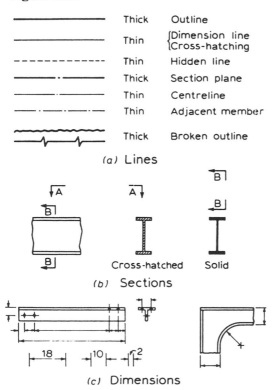

Figure 12.2 Recommendations for lines, sections and dimensions

12.3 Steel sections

Rolled and formed steel sections are represented on steelwork drawings as set out in Figure 12.3. The first two figures indicate the size of section, (for example, depth and breadth). The last figure indicates the weight in kg/m for beams, columns, channels and tees. For angles and hollow sections the last figure gives the thickness of steel. With channels or angles the name may be written or the section symbols used as shown. Built-up sections can be shown either by:

(1) True section for large scale views

or

(2) Diagrammatically by heavy lines with the separate plates and sections separated for clarity for small-scale views. Here only the depth and breadth of the section may be true to scale.

Section		Reference	Example
Universal beam	I	UB	610 x178 x 91 kg/m UB
Universal column	I	UC	203 x 203 x89 kg/m UC
Joist	I	RSJ Joist	203 x 102 x 25.33 kg/m Joist
Channel	C	Channel	254 x 76 x 28.29 kg/m C
Angle	L	Angle	150 x 75 x 10 L
Tee	T	Tee	178 x 203 x 37 kg/m Struct.Tee
Rectangular hollow section	⬚	RHS	150.4 x100 x 6.3 RHS
Circular hollow section	◯	CHS	76.1 O.D. x 5 CHS

Figure 12.3 Representation of rolled and formed steel sections

These two cases are shown in Figure 12.4. The section is often shown in the middle of a member inside the break lines in the length, as shown in (c). This saves having to draw a separate section.

Beams may be represented by lines, and columns by small-scale sections in heavy lines, as shown in Figure 12.5. The mark numbers and sizes are written on the respective members. This system is used for marking plans.

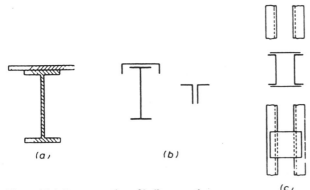

(a)   (b)

(c)

Figure 12.4 Representation of built-up sections

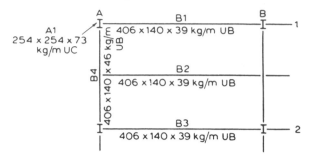

Figure 12.5 Representation of beams and columns

12.4 Grids and marking plans

Marking plans for single-storey buildings present no difficulty. Members are marked in sequence as follows:

Columns A1, A2, ... (See grid referencing below)
Trusses T1, T2, ...
Crane girders CG1, CG2, ...
Purlins P1, P2, ...
Sheeting rails SR1, SR2, ...
Bracing B1, B2, ...
Gable columns GS1, GS2, ...
etc.

Various numbering systems are used to locate beams and columns in multistorey buildings. Two schemes are outlined below:

(1) In plan, the column grid is marked A, B, C, ... in one direction and 1, 2, 3, ... in the direction at right angles. Columns are located A1, C2, ...
Floors are numbered A, B, C, ... for ground, first, second, ..., respectively.
Floor beams (for example, on the second floor) are numbered B1, B2, ...
Column lengths are identified: for example, A4–B is the column on grid intersection A4, length between second and third floors.

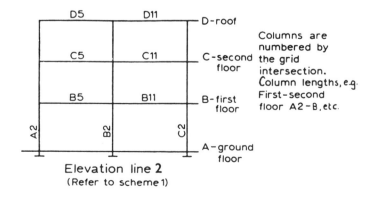

Elevation line 2
(Refer to scheme 1)

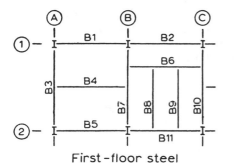

Scheme 1

Beams are numbered
consecutively
Prefix indicates floor

First-floor steel

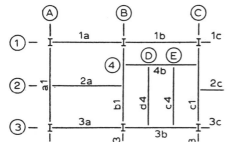

Scheme 2

Beams are numbered
on grid lines
e.g. line 1 – 1a,1b, etc.
First floor requires
prefix B, e.g. B-1a,etc.

Floor steel

Figure 12.6

(2) A grid line is required for each beam.
    The columns are numbered by grid intersections as above.
    The beams are numbered on the grid lines with a prefix letter to give the floor if required. For example,
    Second floor——grid line 1——C-1a, C-1b,——
    Second floor——grid line B——C-b1, C-b2,——

The systems are shown in Figure 12.6. The section size may be written on the marking plan.

12.5 Bolts

12.5.1. Specification

The types of bolts used in steel construction are:

Ordinary or black bolts
Friction-grip bolts

The British standards covering these bolts are:

BS 4190: ISO Metric Black Hexagon Bolts, Screws and Nuts
BS 4395: Part 1: High Strength Friction-Grip Bolts: General Grade
BS 4604: Part 1: The Use of High Strength Friction-Grip Bolts in Structural
              Steelwork: General Grade

The strength grade designation should be specified. BS 5950 gives the strengths for ordinary bolts for grades 4.6 and 8.8. Minimum shank tensions for friction-grip bolts are given in BS 4604. The nominal diameter is given in millimetres. Bolts are designated as M12, M16, M20, M22, M24, M27, M30, etc., where 12, 16, etc. is the diameter in millimetres. The length under the head in millimetres should also be given.

Examples in specifying bolts are as follows:

4 No. 16-mm dia. (or M16) black hex. hd. (hexagon head) bolts, strength grade 4.6 × 40 mm length
20 No. 24-mm dia. friction-grip bolts × 75 mm length

The friction-grip bolts may be abbreviated FG. The majority of bolts may be covered by a blanket note. For example:

All bolts M20 black hex. hd.
All bolts 24-mm dia. FG unless otherwise noted.

12.5.2 Drilling

The following tolerances for drilling ⌐re used:

Ordinary bolts—holes to have a maximum of 2-mm clearance for bolt diameters up to 24 mm and 3 mm for bolts of 24-mm diameter and over.

For friction-grip bolts, holes are drilled the same as set out above for ordinary bolts. Hole diameters are given in Table 35 of BS 5950: Part 1.

Drilling may be specified on the drawing by notes as follows:

All holes drilled 22-mm dia. for 20-mm dia. ordinary bolts
All holes drilled 26-mm dia., unless otherwise noted.

12.5.3 Designating and dimensioning

The representation for bolts and holes in plan and elevation on steelwork drawings is shown in Figure 12.7(a). Some firms adopt different symbols for showing different types of bolts and to differentiate between shop and field bolts. If this system is used, a key to the symbols must be given on the drawing.

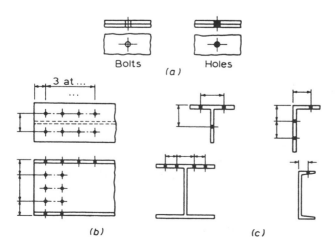

Bolts   Holes
(a)

(b)   (c)

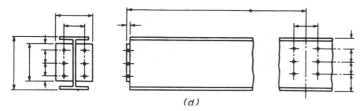

(d)

Figure 12.7

Gauge lines for drilling for rolled sections are given in the *Structural Steelwork Handbook*. Dimensions are given for various sizes of section, as shown in Figure 12.7(c). Minimum edge and end distances were discussed in Section 4.2.2 above.

Details must show all dimensions for drilling, as shown in Figures 12.7(b) and (d). The holes must be dimensioned off a finished edge of a plate or the back or end of a member. Holes are placed equally about centre lines. Sufficient end views and sections as well as plans and elevations of the member or joint must be given to show the location of all holes, gussets and plates.

12.6 Welds

As set out in Chapter 4, the two types of weld are butt and fillet.

12.6.1 Butt welds

The types of butt weld are shown in Figure 12.8 with the plate edge preparation and the fit-up for making the weld. The following terms are defined:

T = thickness of plate
g = gap between the plates
R = root face
a = minimum angle

Values of the gap and root face vary with the plate thickness, but are of the order of 1–4 mm. The minimum angle between prepared faces is generally 50–60 degrees for V preparation and 30–40 degrees for U preparation. For thicker plates the U preparation gives a considerable saving in the amount of weld metal required. The student should consult the complete details given in BS 499: 1965: Part 2, *Welding Terms and Symbols*.

Welds may be indicated on drawings by symbols from Table 1 of BS 499 and these are shown in Figure 12.9. Using these symbols, butt welds are indicated on a drawing as shown in Figure 12.10(a). Reference should be made to BS 499 for a complete set of examples using these symbols.

The weld name may be abbreviated: for example, DVBW for double V butt weld. An example of this is shown in Figure 12.8(b). Finally, the weld may be listed by its full description and an enlarged detail given to show the edge preparation and fit-up for the plates. This method is shown in Figure 12.8(c). Enlarged details should be given in cases where complicated welding is required. Here, detailed instructions from a welding engineer may be required and these should be noted on the drawing.

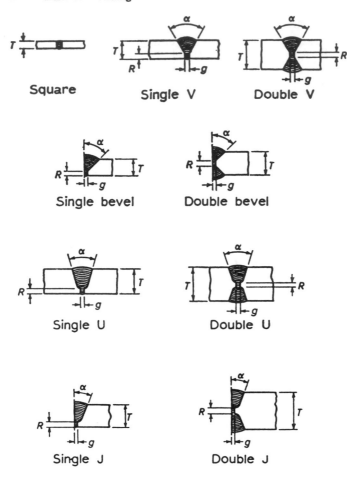

Figure 12.8 Butt welds

Fillet	◁
Square butt weld	∏
Single V butt weld	▽
Double V butt weld	✕
Single U butt weld	∪
Double U butt weld	𝟠
Single bevel weld	⌐
Double bevel weld	⌐
Single J weld	℧
Double J weld	𝔅

Figure 12.9 Symbols for welds

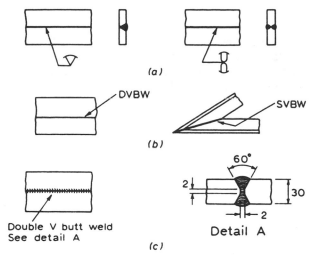

Double V butt weld
See detail A

Detail A

Figure 12.10

12.6.2 Fillet welds

These welds are triangular in shape. As set out in Chapter 4, the size of the weld is specified by the leg length. Welds may be indicated symbolically, as shown in Figure 12.11(a). BS 499 should be consulted for further examples. The weld size and type may be written out in full or the words 'fillet weld' abbreviated, (for example, 6 mm FW). If the weld is of limited length its exact location should be shown and dimensioned. Intermittent weld can be shown by writing the weld size, then two figures which indicate length and space between welds. These methods are shown in Figure 12.11(b). The welds can also be shown and specified by notes, as on the plan view in Figure 12.11(b). Finally, a common method of showing fillet welds is given in Figure 12.11(c), where thickened lines are used to show the weld.

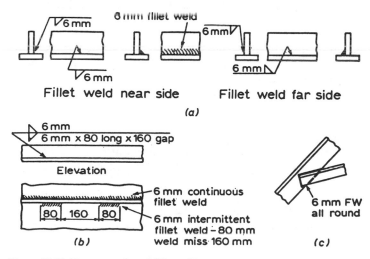

Figure 12.11 Representation of fillet welds

12.7 Beams

Detailing of beams, purlins and sheeting rails is largely concerned with showing the length, end joints, welding and drilling required. A typical example is shown in Figure 12.12. Sometimes it is necessary to show the connecting member and this may be shown by chain dash lines, as shown in Figure 12.12(b).

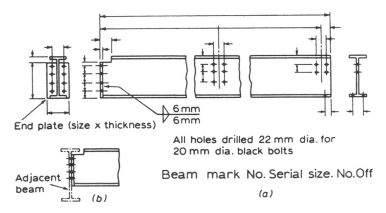

End plate (size x thickness)

6 mm
6 mm

Adjacent beam

(b)

All holes drilled 22 mm dia. for 20 mm dia. black bolts

Beam mark No. Serial size. No.Off

(a)

Figure 12.12 Beam details

12.8 Plate girders

The detail drawing of a plate girder shows the girder dimensions, flange and web plate sizes, sizes of stiffeners and end plates, their location and the details for drilling and welding. Any special instructions regarding fabrication should be given on the drawing. For example, preheating may be required when welding thick plates or 2 mm may require machining off flame-cut edges to reduce the likelihood of failure by brittle fracture or fatigue on high yield-strength steels.

Generally all information may be shown on an elevation of the girder together with sufficient sections to show all types of end plates, intermediate and load-bearing stiffeners. The elevation would show the location of stiffeners, brackets and location of holes and the sections complete this information. Plan views on the top and bottom flange are used if there is a lot of drilling or other features best shown on such a view. The draughtsman decides whether such views are necessary.

Part sectional plans are often used to show stiffeners in plan view. Enlarged details are frequently made to give plate weld edge preparation for flange and web plates, splices and for load-bearing stiffeners that require full-strength welds.

Notes are added to cover drilling, welding and special fabrication procedures, as stated above, Finally, the drawing may contain a material list giving all plate sizes required in the girder. Typical details for a plate girder are shown in Figure 12.13.

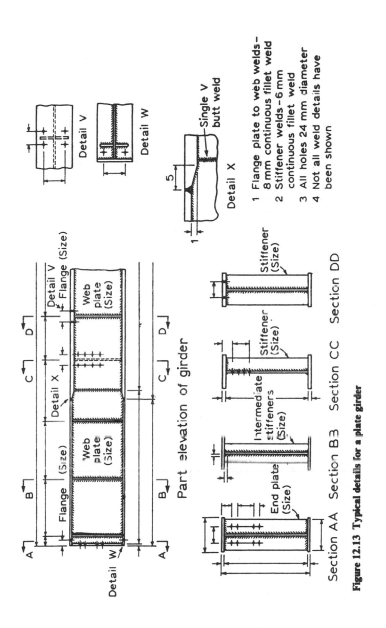

Figure 12.13 Typical details for a plate girder

12.9 Columns and bases

A typical detail of a column for a multi-storey building is shown in Figure 12.14. The bottom column length with base slab, drilling for floor beams and splice details is shown.

A compound crane column for a single-storey industrial building is shown in Figure 12.15. The crane column is a built-up section and the roof portion is a universal beam. Details at the column cap, crane girder level and base are shown.

12.10 Trusses and lattice girders

The rolled sections used in trusses are small in relation to the length of the members. Several methods are adopted to show the details at the joints. These are:

(1) If the truss can be drawn to a scale of 1 in 10, then all major details can be shown on the drawing of the truss.
(2) The truss is drawn to a small scale, 1 in 25, and then separate enlarged details are drawn for the joints to a scale of 1 in 10 or 1 in 5.

Members should be designated by size and length; for example, where all dimensions are in millimetres:

100 × 75 × 10 Angle × 2312 long
100 × 50 × 4 RHS × 1310 long

On sloping members it is of assistance in fabrication to show the slope of the member from the vertical and horizontal by a small triangle adjacent to the member.

The centroidal axes of the members are used to set out the frame and the members should be arranged so that these axes are coincident at the nodes of the truss. If this is not the case, the eccentricity causes secondary stresses in the truss.

Full details are required for bolted or welded splices, end plates, column caps, etc. The positions of gauge lines for drilling holes are given in the *Structural Steelwork Handbook*. Dimensions should be given for edge distances, spacing and distance of holes from the adjacent node of the truss. Splice plates may be detailed separately.

Sometimes the individual members of a truss are itemized. In these cases, the separate members, splice plates, cap plates, etc. are given an item number. These numbers are used to identify the member or part on a material list.

The following figures are given to show typical truss detailing:

Figure 12.16 shows a portion of a flat roof truss with all the major details shown on the elevation of the truss. Each part is given an item number for listing.
Figure 12.17 shows a roof truss drawn to small scale where the truss dimensions, member sizes and lengths are shown. Enlarged details are given for some of the joints.

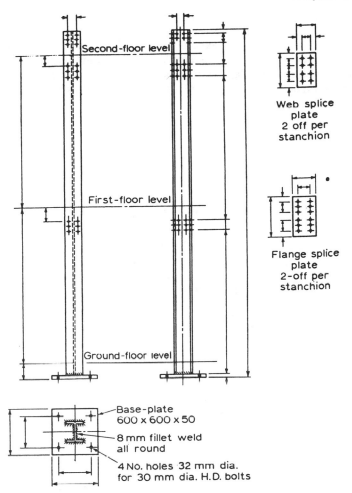

All holes 22 mm dia. for 20 mm dia. black bolts except as noted

Figure 12.14 Stanchion in a multi-storey building

12.11 Computer-aided drafting

Computer-aided drafting (CAD) is now being introduced into civil and structural drafting practice. In the near future it will replace much of the manual drafting work. CAD can save considerable man-hours in drawing preparation, especially where standard details are used extensively.

In computerized graphical systems the drawing is built up on the screen, which is divided into a grid. A menu gives commands (for example, line, circle, arc, text, dimensions, etc.). Data are inputted through the digitizer and keyboard. All drawings can be stored in mass storage devices from which they can be retrieved for subsequent additions or alterations. Updating of drawings using the CAD system can be accomplished with little effort. Standard details used frequently can either be drawn from a program library or created and stored in the user's library.

Overlay of drawings is a common feature in most CAD software. This allows repeated usage of common drawing templates (for example, grid lines, floor plans, etc.) and results in consierable time saving by reducing input time required. Texts and dimensions can be typed from the keyboard using appropriate character sizes.

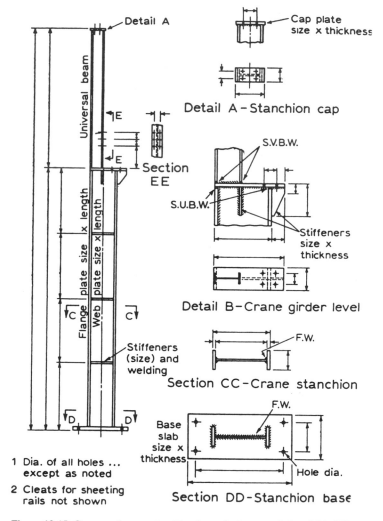

Figure 12.15 Compound crane stanchion for a single-storey industrial building

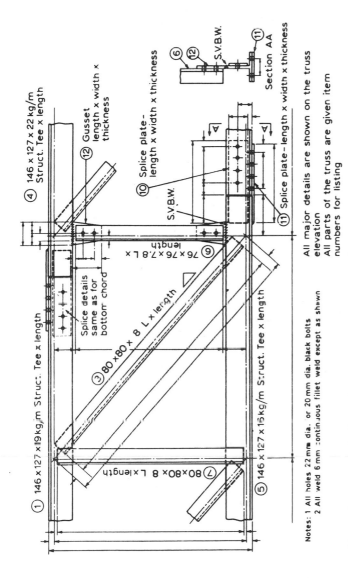

Figure 12.16 Portion of a flat roof truss

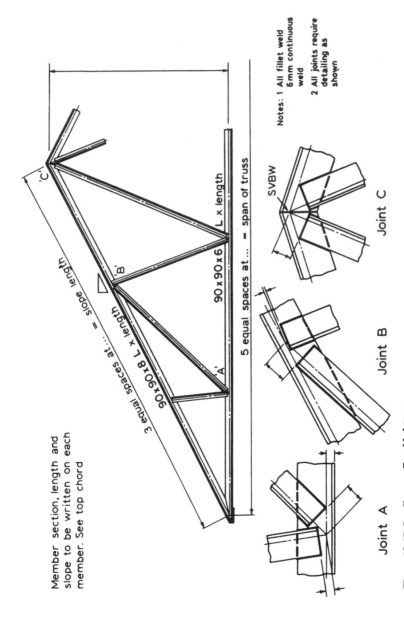

Member section, length and slope to be written on each member. See top chord

= slope length

3 equal spaces at.... L × length

90×90×8

90×90×6

5 equal spaces at.... = span of truss

L × length

SVBW

Notes: 1 All fillet weld 6 mm continuous weld

2 All joints require detailing as shown

Joint A

Joint B

Joint C

Figure 12.17 Small-span all-welded truss

13

Portal design

13.1 Design and construction

13.1.1 Portal type and structural action

The single-storey clear-span building is in constant demand for warehouses, factories and many other purposes. The clear internal appearance makes it much more appealing than a trussed roof building and it also requires less maintenance and heating. The portal may be of three-pinned, pinned-base or fixed-base construction as shown in Figure 13.1(a). The pinned-base portal is the most common type adopted because of the greater economy in foundation design over the fixed-base type. Only pinned-base portals are discussed here.

In-plane the portal resists the following loads by rigid frame action (see Figure 13.1(b)):

Dead and imposed loads acting vertically
Wind causing horizontal loads on the walls and generally
uplift loads on the roof slopes

In the longitudinal direction the building is of simple design and diagonal bracing is provided in the end bays to provide stability and resist wind load on the gable ends and wind friction on sides and roof (see Figure 13.1(d)).

13.1.2 Construction

The main features in modern portal construction shown in Figure 13.1 (c) are:

Columns — uniform universal beam section;

Rafters — universal beams with haunched ends usually of sections 30 – 40% lighter than the columns;

Eaves and ridge joints — site-bolted joints using grade 8.8 bolts where the haunched ends of the rafters provide the necessary lever arm for design. Local joint stiffening is required;

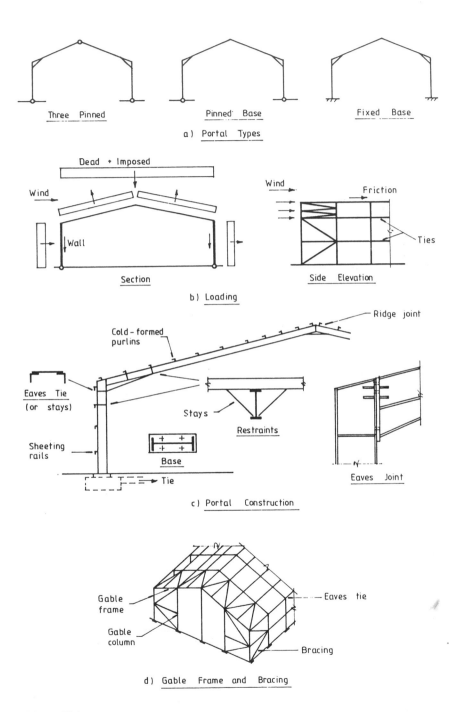

a) Portal Types

b) Loading

c) Portal Construction

d) Gable Frame and Bracing

Figure 13.1

Base — nominally pinned with two or four holding down bolts;

Purlins and sheeting rails — cold-rolled sections spaced at not greater than 1.75–2 m centres;

Stays from purlins and rails — these provide lateral support to the inside flange of portal members;

Gable frame — a braced (not a rigid) frame at the gable ends of the buildings;

Bracing — provided in the end bay in roof and walls;

Eaves and ridge ties — may be provided in larger-span portals though now replaced by stays from purlins or sheeting rails.

13.1.3 Design outline

The code states that either elastic or plastic design may be used. Plastic design gives the more economical solution and is almost universally adopted. The design process for the portal consists of:

Analysis — elastic or plastic;

Design of members taking account of flexural and lateral torsional buckling with provision of restraints to limit out-of-plane buckling;

Sway stability check in the plane of the portal;

Joint design with provision of stiffeners to ensure all parts are capable of transmitting design actions;

Serviceability check for deflection at eaves.

Procedures for elastic and plastic design are set out in BS 5950 and are discussed below.

13.1.4 Foundations

The pinned-base portal is generally adopted because it is difficult to ensure fixity without piling and it is more economical to construct. It is also advantageous to provide a tie through the ground slab to resist horizontal thrust due to dead and imposed load as shown in Figure 13.1(c).

13.2 Elastic design

13.2.1 Code provision

The provisions from BS 5950 are summarized as follows:

(1) Clause 5.2: Analysis is made using factored loads

(2) Clause 5.4.1: The capacity and buckling resistance of members are to be checked using Section 4 of the code.

(3) Clause 5.5.2: The stability of the frame should be checked using Section 2.4.2. of the code. This states that, in addition to designing for wind loads, a separate check is to be made for notional horizontal loads of either:
1% of the factored dead load, or
0.5% of the factored dead + imposed load
These loads are applied at eaves level and act with 1.4 × dead load + 1.3 × imposed load. They are not to be combined with wind loads.

13.2.2 Portal analysis

The most convenient manual method of analysis is to use formulae from the *Steel Designers Manual* (see Further reading at the end of this chapter). A general load case can be broken down into separate cases for which solutions are given and then these results are recombined. Computer analysis is the most convenient method to use, particularly for wind loads and load combinations. The output gives design actions and deflections.

The critical load combination for design is 1.4 × dead load + 1.6 × imposed load. The wind loads mainly cause uplift and the moments are generally in the opposite direction to those caused by the dead and imposed loads. The bending moment diagram for the dead and imposed load case is given in Figure 13.2. This shows the inside flange of the column and rafter near the eaves to be in compression and hence the need for lateral restraints in those areas.

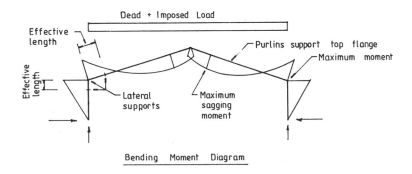

Figure 13.2 Bending moment diagram

As noted in Section 13.1.2, portals are constructed with haunched joints at the eaves. The primary purpose of the haunch is to provide the lever arm to enable the bolted site joints to be made. It is customary to neglect the haunch in elastic analysis. If the haunch is large in comparison with the rafter length more moment is attracted to the eaves and a more accurate solution is obtained if it is taken into consideration. This can be done by dividing the haunch into a number of parts and using the properties of the centre of each part in a computer analysis.

13.2.3 Effective lengths

Member design depends on the effective lengths of members in the plane and normal to the plane of the portal. Effective lengths control the design strengths for axial load and bending.

(1) In-plane effective lengths

The method for estimating the in-plane effective lengths for portal members was developed by Fraser and is reproduced with his kind permission. Computer programs for matrix stability analysis are also available for determining buckling loads and effective lengths. Only the pinned-base portal with symmetrical uniform loads (that is, the critical load case dead + imposed load on the roof) is considered. Reference should be made to Fraser for unsymmetrical load cases (see Further reading at the end of this chapter).

If the roof pitch is less than 10° the frame may be treated as a rectangular portal and the effective length of the column found from a code limited frame sway chart such as Figure 24 from Appendix E of BS 5950. For pinned-base portals with a roof slope greater than 10° Fraser gives the following equation for obtaining the column effective length which is a close fit to results from matrix stability analyses (see Figure 13.3):

$$k_c = \frac{\text{Effective length of column } L_E}{\text{Actual length of column } L_C}$$
$$= 2 + 0.45 G_R$$

where
$$G_R = L_R I_c / I_R L_c$$

L_R, L_c = rafter and column lengths
I_R, I_c = rafter and column moments of inertia

Similar results can be obtained using the limited frame sway chart (BS 5950, Figure 24) if, as Fraser suggests, the pitched roof portal is converted to an equivalent rectangular frame. The beam length is made equal to the total rafter length as shown in Figure 13.3. To apply Figure 24 of the code:

Column top
$$K_1 = \frac{I_c/L_c}{(I_c/L_c) + (I_R/L_R)}$$

Column base
$$K_2 = 1.0 \text{ pinned}$$

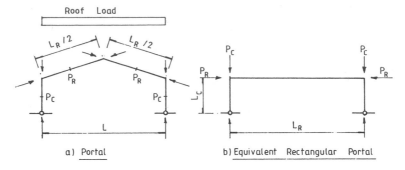

Figure 13.3

Then L_E/L_c for the column is read from Figure 24 of BS 5950.

The effective length of the rafter can be obtained from the effective length of the column because the entire frame collapses as a unit at the critical load, i.e.

Column $$P_{cc} = \frac{\pi^2 E I_c}{K_c^2 L_c^2}$$

Rafter $$P_{RC} = \frac{\pi^2 E I_R}{K_R^2 (L_R/2)^2}$$

whence $$K_R = K_c \frac{2L_c}{L_R} \sqrt{\frac{P_c I_R}{P_R I_c}}$$

where E = Young's modulus

$$K_R = \frac{\text{Effective length of rafter}}{\text{Actual length of rafter}}$$

$P_{cc} \cdot P_{RC}$ = critical loads for column and rafter
            = load factor × design loads

$P_c \cdot P_R$ = average design axial load in the column and rafter

(2) Out-of-plane effective length

The purlins and sheeting rails restrain the outside flange of the portal members. It is essential to provide lateral support to the inside flange, i.e. a torsional restraint to both flanges at critical points to prevent out-of-plane buckling. For the pinned-base portal with the bending moment diagram for dead and imposed load shown in Figure 13.2, supports to the inside flange are required at:

Eaves (may be stays from sheeting rails or a tie)
Within the top half of the column (more than one support may be necessary)
Near the point of contraflexure in the rafter (more than one support may be necessary)

The lateral supports and effective lengths about the minor axis for flexural buckling and lateral torsional buckling are shown in Figure 13.2.

13.2.4 Column Design

The column is a uniform member subjected to axial load and moment with moment predominant. The design procedure for the critical load case of dead + imposed load is as follows.

(1) Compressive resistance

Estimate slenderness ratios L_{EX}/r_x, $L_{EY}r_y \not> 180$

where L_{EX}, L_{EY} = effective lengths for XX and YY axes
r_x, r_y = radii of gyration for XX and YY axes

Compressive strength p_c = Table 27(a) or (b) of the code

Compressive resistance $p_c = P_c A_g$ (A_g = gross area)

(2) Moment capacity

$M_c = p_y S$
where p_y = design strength
S = plastic modulus

(3) Buckling resistance moment

The bending moment diagram and effective length are shown in Figure 13.2.

Equivalent moment factor m – Table 18 of the code:

Slenderness $\lambda = L_{EY}/r_y$
Equivalent slenderness $\lambda_{LT} = uv\lambda$

where u = Handbook
v = Table 14 of the code for λ/x
x = Handbook

Bending strength p_b — Table 11 of the code
Buckling resistance moment $M_b = p_b S$

(4) Interaction expressions

Local capacity check $F/A_g p_y + M/M_c \leqslant 1$
Overall buckling check $F/P_c + mM/M_b \leqslant 1$

13.2.5 Rafter design

The rafter is a member haunched at both ends with the moment distribution shown in Figure 13.2. The portion near the eaves has compression on the inside. Beyond the point of contraflexure the inside flange is in tension and is stable. The top flange is fully supported by the purlins.

    If the haunch length is small it may conservatively be neglected and the design made for the maximum moment at the eaves. A torsional restraint is provided at the first or second purlin point away from the eaves and the design is made in the same way as for the column. In this case the factor $m = 1.0$ and the slenderness correction factor n should be evaluated from Table 16 of the code. Rafter design taking the haunch into account is considered under plastic design in Section 13.3.

13.2.6 Example elastic design of a portal

(1) Specification

The pinned-base portal for an industrial building is shown in Figure 13.4(a). The portals are at 5 m centres and the length of the building is 40 m. The building loads are:

Roof — Dead load measured on slope

     Sheeting 20 gauge $= 0.1 \ kN/m^2$
     Insulation $= 0.15 \ kN/m^2$
     Purlins (Table 5.2, A140/165)
        4.0 kg/m at 1.5 m c/c $= 0.03 \ kN/m^2$

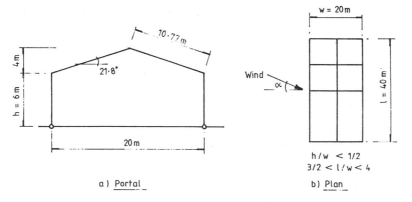

a) Portal   b) Plan

Figure 13.4

Rafter 457 × 191 UB 67 $= 0.13 \text{ kN/m}^2$
        Total $= 0.41 \text{ kN/m}^2$

— Imposed load on plan, BS 6399 $= 0.75 \text{ kN/m}^2$
Walls — same as roof $= 0.41 \text{ kN/m}^2$

Wind load — CP3: Chapter V: Part 2 — Location north-east
                          UK on outskirts of city
Carry out the following work:

Estimate the building loads
Analyse the portal using elastic theory with a uniform section throughout
Design the section for the portal using Grade 43 steel

(2) Loading
Roof — dead $= 0.41 \times 5 \times 10.77/10$ $= 2.21 \text{ kN/m}$
        imposed $= 0.75 \times 5$ $= 3.75 \text{ kN/m}$
        design $= (1.4 \times 2.21) + (1.6 \times 3.75)$ $= 9.09 \text{ kN/m}$
Wall — dead $= 0.41 \times 5 \times 6 \times 1.4$ $= 17.22 \text{ kN}$

Notional horizontal load at each column top
    $= 0.5 \times 0.005 \times 20 \times 9.09 = 0.45 \text{ kN}$

Wind — Basic wind speed $V = 45 \text{ m/s}$
        Ground roughness—category $= 3$
        Building size — class $= B$

The wind load factors S_2, design wind speeds V_s, dynamic pressures q, and external pressure coefficients C_{pe} for the portal are shown in Table 13.1. The internal pressure coefficients C_{pi} from Appendix E of the wind code are $+ 0.2$ or -0.3 for the case where there is a negligible probability of a dominant opening occurring during a severe storm.

Dynamic pressure $q = 0.613 V_s^2/10^3 \text{ kN/m}^2$
Wind load $w = 5q(C_{pe} - C_{pi})$

The diagrams for the characteristic loads are shown in Figure 13.5.

Table 13.1 Wind loads

Element	Height (m)	Factor S_2	Design speed V_s	Dynamic pressure q	External pressure coeff. C_{pe}		
					$\alpha = 0°$		$\alpha = 90°$
					Windward	Leeward	
+							
Roof	10	0.74	33.3	0.68	− 0.33	− 0.4	− 0.7
Walls	6	0.67	30.2	0.56	+ 0.7	− 0.25	− 0.5

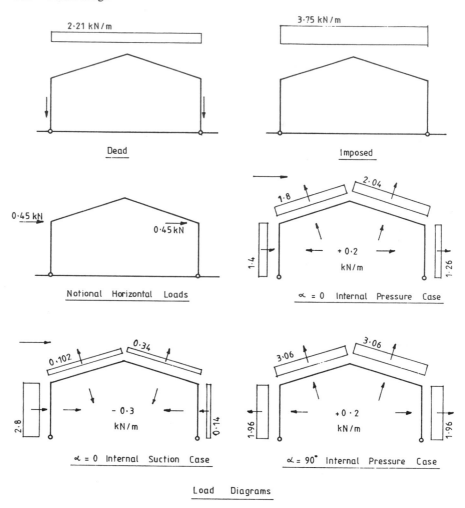

Figure 13.5 Load diagrams

(3) Analysis

The manual analysis for the design dead + imposed load case is set out here. Bending moment diagrams are given for the separate load cases in Figure 13.7.

Frame constants. Refer to Figure 13.6(a)

$$
\begin{aligned}
K &= h/s & &= 0.557 \\
\phi &= f/h & &= 0.667 \\
m &= 1 + \phi & &= 1.667 \\
B &= 2(K + 1) + m & &= 4.781 \\
C &= 1 + 2m & &= 4.334 \\
N &= B + me & &= 12.006
\end{aligned}
$$

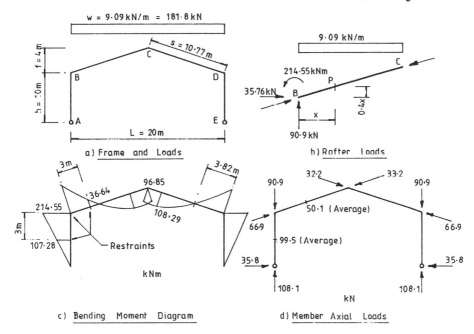

a) Frame and Loads

b) Rafter loads

c) Bending Moment Diagram

d) Member Axial Loads

Figure 13.6

Moments and reactions

$$M_B = wL^2 (3 + 5m)/16N = -214.55 \text{ kNm}$$
$$M_c = (wL^2/8) + mM_B = 96.85 \text{ kNm}$$
$$H = 214.55/6 = 35.76 \text{ kN}$$

Referring to Figure 13.6(b), the moment at any point P in the rafter is

$$M_p = 90.9x - 214.55 - 35.76 \times 0.4x - 9.09x^2/2$$

Put $M_p = 0$ and solve to give x = 3.55 m
Put $dM_p/dx = 0$ and solve to give x = 8.43 m
and maximum sagging moment = 108.29 kNm

Thrust at
$$B = 90.9 \times 4/10.77 + 35.76 \times 10/10.77 = 66.9 \text{ kN}$$

The bending moment diagram and member axial loads are shown in Figure 13.6(c) and (d), respectively. Other values required for design are given in the appropriate figures.

The maximum design moments for the separate load cases shown in Figure 13.5 and moment diagram in Figure 13.7 are given for comparison:

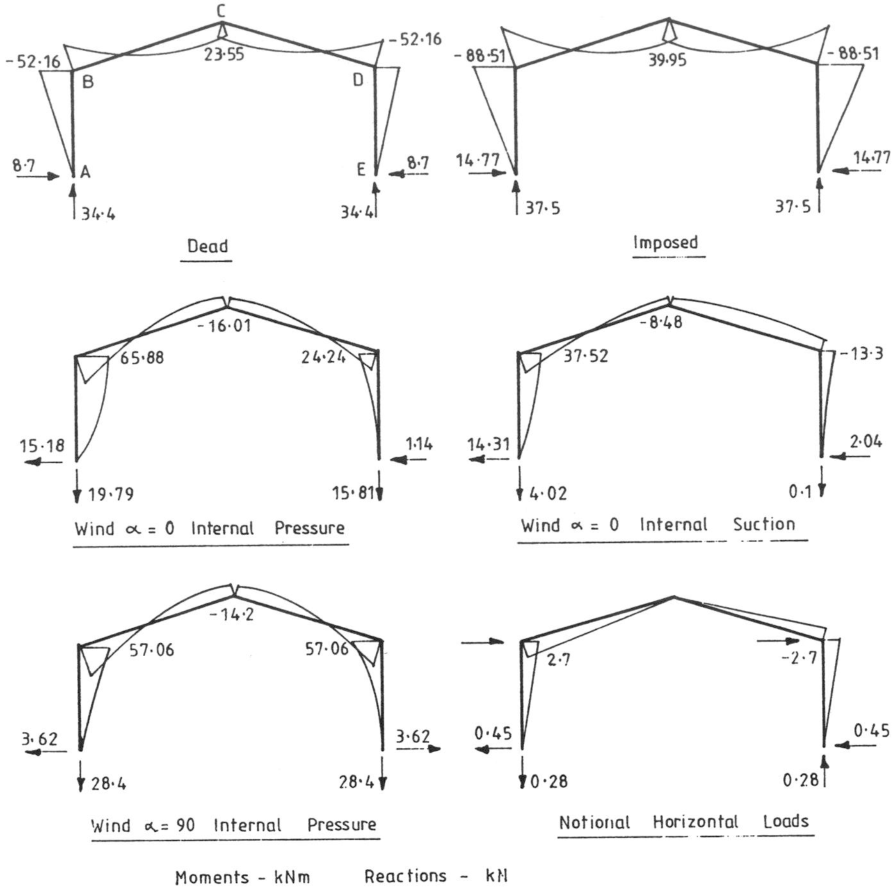

Figure 13.7

Maximum negative moment at the eaves

(1) 1.0 × dead + 1.6 × imposed
$$M_B = - [1.4 \times 52.16 + 1.6 \times 88.51]$$
$$= - 214.55 \text{ kNm}$$

(2) 1.2 [dead + imposed + wind internal suction]
$$M_D = - 1.2 [52.16 + 88.51 + 13.3]$$
$$= - 184.76 \text{ kNm}$$

(3) 1.4 × dead + 1.3 × imposed + notional horizontal load
$$M_D = - [1.4 \times 52.16 + 1.6 \times 88.51 + 2.7]$$
$$= 190.78 \text{ kNm}$$

Maximum reversed moment at the eaves

$-1.0 \times$ dead $- 1.4 \times$ wind internal pressure
$M_B = -52.16 + 1.4 \times 65.88 = 39.1 \text{kNm}$

4 Column design

Trial section

Try 457×152 UB 60, Grade 43, uniform throughout:

$A = 75.9 \text{ cm}^2,$ $r_x = 18.3 \text{ cm},$ $r_y = 3.23 \text{ cm},$
$S = 1280 \text{ cm}^3,$ $u = 0.869,$ $x = 37.5$

Compression resistance – Applied load $F = 90.9$ kN
In-plane slenderness
Fraser's formula

$$L_E/L = 2 + 0.45 \,(2 \times 10.77/6) = 3.62$$

Using BS 5950, Figure 24:

Top $K_1 = \dfrac{1/6}{1/6 + 1/21.54} = 0.782$

Base $K_2 \quad -1.0$
$\quad\quad L_E/L = 3.5$

Slenderness $L_{EX}/r_x = 3.62 \times 6000/183 = 118.7$

Out-of-plane slenderness with restraint at the mid-height of the column as shown in Figure 13.6(c):

Slenderness $L_{EY}/r_y = 3000/32.3 = 92.9 < L_{EX}/r_x$

Compressive strength $p_c = 121 \text{ N/mm}^2$ — Table 27(a) of the code

Compressive resistance $P_c = 121 \times 75.9/10 = 918.4$ kN

Moment capacity $M_c = 275 \times 1280/10^3 = 352$ kNm

Buckling resistance moment   Applied moment $M = 214.55$ kN.
The bending moment diagram is shown in Figure 13.6(c).
Equivalent moment factor $m = 0.76$ for $\beta = 0.5$ — Table 18 of the code.

Slenderness factor for $\lambda/x = 92.9/37.5 = 2.48$
$\quad\quad\quad\quad\quad\quad\quad\quad v = 0.931$ — Table 14 of the code

Equivalent slenderness

$$\lambda_{LT} = 0.869 \times 0.931 \times 92.9$$
$$= 75.2$$

Bending strength
$$p_b = 178.6 \text{ N/mm}^2 - \text{Table 11 of the code}$$

Buckling resistance moment:

$$M_b = 175.6 \times 1280/10^3 = 224.8 \text{ kNm}$$

Interaction expressions

Local capacity $\dfrac{90.9 \times 10}{275 \times 75.9} + \dfrac{214.55}{352} = 0.65$

Overall capacity $\dfrac{90.9}{918.4} + \dfrac{0.76 \times 214.55}{224.8} = 0.83$

The section is satisfactory.

5 Rafter design check

Compression resistance $F = 66.9 \text{ kN}$

The average compressive forces for the rafter and column are shown in Figure 13.6(d).

In-plane slenderness

$$\frac{L_E}{L_{R/2}} = \frac{3.62 \times 2 \times 6}{21.54} \sqrt{\frac{99.5}{50.1}} = 2.84$$

$L_{EX}/r_x = 2.84 \times 10770/183 = 167$ < 180

Out-plane slenderness: $L_{EY} = 3000 \text{ mm} - \text{Figure 13.6(c)}$

$L_{EY}/r_y = 3000/32.3$ $= 92.9$
$\quad p_c = 65.7 \text{ N/mm}^2 - \text{Table 27(a) of the code}$
$\quad P_c = 65.7 \times 75.9/10 = 498.7 \text{ kN}$

Buckling resistance moment
The slenderness factor n is evaluated using Table 16 of the code:

$\beta \quad = 36.64/214.55 = 0.17 - \text{Figure 13.6(c)}$
$M_o = 9.09 \times (3 \times 10/10.77)^2/8 \qquad = 8.82 \text{ kNm}$

$$\gamma = 214.55/8.82 \qquad\qquad = 24.3$$
$$n = 0.84 \text{ --- Table 16 of the code}$$
$$\lambda_{LT} = 0.84 \times 0.869 \times 0.931 \times 92.9 = 63.1$$
$$p_b = 205.6 \text{ N/mm}^2 \text{ --- Table 11 of the code}$$
$$M_b = 205.6 \times 1289/10^3 \qquad\qquad = 263.2 \text{ kNm}$$

Interaction expressions for overall buckling

$$\frac{66.9}{498.7} + \frac{214.55}{263.2} = 0.95$$

The section is satisfactory.

13.3 Plastic design

13.3.1 Code provisions

BS 5950 states in Section 5.3 that plastic design may be used in the design of structures and elements provided that the following main conditions are met:

(1) The loading is predominantly static.

(2) Structural steels with stress–strain diagrams as shown in Figure 2.1 are used. The plastic plateau permits hinge formation and rotation necessary for moment redistribution at plastic moments to occur.

(3) Member sections are plastic where hinges occur. Members not containing hinges are to be compact.

(4) Torsional restraints are required at hinges and within specified distances from the hinges

Provisions regarding plastic design of portals are given in the code in Section 5.5. The code states that a torsional restraint need not be provided at the last hinge to form. Other important provisions deal with overall sway stability and column and rafter stability and these are discussed below.

13.3.2 Plastic analysis — uniform frame members

Plastic analysis is set out in books such as that by Horne and only plastic analysis for the pinned-base portal is discussed here. The plastic hinge (the formation of which is shown in Figure 5.8) is the central concept. This rotates to redistribute moments from the elastic to the plastic moment distribution.

Referring to Figure 13.8(a), as the load is increased hinges form first at the points of maximum elastic moments at the eaves. Rotation occurs at

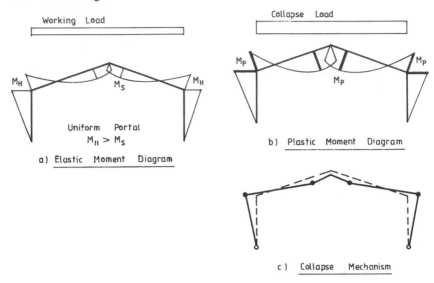

a) Elastic Moment Diagram

b) Plastic Moment Diagram

c) Collapse Mechanism

Figure 13.8

the eaves' hinges with the rafter acting like a simply supported beam, taking more load until two further hinges form near the ridge, when the rafter collapses. The plastic bending moment diagram and collapse mechanism are shown in Figure 13.8. In general, the number of hinges required to convert the portal to a mechanism is one more than the statical indeterminacy. With unsymmetrical loads such as dead + wind load two hinges only form to cause collapse.

For the location of the hinges to be correct, the plastic moment at the hinges must not be exceeded at any point in the structure. That is why, in Figure 13.8 (b), two plastic hinges form at each side of the ridge and not one only at the ridge at collapse. The critical mechanism is the one which gives the lowest value for the collapse load.

The collapse mechanism which occurs depends on the form of loading. Plastic analysis for the pinned-base portal is carried out in the following stages:

(1) The frame is released to a statically determinate state by inserting rollers at one support.

(2) The free bending moment diagram is drawn.

(3) The reactant bending moment diagram due to the redundant horizontal reaction is drawn.

(4) The free and reactant moment diagrams are combined to give the plastic bending moment diagram with sufficient hinges to cause the frame or part of it (e.g. the rafter) to collapse. As mentioned above, the plastic moment must not be exceeded.

The process of plastic analysis for the pinned-base portal is shown for the case of dead and imposed load on the roof in Figure 13.9. The frame is taken to be uniform throughout and the bending moment diagrams are drawn on the opened-out frame. The case of dead + imposed + wind load is treated in the design example.

The exact location of the hinge near the ridge must be found by successive trials or mathematically if the loading is taken to be uniformly distributed. Referring to Figure 13.10, for a uniform frame, hinge X is located by

$$g = h + 2fx/L$$

The free moment at X in the released frame is

$$M_x = wLx/2 - wx^2/2$$

The plastic moments at B and X are equal, i.e.

$$M_p = Hh = M_x - Hg$$

Put $dH/dx = 0$ and solve for x and calculate H and M_p. Symbols used are shown in Figure 13.10.

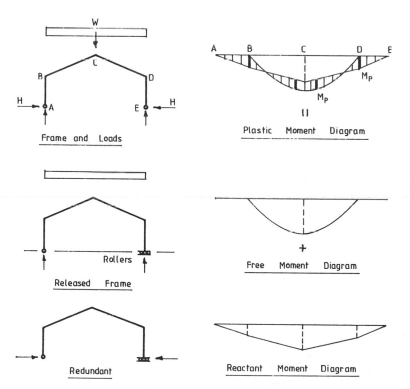

Figure 13.9

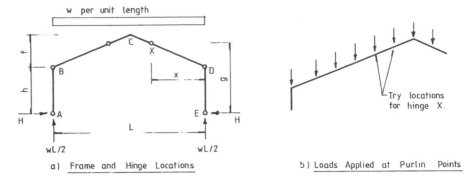

a) Frame and Hinge Locations   b) Loads Applied at Purlin Points

Figure 13.10

If the load is taken to be applied at the purlin points as shown in Figure 13.10(b) the hinge will occur at a purlin location. The purlins may be checked in turn to see which location gives the maximum value of the plastic moment M_p.

13.3.3 Plastic analysis — haunches and non-uniform frame

Haunches are provided at the eaves and ridge primarily to give a sufficient lever arm to form the bolted joints. The haunch at the eaves causes the hinge to form in the column at the bottom of the haunch. This reduces the value of the plastic moment when compared with the analyses for a hinge at the eaves intersection. The haunch at the ridge will not affect the analyses because the hinge on the rafter forms away from the ridge. The haunch at the eaves is cut from the same UB as the rafter. The depth is about twice the rafter depths and the length is often made equal to span/10.

It is also more economical to use a lighter section for the rafter than for the column. The non-uniform frame can be readily analysed as discussed below. It is also essential to ensure that the haunched section of the rafter at the eaves remains elastic. That is, the maximum stress at the end of the haunch must not exceed the design strength p_y:

$$p_y > F/A + M/Z$$

where

F = axial force
M = moment
A = area
Z = elastic modulus

The analysis of a frame with haunched rafter and lighter rafter than column section is demonstrated with reference to Figure 13.11(a). Let

M_p = plastic moment of resistance of the column
qM_p = plastic moment of resistance of the rafter
q = normally $0.6 - 0.7$

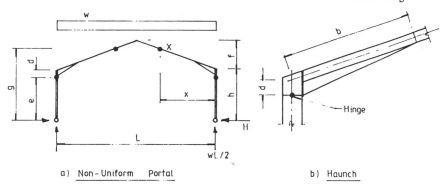

a) Non-Uniform   Portal   b) Haunch

Figure 13.11

Frame dimensions are shown in Figure 13.11.

Hinge in column $M_p = He$
Hinge in rafter $qM_p = M_x - Hg = qHe$

where M_x = free moment at X in the released frame
$g = h + 2fx/L$.

Put $dH/dx = 0$ and solve to give x and so obtain H and M_p.

13.3.4 Section design

At hinge locations design is made for axial load and plastic moment. The following two design procedures can be used:

(1) Simplified method

Local capacity check:

$$\frac{F}{A_g p_y} + \frac{M}{M_c} \leqslant 1$$

where F, M = applied load and plastic moment
M_c = moment capacity
A_g = gross area

(2) Exact method

Axial load reduces the plastic moment of resistance of a section. The bending moment is resisted by two equal areas extending inwards from the edges. The central area resists axial load and this area may be confined to the web or extend into the flange under heavy load. The stress diagrams are shown in Figure 13.12.

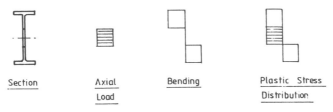

Section   Axial   Bending   Plastic Stress
                   Load   Distribution

Figure 13.12

The formulae for calculating the reduced plastic moduli of sections subjected to axial load and moment are given in *Steelwork Design to BS 5950*, Volume 1 and the calculation procedure is as follows:

$$n = F/A_g p_y$$

where F = applied axial load
          A_g = gross area

Change values of n are given for each section. For lower values of n the neutral axis lies in the web and the reduced plastic modulus:

$$S_R = k_1 - k_2 n^2$$

where values of k_1 and k_2 are given in the Handbook. A formula is also given for upper values of n. Reduced moment capacity, $M_R = S_R p_y > M$ applied moment.

13.3.5 Sway stability

The code gives two procedures for checking the overall sway stability of a portal. These ensure that the elastic buckling load is not reached and that the effects of additional moments due to deflection of the portal are taken into account.

(1) Limiting horizontal deflection at eaves

The horizontal deflection calculated by linear elastic analysis at the top of the columns due to the notional horizontal loads should not exceed $h/1000$, where h is the column height. The notional loads are 0.5% of the factored roof dead and imposed loads applied at the column tops.

(2) Limiting span/depth ratio of the rafter

The following equation must be satisfied:

$$\frac{L_b}{D} \leqslant \frac{44}{\Omega} \frac{L}{h} \left(\frac{\rho}{4 + \rho\, L_r/L} \right) \left(\frac{275}{p_{yr}} \right)$$

where $\rho = (2I_c/I_r)\ (L/h)$
L = portal span
D = depth of rafter
h = column height
L_r = total developed length of rafter
p_{yr} = design strength of rafter
$I_c,\ I_r$ = moments of inertia of column and rafter
$\Omega = W_r/W_o$
W_r = factored vertical load on the rafter
W_o = maximum value of W_r which could cause failure of the rafter treated as a fixed end beam of span L
L_b = effective span of the bay
When the depth of haunch is not less than $2D$,
$L_b = L - L_h$.
Otherwise $L_b = L$
L_h = length of haunch

13.3.6 Restraints and member stability

(1) The need for restraints

Restraints are required to ensure that:

(1) Plastic hinges can form in the deep I-sections used.
(2) Overall flexural buckling of the column and rafter about the minor axis does not occur.
(3) There is no lateral torsional buckling of an unrestrained compression flange on the inside of members.

The code requirements regarding restraints and member stability are set out below. A restraint should be capable of resisting 2.5% of the compressive force in the member or part being restrained.

(2) Column stability

The column contains a plastic hinge near the top at the bottom of the haunch. Below the hinge it is subjected to axial load and moment with the inside flange in compression. The code states in Section 5.3.5 that torsional restraints (i.e. restraints to both flanges) must be provided at or within member depth $D/2$ from a plastic hinge.
  In a member containing a plastic hinge the maximum distance from the restraint at the hinge to the adjacent restraint depends on whether or not restraint to the tension flange is taken into account. The following procedures apply:

(i) *Restraint to tension flange not taken into account*
  This is the conservative method where the distance from the hinge restraint to the next restraint is given by:

$$L_m = \frac{38\ r_y}{[(f_c/130) + (p_y/275)^2\ (x/36)^2]^{0.5}}$$

where f_c = compressive stress due to axial load

r_y = radius of gyration about YY axis using the minimum value if the section varies

x = torsional index using the maximum value if the section varies. See Appendix B.2.5 in the code

When this method is used no further checks are required. The locations of restraints at hinge H and at G, L_m below H are shown in Figure 13.13 (a). It may be necessary to introduce a further restraint at F below G, in which case column lengths GF and FA would be checked for buckling resistance for axial load and moment. The effective length for the XX axis may be estimated for the portal as set out in Section 13.2.3. The *Steelwork Design Guide to BS 5950* takes the effective length for the XX axis as HA in Figure 13.13, the distance between the plastic hinge and base. The effective lengths for the YY axis are GF and FA. Note that compliance with the sway stability check ensures that the portal can safely resist in-plane buckling and additional moments due to frame deflections.

(ii) *Restraint to tension flange taken into account*
A method for determining spacing of lateral restraints taking account of restraint to the tension flange is given in Clause 5.5.3.5 of BS 5950. A more rigorous method is given in Appendix G. The formula from Clause 5.5.3.5 gives the following limit for Grade 43 steel:

$$L_t \leqslant \frac{620 r_y x}{[72x^2 - 10^4]^{0.5}}$$

where K_1 = 620 in this case where the ratio of haunch depth to rafter depth is taken as 1.0. The code specifies that the buckling resistance moment M_B calculated using an effective length L_E equal to the spacing of the tension flange restraints must exceed the equivalent uniform moment for that column length. The restraints are shown in Figure 13.13(b). The column length AF is checked as set out above.

(3) Rafter stability near ridge

The tension flange at the hinge in the rafter near the ridge is on the inside and no restraints are provided. Clause 5.5.3.1 states that a torsional restraint is not needed at the last hinge to form. In the portal two hinges form last near the ridge. A purlin is required at or near the hinge and purlins should be placed at a distance not exceeding L_m on each side of the hinge. The purlin arrangement is shown in Figure 13.13 (c). In some cases a restraint to the inside flange is needed here when stresses reverse to compression under wind uplift loads.

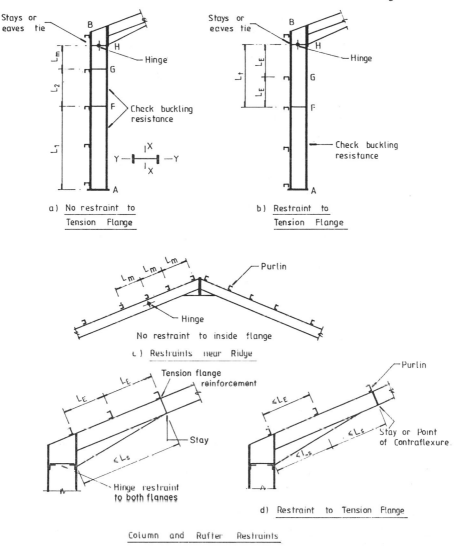

a) No restraint to
   Tension Flange

b) Restraint to
   Tension Flange

c) Restraints near Ridge

d) Restraint   to Tension Flange

Column and Rafter   Restraints

Figure 13.13 Column and rafter restraints

(4) Rafter stability at haunch

The requirements for rafter stability are set out in Clause 5.5.3.5 of the code. The haunch and rafter are in compression on the inside from the eaves for a distance to the second or third purlin up from the eaves where the point of contraflexure is usually located:

(i) *Restraint to tension flange not taken into account*
The maximum distance between restraints to the compression flange must not exceed L_m as set out in (2) above or that to satisfy the overall

buckling expression given in Section 4.8.3.3.1 of the code. For the tapered member the minimum value of r_y and maximum value of x are used. No restraint is to be assumed at the point of contraflexure.

(ii) *Restraint to tension flange taken into account*

When the tension flange is restrained at intervals by the purlins possible restraint locations for the compression flange are shown in Figure 13.13(d). The code requirements are:

(a) The distance between restraints to the tension flange L_E must not exceed L_m or, alternatively, the interaction expression for overall buckling given in Section 4.8.3.3.1 of the code based on an effective length L_E must be satisfied.

(b) The distance between restraints to the compression flange must not exceed

$$L_s = \frac{K_1 r_y x}{(72x^2 - 10^4)^{1/2}} \text{ for Grade 43 steel}$$

where K_1 is given by
Haunch depth/rafter depth 1 $K_1 = 620$
Haunch depth/rafter depth 2 $K_1 = 495$

The following provisions are set out in the code:

(1) The rafter must be a universal beam.
(2) The depth of haunch must not exceed three times the rafter depth.
(3) The haunch flange must not be smaller than the rafter flange.

If these conditions are not met the rigorous method given in Appendix G of the code should be used.

The code also specifies that for the purlins to provide restraint to the top flange, they must be connected by two bolts and have a depth not less than one quarter of the rafter depth. In this case, a vertical lateral restraint can be assumed at the point of contraflexure without provision of stays.

13.3.7 Serviceability check for eaves deflection

BS 5950 specifies in Table 5 that the deflection at the column top in a single-storey building is not to exceed height/300 unless such deflection does not damage the cladding. The deflections to be considered are due to the unfactored imposed and wind loads. If necessary, an allowance can be made for dead load deflections in the fabrication.

A formula for horizontal deflection at the eaves due to uniform vertical load on the roof is derived. Deflections for wind load should be taken from a computer analysis.

Referring to Figure 13.14, because of symmetry of the frame and loading the slope at the ridge does not change. The slope θ at the base is equal to the area of the M/EI diagram between the ridge and the base given by

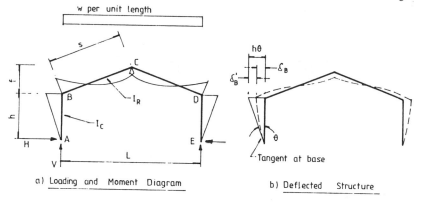

Figure 13.14

$$\theta = \frac{1}{EI_R} [\frac{wL^2s}{24} + Hs (h + \frac{f}{2}) - \frac{VLs}{4}] + \frac{Hh^2}{2EI_c}$$

The deflection at the eaves is then

$$\delta_B = h\theta - \frac{Hh^3}{6EI_c}$$

where H = horizontal reaction at base
V = vertical reaction at base

w = characteristic imposed load on the roof
I_c, I_R = moments of inertia of the column and rafter

Frame dimensions are shown in Figure 13.14.

13.3.8 Design of joints

(1) Eaves joint

The eaves joint arrangement is shown in Figure 13.15(a). The steps in the joint design check are:

(a) *Joint forces*
Take moment about X:

$$Vd/2 - M_p + Ta = 0$$

Bolt tension $T = (M_p - Vd/2)/a$
Compression $C = T + H$
Haunch flange force $F = C \sec \phi$

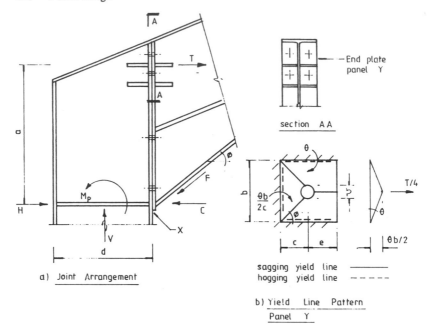

Figure 13.15

(b) *Bolt design for joint shown*

Tension $F_T = T/4$
Shear $F_s = V/8$

for a given bolt size with capacities P_T in tension and P_s in shear. The interaction expression is

$$F_s/P_s + F_T/P_T \leqslant 1.4$$

(c) *Column flange check and end plate design*

Adopt a yield line analysis (see Horne and Morris and MacGinley in the Further reading at the end of this chapter). The yield line pattern is shown in Figure 13.15(b) for one panel of the end plate or column flange. The hole diameter is v.

Work done by the load $= T\theta\, b/8$

Work done in the yield lines

$$= 4m\theta(c + e) - mv\theta\,(1 + \cos \phi)$$

$$+ \frac{mb}{c}\,[b - \frac{v}{2}\sin \phi]$$

Equate the expressions and solve for *m*. The plate thickness required

$$t = (4m/p_y)^{0.5}$$

(d) *Haunch flange*

Flange force $F < p_y \times$ flange area. The haunch section is checked for axial load and moment in the haunch stability check.

(e) *Column stiffener*

Design the stiffener for force *C*. See Section 6.3.7.

(f) *Column web*

Check for shear *T*. See Section 5.6.2.

(g) *Column and rafter webs*

These webs are checked for tension *T*. The small stiffeners distribute the load.

(h) *Welds*

The fillet welds from the end plate to rafter and on the various stiffeners must be designed.

The main check calculations are shown in the example in Section 13.3.9

(2) *Ridge joint*

The ridge joint is shown in Figure 13.16. The bolt forces can be found by taking moments about Z:

$$T = (M - Hh)/g$$

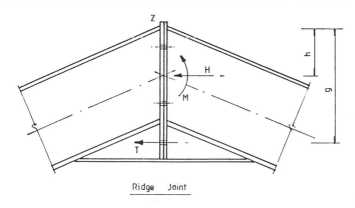

Ridge   Joint

Figure 13.16

where H = horizontal reaction in the portal
                                    M = ridge moment

Joint dimensions are shown in the figure. Other checks such as for end plate thicknesses and weld sizes are made in the same way as for the eaves joint.

13.3.9 Example plastic design of a portal

(1) Specification

Redesign the pinned-base portal specified in Section 13.2.7 using plastic design. The portal shown in Figure 13.17(a) has haunches at the eaves 1.5 m long and 0.45 m deep. The rafter moment capacity is to be approximately 75% of that of the column.

(2) Analysis — dead and imposed load

The frame dimensions, loading and plastic hinge locations are shown in Figure 13.17(a). The plastic moments in the column at the bottom of the haunch and in the rafter at x from the eaves are given by the following expressions:

Column $M_p = 5.55H$
Rafter $0.75M_p = 90.9x - 9.09x^2/2 - H(6 + 0.4x)$

Reduce to give $H = \dfrac{90.9x - 4.55x^2}{10.16 + 0.4x}$

Put $dH/dx = 0$ and collect terms to give

$$x^2 + 50.7x - 507.4 = 0$$

Solve $x = 8.56$ m
      $H = 32.7$ kN
      $M_p = 181.5$ kNm—column
$0.75M_p = 136.1$ kNm—rafter

The plastic bending moment diagram with moments needed for design is shown in Figure 13.17(b) and the co-existant thrusts in (c).

(3) Analysis—dead + imposed + wind load

The frame is analysed for the load case:

1.2 [dead + imposed + wind internal suction]

The roof wind loads acting normally are resolved vertically and horizontally and added to the dead and imposed load. The frame and hinge locations

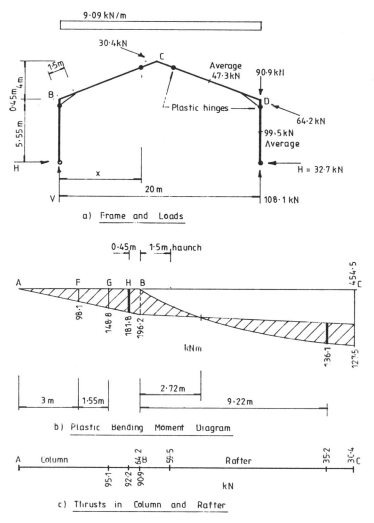

Figure 13.17

are shown in Figure 13.18(a) and the released frame, loads and reactions in (b).

The free moment in the rafter at distance x from the eaves point can be expressed by

$$M_x = 66.5x - 57.24 - 3.52x^2$$

The equations for the plastic moments at the hinges are:

Columns $M_p = 5.55H + 2.59$
Rafter $0.75M_p = 66.5x - 57.24 - 3.52x^2 - H(6 + 0.4x)$

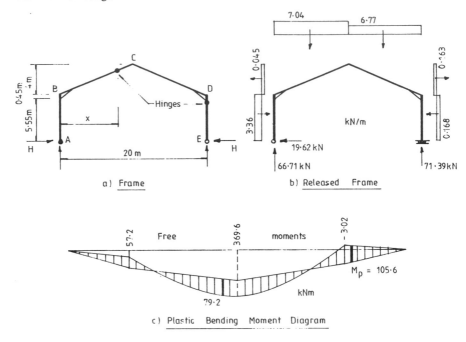

Figure 13.18

This gives

$$H = \frac{66.5x - 59.18 - 3.52x^2}{10.16 + 0.4x}$$

Put $dH/dx = 0$ and solve to give $x = 8.39$m.

$$H = 18.6 \text{ kN}$$
$$M_p = 105.1 \text{ kN}$$

The plastic bending moment diagram is shown in Figure 13.18(c). The moments are less than for the dead + imposed load case.

(4) Column section

The design actions at the column hinge are

$$M_p = 181.8 \text{ kNm}$$
$$F = 92.2 \text{ kN}$$
$$S = 181.8 \times 10^3/275 = 661.1 \text{ cm}^3$$

Try 406×140 UB 46, $S = 888 \text{ cm}^3$, $A = 59 \text{ cm}^2$:

$r_x = 16.3$ cm, $r_y = 3.02$ cm, $u = 0.87$, $x = 38.8$
$I_x = 15\ 600$ cm^4
$n = 92.2 \times 10/(59 \times 275) = 0.056 < 0.444$

Reduced plastic modulus:

$S_R = 888 - 1260n^2 = 883.9$ cm^3

The section is satisfactory.

(5) Rafter section

The design actions at the rafter hinge are:

$M_p = 136.1$ kNm
$F = 35.2$ kN
$S = 136.1 \times 10^3/275 = 494.9$ cm^3

Try 356×127 UB 39, $S = 654$ cm^3, $A = 49.4$ cm^2:

$r_x = 14.3$ cm, $r_y = 2.69$ cm, $Z = 572$, $I_x = 10\ 100$ cm^4
$x = 34.3$
$n - 35.5 \times 10/(49.4x\ 275) = 0.026 < 0.437$
$S_R = 654 - 0.43 = 653.6$ cm^3

The section is satisfactory.

Check that the rafter section at the end of the haunch remains elastic under factored loads. The actions are

$M - 96.9$ kNm
$F = 59.5$ kN

$$\text{Maximum stress} = \frac{59.5 \times 10}{49.4} + \frac{96.9 \times 10^3}{572}$$
$$= 181.4\ \text{N/mm}^2$$

This is less than $p_y = 275$ N/mm^2. The section remains elastic.

(6) Column restraints and stability

A tie is provided to restrain the hinge section at the eaves. The distance to the adjacent restraint using the conservative method is:

$$L_m \leqslant \frac{38 \times 30.2}{\{[(92.2 \times 10)/(59 \times 130)] + (38.8/36)^2\}}$$

$$= 1013.7 \text{ mm}$$

The arrangement for the column restraints is shown in Figure 13.19(a). The column is checked between the second and third restraints G and F over a length of 1.55 m. The moments and thrusts are shown in Figure 13.17:

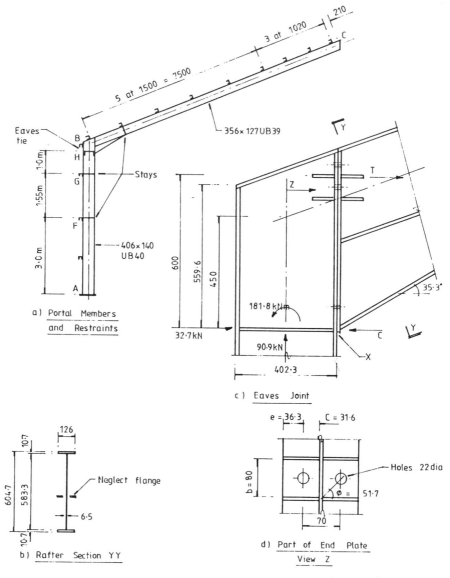

a) Portal Members and Restraints

c) Eaves Joint

b) Rafter Section YY

d) Part of End Plate View Z

Figure 13.19

M_F = 98.1 kNm
M_G = 148.8 kNm
F_g = 95.1 kN

The effective lengths and slenderness ratios are

$$k_c = [2 + \frac{0.45 \times 21.54 \times 15\ 600}{6 \times 10\ 100}] = 4.49$$

L_{EX}/r_x = 4.49 × 6000/163 = 165.3

Note that the *Steelwork Design Guide to BS 5950* would use L_{EX}/r_x = 5550/163 = 34 in this check

L_{EY}/r_y = 1700/30.2   = 56.3
p_c = 66.7 N/mm²—Table 27(a) in the code
P_c = 66.7 × 59/10   = 393.5 kN
β = 98.1/148.8   = 0.66
m = 0.83—Table 18 in the code
λ/x = 56.38.8   = 1.45
v = 0.972—Table 14 in the code
λ_{LT} = 0.87 × 0.972 × 56.3 =47.6
p_b = 243.8 N/mm²—Table 11 in the code
M_b = 243.8 × 888/10³   = 216.5 kNm

Overall buckling check:

$$\frac{95.1}{393.5} + \frac{0.83 \times 148.8}{216.5} = 0.81 \quad \text{Satisfactory}$$

Note that the in-plane stability or sway stability is checked by the code procedure in (8) below.

(7) Rafter restraints and stability

The rafter section at the eaves is shown in Figure 13.19(b). The centre flange is neglected and the section properties are calculated to give

A = 64.9 cm²
S = 1353 cm²
r_y = 2.35 cm
x = 0.566 h_s (A/J)$^{0.5}$—Appendix B2.5.1 in the code
J = 1/3 [2 × 1.07³ × 12.6 + 0.65³ × 58.33]
   = 15.6 cm⁴
x = 0.566 × 59.4 (64.9/15.6)$^{0.5}$ = 68.6

The rafter stability is checked taking account of restraint to the tension flange using Clause 5.5.3.5 in the code.

Depth of haunch/depth of rafter

$$= 559.5/352.8 = 1.59$$

$$K_1 = 546$$

$$L_s = \frac{546 \times 23.5 \times 68.6}{[72 \times 68.6^2 - 10^4]^{0.5}} = 1535.0 \text{ mm}$$

Provide stays at the end of the haunch.

Check overall buckling in accordance with Clause 4.8.3.3.1 using an effective length L_E of 1500 mm for the haunch:

$$
\begin{aligned}
L_{EY}/r_Y &= 1500/23.5 &&= 63.8 \\
\lambda/x &= 63.8/68.6 &&= 0.93 \\
v &= 1.0\text{—Table 14 in the code} \\
u &= 1.0 \\
\lambda_{LT} &= 63.8
\end{aligned}
$$

The smallest value of r_y and the largest value of x have been used.

At the large end of the haunch, at the eaves, consider as a welded section:

$$
\begin{aligned}
p_b &= 169.4 \text{ N/mm}^2 \text{ from Table 12 of the code} \\
p_c &= 193.4 \text{ N/mm}^2 \text{ from Table 27(c) of the code} \\
M_b &= 169.4 \times 1353/10^3 = 229.2 \text{kNm} \\
F &= 64.2 \text{ kN from Figure 13.17} \\
M_x &= 196.2 \text{ kNm from Figure 13.17}
\end{aligned}
$$

Interaction expression

$$\frac{64.2 \times 10}{193.4 \times 64.9} + \frac{196.2}{229.2} = 0.91$$

At the small end of the haunch, consider as a rolled section 356 × 127 UB 39 for which $A = 49.4$ cm^2, $S_x = 654$ cm^3. The actions at the end of the haunch are:

$$
\begin{aligned}
F &= 59.5 \text{ kN} &&\text{See section (5) above} \\
M_x &= 96.9 \text{ kNm} \\
p_b &= 203.9 \text{ N/mm}^2 \text{ from Table 11 of the code} \\
p_c &= 213.4 \text{ N/mm}^2 \text{ from Table 27(b) of the code} \\
M_b &= 203.9 \times 654/10\ 3 = 133.4 \text{ kNm}
\end{aligned}
$$

Interaction expression

$$\frac{59.5 \times 10}{213.4 \times 49.4} + \frac{96.9}{133.4} = 0.79$$

The haunch is satisfactory.
  If restraint to the tension flange is not considered:

At the start of the haunch

$$L_m = \frac{38 \times 23.5}{[(64.2 \times 10)/(64.9 \times 130) + (68.6/36)^2]^{0.5}}$$

$$= 463.8 \text{ mm}$$

At the end of the haunch

$$L_m = \frac{38 \times 26.9}{[(35.2 \times 10)/(49.4 \times 130) + (35.3/36)^2]^{0.5}}$$

$$= 1030.5 \text{ mm}$$

An additional purlin must be located at, say, 450 mm from the eaves with stays to the compression flange.
  The above calculations follow the code procedure and the rafter section complies with the sway stability check made in (8) below. If the rafter is checked using the in-plane effective length derived in the elastic design in Section 13.2.3 the slenderness L_{EX}/r_x exceeds 180, and its depth would need to be increased if this more conservative design procedure is used: However, the rafter is primarily a flexural member carrying small axial load and the restriction $l/r < 180$ intended for struts should not, in fact, apply.
  At the hinge location near the ridge the purlins will be spaced at 1020 mm as shown in Figure 13.19(a).

(8) Sway stability

Use limiting span/depth of rafter ratio. The haunch depth 604.7 mm (Figure 13.19(b)) is less than twice the rafter depth, the effective span $L_b = 20$ m:

$$\rho = \frac{2 \times 15\ 600 \times 20}{10\ 100 \times 6} = 10.29$$

$$M_p = 654 \times 275/10^3 = 179.9 \text{ kNm—rafter}$$
$$= W_o L/16$$
$$W_o = 143.9 \text{ kN}$$
$$\Omega = 9.09 \times 20/143.9 = 1.26$$

$$\frac{L}{D} = \frac{20000}{352.8} = 56.7 \leqslant \frac{44 \times 20}{1.26 \times 6} \left(\frac{10.29}{4 + 10.29 \times 21.54/20} \right)$$

$$\leqslant 79.4 \quad \text{Satisfactory}$$

(9) Serviceability—Deflection at eaves

An elastic analysis for the imposed load on the roof of 3.75 kN/m gives $H = 15.26$ kN and $V = 37.5$ kN. See sections 13.2.2. and 13.2.6. From Section 13.3.7.

$$\theta = \frac{1}{205 \times 10^3 \times 10\,100 \times 10^4} \left[\frac{3.75 \times 20\,000^2 \times 10\,770}{24} \right.$$

$$\left. + 15\,260 \times 10\,770 \left(600 + \frac{4000}{2} \right) - \frac{37\,500 \times 20\,000 \times 10\,770}{4} \right]$$

$$+ \frac{15\,260 \times 6000^2}{2 \times 205 \times 10^3 \times 15\,600 \times 10^4} = 7.068 \times 10^{-3} \text{ radians}$$

Deflections at eaves

$$S_B = 6000 \times 7.068 \times 10^{-3} - \frac{15\,260 \times 6000^3}{6 \times 205 \times 10^3 \times 15\,600 \times 10^4}$$

$$= 25.2 \text{ mm} = \text{height}/238$$

This exceeds $h/300$ but metal sheeting will accommodate the deflection.

(10) Design of joints

The arrangement for the eaves joint is shown in Figure 13.19(c). Selected check calculations only are given.

(a) *Joint forces* Take moment about X
$$T = [181.8 - 90.9 \times 0.402/2]/0.6 = 272.5 \text{ kN}$$

(b) *Bolts* $F_T = 272/4 = 68.13$ kN
$F_s = 90.0/6 = 15.15$ kN

Try 20 mm Grade 8.8 bolt. From the *Steel Design Guide to BS 5950*, Volume 1:

$$P_T = 110 \text{ kN}$$
$$P_s = 91.9 \text{ kN}$$

$$(15.15/91.9) + (68.13/110) = 0.78 < 1.4$$

The bolts are satisfactory.

(c) *Column flange*

See Figure 13.19(d) and Section 13.3.8. The yield line analysis gives:

$$\frac{272.5 \times 10^3 \times 80}{8} = 4 \text{ m} \times 67.9 - \text{m} \times 22 (1 + 0.62)$$

$$+ \frac{\text{m} \times 80}{31.6} \left(80 - \frac{22 \times 0.78}{2} \right)$$

$$= 416.8 \text{ m}$$

$m = 6537.9$ Nmm/mm

$$t = \left(\frac{4 \times 6537.9}{275} \right)^{0.5} = 9.75 \text{ mm}$$

The flange thickness 11.2 mm is adequate. The rafter end plate can be made 12 mm thick.

(d) *Column web shear*

$F_v = 0.6 \times 275 \times 402.3 \times 6.9/10^3 = 458.8$ kN
> 272.8 kN   Satisfactory

A similar design procedure is carried out for the ridge joint.

(11) Comments

(a) *Wind uplift*

The rafters on portals with low roof angles and light roof dead loads require checking for reverse bending due to wind uplift. Restraints to inside flanges near the ridge are needed to stabilize the rafter. Joints must also be checked for reverse moments.

The portal designed above has a relatively high roof angle and heavy roof dead load. Checks for load, $-1.4 \times$ wind $+ 1.0 \times$ dead, for wind angles 0° and 90° show that the frame remains in the elastic range and the rafter is stable without adding further restraints. See Figure 13.13.

(b)   *Eaves joint design*

In designing the joint, the effect of axial load is usually ignored, and the bolts are sized to resist moment only. In the above case $M = 180.8$ kN, and the bolt tension $T = 75.3$ kN<110 kN. The shear is taken by the bottom two bolts.

13.4 Further reading for portal design

FRASER, D. J., 'Effective Lengths in Gable Frames, Sway not Prevented', *Civil Engineering Transactions, Institution of Engineers, Australia* **CE 22**, No. 3, 1980

FRASER, D. J., 'Stability of Pitched Roof Frames', *Civil Engineering Transactions, Institution of Engineers, Australia* **CE28**, No. 1, 1986

HORNE, M. R., *Plastic Theory of Structures*, Nelson, London, 1971

HORNE, M. R. and MORRIS, L. J., *Plastic Design of Low Rise Frames*, Collins, London, 1981

MACGINLEY, T. J., *Reinforced Concrete — Design Theory and Examples*, E & FN Spon, London, 1978

STEEL DESIGNERS MANUAL, Crosby Lockwood, London, 1972

Steelwork Design Guide to BS 5950: Part 1, Volume 1, *Section Properties. Member Capacities*, The Steel Construction Institute, Ascot, 1987

Steelwork Design Guide to BS 5950: Part 1, volume 2, *Worked Examples*, The Steel Construction Institute, Ascot, 1986

References

1 RICHARDS, K.G., *Fatigue Strength of Welded Structures*, The Welding Institute, Cambridge, 1969
2 BOYD, G.M., *Brittle Fracture in Steel Structures*, Butterworths, London, 1970
3 *Building Regulations and Associated Approved Documents*, HMSO, London, 1985
4 ELLIOTT, D.A., *Fire and Steel Construction. Protection of Structural Steelwork*, Constrado, London, 1981
5 WARD BUILDINGS COMPONENTS, *Multibeam Purlin and Cladding Rail System*, Ward Brothers (Sherburn) Ltd, 1984
6 CASE, J. and CHILER, A.H., *Strength of Materials*, Edward Arnold, London, 1964
7 HOLMES, M and MARTIN, L.H., *Analysis and Design of Structural Connections—Reinforced Concrete and Steel*, Ellis Horwood, London, 1983
8 PRATT, J.L., *Introduction to the Welding of Structural Steelwork*, Constrado, London, 1979
9 SALMON, C.G. *et al., Laboratory Investigation of Unstiffened Triangular Bracket Plates*, ASCE Structural Division, April 1964
10 PASK, J.W., *Manual on Connections for Beam and Column Construction*, BCSA, London, 1982
11 HORNE, M.R. and MORRIS, L.J., *Plastic Design of Low Rise Frames*, Collins, London, 1981
12 TRAHAIR, N.S., *The Behaviour and Design of Steel Structures*, Chapman and Hall, London, 1977
13 JOHNSON, B.G. [ed.], *Guide to Stability Design Criteria for Metal Structures*, 3rd edn. Wiley, New York, 1976
14 HORNE, M.R., *Plastic Theory of Structures*, Pergamon Press, Oxford, 1979
15 STEEL DESIGNERS MANUAL, Crosby Lockwood, London, 1972
16 GHALI, A. and NEVILLE, A.M., *Structural Analysis*, Chapman and Hall, London, 1978
17 TIMOSHENKO, S.P. and GERE, J.M., *Theory of Elastic Stability*, McGraw-Hill, New York, 1961
18 BRESLER, B. and LIN, T.Y., *Design of Steel Structures*, Wiley, New York, 1964
19 PORTER, D.M., ROCKEY, K.C. and EVANS, H.R., 'The Collapse Behaviour of Plate Girders Loaded in Shear', *The Structural Engineer* August 1975
20 JENKINS, W.M., *Structural Mechanics and Analysis*, Van Nostrand-Reinhold, New York, 1982
21 BATES, W., *Design of Structural Steelwork: Lattice Framed Industrial Buildings*, Constrado, London, 1983
22 COATES R.C., COUTIE, M.G. *and* KONG, F.K., *Structural Analysis*, 2nd edn, ELBS/Van Nostrand-Reinhold, London, 1980

Computer References

23 IBM Disk Operating System, Version 3.1, Microsoft Corp., 1985
24 BASIC, Version 3.0, Microsoft Corp., 1984
25 AUTOCAD, Version 2.18, *User's Manual*, Autodesk Inc., 1985
26 HARRISON, H.B., *Structural Analysis and Design: Some Minicomputer Applications*, Parts I and II, Pergamon Press, Oxford, 1980
27 MACGINLEY, T.J. and ANG, T.C., Limit State Design and Microcomputers', paper presented at the International Symposium on Structural Steel Design and Construction, Singapore, 1985

British Standards

The following British Standards are referred to in this book.

BS 5950: *Structural Use of Steelwork in Building*

> Part 1: Code of Practice for Design in Simple and Continuous Construction: Hot Rolled Section.   Part 5: Code of Practice for Design in Cold Formed Sections

BS 6399: *Design Loading for Buildings*

> Part 1: Code of Practice for Dead and Imposed Loads

CP3:   Chapter V: Loading. Part 2: Wind Loads

BS 4360: Specification for Weldable Structural Steels

BS 5493: Code of Practice for Protective Coating of Iron and Steel Structures against Corrosion

BS 4190: 150 Metric Black Hexagon, Bolts, Screws and Nuts

BS 4395: High Strength Friction Grip Bolts and Associated Nuts and Washers for Structural Engineering. Part 1: General Grade

BS 4604: The Use of High Strength Friction Grip Bolts in Structural Steelwork, Metric Series. Part 1: General Grade

BS 8110: Structural Use of Concrete. Part 1: Code of Practice for Design, Materials and Workmanship

Handbooks

Structural Steelwork Handbook

Properties and Safe Load Tables, BCSA, Constrado, London, 1982

Steelwork Design: Guide to BS 5950: Part 1: 1985. Volume 1, Section Properties, Member Capacities, Constrado, 1985

Index